PSA 1982

VOLUME ONE

PSA 1982

PROCEEDINGS OF THE 1982
BIENNIAL MEETING
OF THE
PHILOSOPHY OF SCIENCE
ASSOCIATION

volume one

Contributed Papers

edited by

PETER D. ASQUITH
&
THOMAS NICKLES

1982
Philosophy of Science Association
East Lansing, Michigan

Library of Congress Catalog Card Number 72-624169

Cloth Edition: ISBN 0-917586-18-2

Paperback Edition: ISBN 0-917586-17-4

ISSN: 0270-8647

Manufactured in the United States of America
by Edwards Brothers, Ann Arbor, Michigan

CONTENTS

PART I. DISCOVERY, RATIONALITY AND HISTORY OF SCIENCE

PART II. CAUSATION

PART III. SCIENTIFIC REALISM AND OBSERVATION

Part IV. Probability and Statistical Inference

Part V. Measurement, Verisimilitude and Decision

PREFACE

Among the charges of the By-Laws of the Philosophy of Science Association is furthering "free discussion from diverse standpoints in the field of philosophy of science." To achieve this end PSA engages in a variety of activities including the holding of Biennial Meetings and arranging for the publication of the Proceedings of these meetings. Since 1976 PSA itself has published the Proceedings utilizing "camera-ready typewritten copy supplied by the authors" in an effort to make the Proceedings available at an affordable price. Proceedings have been published in two volumes--Volume 1 containing the contributed papers and Volume 2 containing the symposia papers.

The practice, which this volume continues, has been to publish the contributed papers in advance of the meeting. This eliminates the need for a full presentation of the paper at the meeting and instead allows that the time be primarily devoted to discussion with an audience that has been able to inform itself of the author's views in advance of the session.

The labor in preparing the volume was divided. The Program Committee selected the papers from those submitted. The PSA business office determined the format for camera-ready copy and the production schedule. The contributors produced the papers in the form of camera-ready copy. The Executive Secretary and his staff proofread the papers, made necessary corrections in the copy, and prepared the whole volume for the printer.

Our debts are numerous. Thanks go to the 1982 Program Committee--Ned Block, Jeffrey Bub, Allan Gibbard, Gary Gutting, Nancy Maull, and Teddy Seidenfeld and to the contributors to this volume. Christine Rosenberg had the responsibility for the mechanics of taking a collection of individual papers and organizing them in a format that would enable them to become a book. Phil West assisted in the proofreading and in the necessary library work.

Peter D. Asquith
Department of Philosophy
Michigan State University

Thomas Nickles
Department of Philosophy
University of Nevada-Reno

July 16, 1982

SYNOPSIS

The following brief summaries provide an introduction to each of the papers in this volume.

1. *The Rationality of the Copernican Revolution*. **Martin V. Curd.** The claim that even in 1543 the Copernican theory was objectively superior to the Ptolemaic theory is explained and defended. The question is then raised concerning the relevance of this insight for our understanding of the rationality of the Copernican revolution. It is proposed that (a) the decision to reject the Ptolemaic theory first became clearly rational early in the 17th century as a result of Galileo's observations of the phases of Venus, and (b) the decision to accept the Copernican theory only became clearly rational when Newtonian gravitational theory provided reasonable physical grounds for rejecting the Tychonic theory towards the end of the 17th century.

2. *Explaining Scientific Discovery*. **Noretta Koertge.** Since philosophers of science have shown that discoveries cannot be predicted, how can historians of science explain them? The concept of discovery is explicated and what is required in order to provide a covering law explanation of past scientific discoveries is analyzed. The account relies on Hempel's model of genetic explanation, Popper's situational logic and Salmon's theory of statistical relevance. The <u>Verstehen</u> approach also plays an important role.

3. *Curried Lakatos, or, How Not to Spice Up the Norm-Ladenness Thesis*. **Stephen J. Wykstra.** Using Currie's critique as a foil, this paper reconstructs Lakatos's thesis that historiography of science is laden with normative assumptions about scientific rationality. It is argued that this thesis comprises both a heuristic claim and a constitutive claim. The Received Critique of Lakatos fails to see that "internal history" and "rational reconstruction" receive a special meaning (by which they designate "rational preconstructions") when used in the context of the heuristic claim. Currie avoids this mistake, but attributes to Lakatos an "investigation-surrogate claim" which misrepresents the heuristic claim, oversimplifying the relation Lakatos envisions between hard cores and the solutions they generate.

4. *Causal and Explanatory Asymmetry*. **Daniel M. Hausman.** This paper asks why causal asymmetries should give rise to explanatory asymmetries. One way to give some rationale for the asymmetries of causal explanation is to adopt a pragmatic view of explanation and to stress the fact that causes can be used to manipulate their effects. This paper argues, however, that when one recognizes that causal

asymmetry is fundamentally an asymmetry of "connectedness", one can see how causal asymmetry leads to an objective difference between explanations in terms of causes and explanations in terms of effects. The pragmatic differences are subsidiary.

5. *Mackie's Singular Causality and Linked Overdetermination*. Robert H. Ennis. Necessary-condition analyses of singular causal claims are particularly vulnerable to cases of linked overdetermination, so named because the nonoperation of the back-up factor (in fail-safe cases) or the preempted factor (in preemptive cases) is linked to the operation of the actual cause. As an example J. L. Mackie's analysis is here challenged with a simple switch-light case. Three replies are considered, a facts-vs.-events reply, a different-effect reply, and an in-the-circumstances reply. All are found deficient.

6. *The "Established Maxim" and Causal Chains*. A. David Kline. There is a widely accepted ancient principle which holds that effects must occur as soon as possible after their causes. This paper takes this principle seriously and shows that if one believes there are causal chains then one is forced to accept the view that the temporal order is discrete or that some causally related events form a dense sequence.

7. *Hermeneutical Realism and Scientific Observation*. Patrick A. Heelan. Using the methods of hermeneutic phenomenology, and against the background of the principle that the real is what is or can be given in a public way in perception as a state of the World, and of the thesis established elsewhere that acts of perception are always epistemic, contextual, and hermeneutical, the writer proposes that objects of scientific observation are perceptual objects, states of the World described by theoretical scientific terms and, therefore, real. This thesis of Hermeneutical Realism is proved by showing how the response of a standard instrument is 'read' as if it were a 'text'. Conclusions are then drawn about a number of topics, including Scientific Realism, Conventionalism, and Cultural Relativism.

8. *The Historical Objection to Scientific Realism*. Jarrett Leplin. A realist interpretation of successful science is defended against a historical induction to the ultimate failure of current science from the failure of theories which once excelled by current standards. The defense requires (1) restrictions on the forms of success which realism, by its own lights, must explain, (2) referential stability through theory changes where the rejected theory achieves such success, and (3) degrees of truth for scientific statements.

9. *Realism, Miracles, and the Common Cause*. James Robert Brown. The principle of the common cause, which gets its justification from the miracle arguments, probably constitutes the best reason for being a scientific realist. However, results in quantum mechanics steming from the work of Bell raise difficulties which anti-realists have been quick to seize. The author tries to overcome the problem and save scientific realism by reformulating the principle of the common cause so that a distinction is made between a priori and a posteriori correlations.

xx

10. *Scientific Realism and the Hierarchical Counterfactual Path from Data to Theory.* Ronald Laymon. Using the Schwarzschild calculation of the Relativistic bending of starlight near the sun as an illustration, it is shown that the relationship between theory and data requires a hierarchy of structures of different logical type. An essential feature of this hierarchy is the use of idealizations and approximate truths. On the basis of a counterfactual analysis of these concepts, it is shown that confirmation is possible even though statistical measures of goodness of fit are not satisfied. The consequences of this view of confirmation and hierarchical structure for scientific realism are then considered.

11. *The Explanatory Import of Dispositions: A Defense of Scientific Realism.* Jon D. Ringen. It is widely assumed that disposition predicates do not designate entities which could be causal factors in the production of natural phenomena. Yet, the fact that an object has a given dispositional property is often taken to help explain behavior exhibited by objects to which the disposition is ascribed. Instrumentalist, realist, and rationalist analyses of disposition predicates embody three quite distinct views of how both assumptions could be correct. It is argued that the instrumentalist fails to capture basic intuitions concerning the explanatory import of disposition ascriptions, the rationalist tries unsuccessfully to locate necessary connections in nature, and the realist provides an account which is intuitively satisfying without introducing otiose entities into the ontology of empirical science.

12. *The Generalization of de Finetti's Representation Theorem to Stationary Probabilities.* Jan von Plato. de Finetti's representation theorem of exchangeable probabilities as unique mixtures of Bernoullian probabilities is a special case of a result known as the ergodic decomposition theorem. It says that stationary probability measures are unique mixtures of ergodic measures. Stationarity implies convergence of relative frequencies, and ergodicity the uniqueness of limits. Ergodicity therefore captures exactly the idea of objective probability as a limit of relative frequency (up to a set of measure zero), without the unnecessary restriction to probabilistically independent events as in de Finetti's theorem. The ergodic decomposition has in some applications to dynamical systems a physical content, and de Finetti's reductionist interpretation of his result is not adequate in these cases.

13. *On After-Trial Criticisms of Neyman-Pearson Theory of Statistics.* Deborah G. Mayo. Despite its widespread use in science, the Neyman-Pearson Theory of Statistics (NPT) has been rejected as inadequate by most philosophers of induction and statistics. They base their rejection largely upon what the author refers to as <u>after-trial criticisms of NPT</u>. Such criticisms attempt to show that NPT fails to provide an adequate analysis of specific inferences after the trial is made, and the data is known. In this paper, the key types of after-trial criticisms are considered and it is argued that each fails to demonstrate the inadequacy of NPT because each is based on judging NPT on the grounds

of a criterion that is fundamentally alien to NPT. As such, each may
be seen to either misconstrue the aims of NPT, or to beg the question
against it.

   14. *A Bayesian Argument in Favor of Randomization*. Zeno G.
Swijtink. Randomization is a generally accepted principle of sound
experimental design and common practice among working scientists. But
Bayesian statisticians reject it, most often because of decision
theoretic argument against randomization. I trace it back to Abraham
Wald's Theory of Inductive Behavior and argue that Bayesians should
concur with Ronald Fisher's criticism of Wald's analysis of randomization. The paper ends with a Bayesian argument in favor of randomization: randomization can lead to an increase in expected utility.

   15. *How to Commit the Gambler's Fallacy and Get Away With It*.
Davis Baird and Richard E. Otte. In a recent article Ian Hacking argues
that there can be cases where no probabilities may correctly be ascribed
to individual members of a population, while probabilities are correctly
ascribable to the population as a whole. In this paper a simple artificial coin-flipping model for such probabilities, not 'grounded from
below' is constructed. The inferences licensed by this model and a
consequence of the model for the theory of statistical tests is explored.

   16. *Quantity and Quality: Some Aspects of Measurement*. Arnold
Koslow. A description is given of the quantitative-qualitative distinction for terms in theories of measurable attributes, and, adjoined to
that account, a suggestion is made concerning the sense in which empirical relational systems have an empirical attribute as their topic or
focus. Since this characterization of quantitative terms, relative to
a partition, makes no explicit reference to numbers, concatenation
operations, or ordering relations, we show how our results are related
to some standard theorems in the literature. Analogs of representation
and uniqueness theorems are proved, and the notions of exact quantitative
term and the underlying attribute of a quantitative term, are described
and studied.

   17. *Approximate Generalizations and Their Idealization*. Ernest
W. Adams. Aspects of a formal theory of approximate generalizations,
according to which they have degrees of truth measurable by the propor-
tions of their instances for which they are true, are discussed. The
idealizability of laws in theories of fundamental measurement is considered: given that the laws of these theories are only approximately
true "in the real world", does it follow that slight changes in the
extensions of their predicates would make them exactly true?

   18. *Truthlikeness for Quantitative Statements*. Ilkka Niiniluoto.
The most elaborate recent accounts of truthlikeness (verisimilitude)
apply this notion primarily to generalizations in first-order languages
with qualitative predicates. This paper outlines a new approach to the
definition of truthlikeness for quantitative statements, including
singular statements (point estimation), interval statements (interval
estimation), and quantitative laws. In the case of laws, the basic

issue is reduced to the topological problem of measuring the distance
between two real-valued functions. The solution of this problem makes
it possible to define also the notion of approximate truth for quantitative laws.

19. *Economics, Risk-Cost-Benefit Analysis, and the Linearity
Assumption*. K. S. Shrader-Frechette. An offshoot of decision analysis,
risk-cost-benefit analysis (RCBA) dominates US policymaking regarding
science and technology. In this paper a central normative presupposition of RCBA, called "the linearity assumption" is argued against. This
is that there is a linear relationship between the actual probability
of fatality and the value of avoiding a social risk or the cost of a
social risk. The main object of this essay is to show that the presuppositions underlying the linearity assumption are highly questionable.
It is maintained that assessors ought to give more consideration to
broadening their interpretations of "unit cost" and "societal risk" and
to abandoning their claims about linearity.

20. *Do Virtual Particles Exist?*. Robert Weingard. In this
paper a few facts about Feynman diagrams and the perturbation expansion
of the S-matrix are reviewed and discussed in connection with the question of the ontological status of virtual particles.

21. *The Logic of Experimental Questions*. R. I. G. Hughes. The
pair (A, Δ), where A is a physical quantity (an observable) and Δ a
subset of the reals, may be called an 'experimental question'. The set
Q of experimental questions is, in classical mechanics, a Boolean algebra, and in quantum mechanics an orthomodular lattice (and also a
transitive partial Boolean algebra). The question is raised: can we
specify a priori what algebraic structure Q must have in any theory
whatsoever? Several proposals suggesting that Q must be a lattice are
discussed, and rejected in favor of the weak claim that Q must be a
Boolean atlas.

22. *The Status and Meaning of the Laws of Inertia*. Robert Alan
Coleman and Herbert Korte. The Law of Inertia plays a key role in the
scheme of constructive axioms for the General Theory of Relativity. A
new formulation of this law which avoids the circularity problems in-
herent in previous formulations is presented. The empirical status of
this law and the manner in which it provides a non-conventional foundation for the Law of Motion and the definition of physical forces is
established. First, quite general path structures are discussed which
are not defined at the outset in terms of geodesic paths and which
require for their description only the local differential topological
structure. Secondly, a number of theorems are presented which serve
as purely local differential topological criteria for singling out from
among the general path structures the geodesic ones that represent the
inertial structure of spacetime.

23. *Tachyon Signals, Causal Paradoxes, and the Relativity of
Simultaneity*. Steven F. Savitt. Some elementary properties of tachyons
are described and then it is argued that the claim that (T) Tachyons
exist, is incompatible with the truth of the Special Theory of Relativity
(STR). First it is argued that from T, STR, and the negation of the
principle that (P1) Effect never precedes cause, one can derive a paradoxical conclusion, one of the so-called "causal paradoxes". An
obvious response is to affirm (P1), but then it is argued that (P1) and
(T) entail that STR is false.

24. *Temporality, Secondary Qualities, and the Location of Sensations*. Paul Fitzgerald. Several philosophers have argued that "temporal
becoming" is mind-dependent, a claim they see as analogous to the traditional one about the mind-dependence of secondary qualities. They have
tended to assume that the classical secondary qualities are mind-dependent, and also that the close analogue for time of directly experienced
secondary qualities is an irreducibly indexical *nowness*. In an earlier
article it was argued that we should reject the second assumption. Here
it is shown why there is indeed a genuine problem of the ontological
status of directly experienced temporality and spatiality, a problem
analogous to the traditional one about secondary qualities.

25. *The Apparent Inconsistency of Moulines' Treatment of Equilibrium Thermodynamics*. John H. Harris. Moulines in his "A Logical
Reconstruction of Simple Equilibrium Thermodynamics" shows that Sneedian
constraints play an essential role even in the purely theoretical
development of the mathematical formalism of at least one actual scientific theory. However, Moulines' treatment is apparently inconsistent
because of the way he represents constraints. A very simple non-Sneedian
way of representing constraints is given which removes the difficulty.

26. *The Levels of Selection*. Robert Brandon. In this paper
Wimsatt's analysis of units of selection is taken as defining the units
of selection question. A definition of <u>levels</u> <u>of</u> <u>selection</u> is offered
and it is shown that the levels of selection question is quite different
from the units of selection question. Some of the relations between
units and levels are briefly explored. It is argued that the levels of
selection question is <u>the</u> question relevant to explanatory concerns, and
it is suggested that it is <u>the</u> question relevant to ontological concerns.

27. *Grades of Organization and the Units of Selection Controversy*.
Robert C. Richardson. Much recent work in sociobiology can be understood as designed to demonstrate the sufficiency of selection operating
at lower levels of organization by the development of models at the level
of the gene or the individual. Higher level units are accordingly viewed
as artifacts of selection operating at lower levels. The adequacy of
this latter form of argument is dependent upon issues of the complexity
of the systems under consideration. A taxonomy is proposed elaborating
a series of types, or grades, of hierarchically organized systems. These
range from <u>aggregative</u> systems, in which there is no organization relevant to systemic properties, through several graded variations reflecting
various degrees of functional interdependence of components, to <u>integrated</u>

systems, which manifest component specialization and diversification as well as a subordination of component function to systemic function. It is suggested that the most complex form of organization is plausibly treated as indicative of higher level units of selection.

28. *The Insights and Oversights of Molecular Genetics: The Place of the Evolutionary Perspective*. **John Beatty.** A general case about the insights and oversights of molecular genetics is argued for by considering two specific cases: the first concerns the bearing of molecular genetics on Mendelian genetics, and the second concerns the bearing of molecular genetics on the replicability of the genetic material. As in the first case, it is argued that Mendel's law of segregation cannot be explained wholly in terms of molecular genetics--the law demands evolutionary scrutiny as well. In the second case, it is argued that an account of the replicability of the genetic material in terms of molecular genetics is not entirely independent of evolutionary considerations, in the sense that it raises further evolutionary questions. The limitations of the molecular-genetic approach in these cases point to the limitations of that approach in general.

29. *Can Darwinian Inheritance Be Extended from Biology to Epistemology?* **Carla E. Kary.** An attempt is made to answer this question by first analyzing the structure of Darwinian inheritance as it is exemplified in the current biological theory of evolution. Based on this analysis a generalized framework for Darwinian inheritance containing conditions which must be met by all proper Darwinian evolutionary theories is developed. Subsequently this framework is employed to assess, in part, the adequacy of Toulmin's evolutionary epistemology. Although Toulmin asserts that Darwinian inheritance operates in conceptual development, the framework is used to show that the structure of concepts which he presents is unable to support the requirements for this kind of inheritance. It is concluded that Toulmin's evolutionary theory is not truly Darwinian.

30. *Recovering Philosophy from Rorty.* **Steve Fuller.** This paper considers Richard Rorty's thesis that philosophy has yielded all its subject matter to the sciences so as to no longer qualify as an autonomous discipline. We do not question his controversial historical diagnosis, but instead argue that all it shows is that the practice of philosophy does not depend on any particular subject-matter. The "philosophical turn" is taken whenever a problem is posed or an explanation is needed, for in either case one needs to go beyond the given phenomena in order to account for excluded possibilities, the choice of which makes the explainer's presence integral, in a manner that is often obscured by naive prose canons. Furthermore, the emphasis on subject matter has led philosophers to misconceive the source of difficulty in raising metaphysical questions, which is largely a matter of setting up the right narrative perspective, a point often made by literary critics.

31. *Was Carnap a Complete Verificationist in the Aufbau?* Richard
Creath. It is argued that Carnap was not a complete verificationist in
the Aufbau despite the widespread view that he was. That doctrine would
be intrinsic to constructionalism only if either of two additional
assumptions are made, and there is no reason to believe that Carnap
made these assumptions. Further, in the Aufbau Carnap did not demand
verifiability independently of constructionalism, and his clear rejection of verifiability in Pseudoproblems counts heavily against his ever
having accepted it in the Aufbau.

32. *The Creation of Similarity: A Discussion of Metaphor in
Light of Tversky's Theory of Similarity.* Eva Feder Kittay. The cognitive gain in the use of metaphor and simile is nicely elucidated by
Tversky's theory of similarity. The features of the theory which are
of special importance are the directionality and context-dependency of
similarity judgments. These indicate the extent to which such judgments
are classificatory and that similarity is not only the cause of an
object's classification but is also a derivative of groupings. Metaphor
and simile exploit certain cognitive features involved in the relation
between classification, context and similarity judgments so as to make
possible the creation of similarity, which, from a conceptual standpoint,
is the prime motivation for metaphor.

33. *Science and Play.* Michael Goldman. Gonzalo Munévar has
recently suggested that a criterion for scientific success and scientific progress can be found in the ability of a culture to "get along
better" with the help of that science, and that as a consequence there
is much to be said in favor of a proliferationist approach to scientific
methodology. I argue that there are severe constraints upon the
possibility and desirability of proliferation even under these conditions. I offer some tentative suggestions for defining areas to which
the limited resources available for scientific research may be most
constructively put.

PART I

DISCOVERY, RATIONALITY AND HISTORY OF SCIENCE

<u>The</u> <u>Rationality</u> <u>of</u> <u>the</u> <u>Copernican</u> <u>Revolution</u>

Martin V. Curd

Purdue University

Given the central importance of the Copernican theory to the birth of modern science, it is somewhat surprising to learn that there were remarkably few committed Copernicans prior to 1600. In a recent study, Westman (1980) finds only ten, of whom fewer than five were major scientific figures. The vast majority of scientists in this period continued to accept the Earth-centered astronomy of Ptolemy or later switched to the geoheliocentric system of Tycho Brahe. This widespread reluctance to adopt the heliocentric theory suggests the following questions about the Copernican Revolution: When and why did it become rational to reject the Ptolemaic theory as false? When and why did it become rational to accept the Copernican theory as true? Or, to put these questions in the more familiar language of the epistemologist: When and why did the beliefs that the Ptolemaic theory is false and that the Copernican theory is true become <u>justified</u> beliefs?

The traditional answers to these questions are now widely recognized as being unsatisfactory. The Copernican theory, contrary to popular myth, was not significantly more accurate in its predictions than the Ptolemaic, neither was it in any straightforward sense simpler than its rival (Palter, 1970; Babb, 1977). Indeed, at least one prominent historian and philosopher of science, Thomas Kuhn, has reached the conclusion that the Copernican Revolution turned upon irrational and subjective factors that reduce to "matters of taste." (Kuhn 1957, p. 171).

This extreme and unpalatable conclusion has not gone unchallenged. Lakatos and Zahar (1975) have argued that, from its inception, the Copernican research programme was both heuristically and empirically progressive with respect to its Ptolemaic rival. They base the second half of this conclusion on the claim that the essential geometric structure of the heliocentric theory (as presented, for example in the postulates of the <u>Commentariolus</u>, which for them constitutes the Aristarchan hard-core of the Copernican programme), had <u>excess</u>

PSA <u>1982</u>, Volume 1, pp. 3-13

4

<u>predictive</u> <u>power</u> over the geocentric theory, some of which <u>immediately</u> corroborated the new theory. This claim rests in turn, however, on the adoption of Zahar's revised conception of what constitutes a 'novel fact', according to which it is no longer necessary that a novel fact be previously unknown as long as it was not used in the construction of the theory.

In a similar vein, though on sounder grounds, Glymour (1980a; 1980b) argues that the Copernican theory was objectively superior to the Ptolemaic with respect to its ability to explain and be tested by the then-known facts of positional astronomy. Related claims have been made by Hall (1970), Millman (1976) and Heidelberger (1976).

These claims are important because they attempt to give a clear sense to the judgment that, prior to Galileo's telescopic discoveries, Newton's gravitational theory and the detection of stellar aberration and stellar parallax, the heliocentric theory, though not quantitatively more accurate, was nonetheless objectively better than Ptolemy's. It should be noted however, that neither Lakatos and Zahar nor Glymour wish to draw any direct conclusions from this judgment about the rationality of either accepting the Copernican theory as true or of pursuing the Copernican programme in the 16th and 17th centuries. Glymour is explicit about this. He uses the comparison between the heliocentric and the geocentric theories merely as an illustration of the analysis of theory testing and confirmation presented in his book. Nevertheless, both sets of authors claim to have shed significant light on the objective factors which lie behind the pro-Copernican admiration for the systematic unity, harmony and coherence of the heliocentric theory expressed in the writings of men like Rheticus and Kepler. These considerations are manifestly not subjective, irrational, or merely aesthetic as Kuhn and others have sometimes insisted. Due to limitations of space, I shall briefly explain only the most important of these objective considerations and then comment on their significance for assessing the rationality of the Copernican Revolution.

We first note that Equation (1) is derivable from the essential geometric structure of the Copernican system:

$$(1) \qquad 1/T_p = 1/T_e \pm 1/S_p$$

where T_p is the heliocentric period of the planet P, T_e is the heliocentric period of the Earth (approximately 365.25 days-- the mean solar year) and S_p is the interval of time between successive episodes of the retrograde motion of P as viewed from the Earth. The plus sign holds when P is an inferior planet, the minus sign when P is a superior planet. Since retrograde motion occurs only at inferior conjunction for the inferior planets and at opposition for the superior planets, S_p is the period of, or the time taken to complete, one cycle of anomaly, which is approximately constant for each planet. In the Ptolemaic system, where epicylic rotation is reckoned with respect to the line joining the center of the epicycle to the center of the deferent, S_p is the period of the major epicycle for each planet. The important fact about S_p is that it is an observable quantity. It can be calculated from observations without presupposing the truth of any astronomical theory. Since T_e is also known from observation, Equation (1) can be used in the Copernican theory to calculate the heliocentric periods of the inferior planets Mercury and Venus, yielding values of 88 days and 225 days respectively. When we turn to the superior planets, Mars, Jupiter and Saturn, all three of the quantities T_p, T_e and S_p are already known from independent observations. In the Ptolemaic system T_p represents the period of the deferent for each superior planet. Thus the Copernican theory predicts that Equation (1) holds in the case of the three superior planets and is successfully tested in these three cases. This is an important part of the excess predictive power of the heliocentric theory over the geocentric theory. While Ptolemy's theory is consistent with Equation (1) for the superior planets, that theory does not predict or explain it. The relationship is represented in the Ptolemaic theory by the fact that once the deferential and epicyclic periods are chosen as T_p and S_p respectively, the radius of the major epicycle always remains parallel to the line joining the Earth to the Sun. Thus the solar component in the apparent motion of the superior planets remains an unexplained fact in the Ptolemaic theory but follows as a natural geometric consequence of the heliocentric theory. Similarly, the Copernican theory, unlike the Ptolemaic, provides a simple geometric explanation for the restricted elongation of the inferior planets and for the decrease in the values of S_p as the planets are more distant from the Earth in either direction.

Next we turn to the determination of planetary distances and the order of the planets. In the Ptolemaic theory, planetary distances from the Earth can be determined neither as absolute magnitudes nor as ratios of a common unit from observation alone, if one remains strictly within the limits of that theory. The most that one can calculate from observation is the ratio r_p/R_p, where r_p is the radius of the major epicycle and R_p is the radius of the planetary deferent (which is different for each planet but unknown). Thus the order of the planets cannot be based on the values of R_p. Such order assignments were made, however, on the basis of an additional postulate, P: The period of a planet increases with the size of its orbit, that is, with its distance from the fixed center of revolution. Such a postulate was annunciated and defended by Aristotle in <u>De Caelo</u>, Book II, Chapter X. The fixed center in the Ptolemaic system is, of course, the Earth. Thus for the superior planets, the order Mars, Jupiter, Saturn follows from the known values of the deferential periods with the help of postulate P. But for the inferior planets, Mercury and Venus, the assignment of order remains arbitrary since their deferential periods are the same as that of the Sun. Once an order had been decided upon, it was also possible to estimate the relative sizes of the planetary orbits on the basis of a further postulate which is sometimes referred to as 'the plenitude principle': The greatest lower bound on the distance of one planet from the Earth is the same as the least upper bound on the distance between the Earth and the next interior planet. (Kuhn 1957, pp. 80-81, 174). But, again, this requires going beyond the resources of the Ptolemaic theory itself.

In the Copernican theory, things stand much better. In the heliocentric system all planetary distances from the center of revolution (which is now taken to be the Sun) are determinable from observation as ratios of one basic unit, the Earth-Sun distance. (Kuhn 1957, pp. 174-5; see Grafton (1973) for a more sophisticated discussion). Order assignments then follow directly from these distance determinations and thus test and confirm postulate P. Of course this is not an empirical confirmation of the heliocentric theory since P is not known to be true, but it clearly demonstrates the unity and systematic integrity of the Copernican theory, a feature noticeably lacking in the Ptolemaic and which played a major role in the arguments of Copernicus, Rheticus and Kepler.[3]

Given that the Copernican theory is superior to the Ptolemaic theory in the sense just explained (being better tested by the same positional data and possessing greater systematic unity and explanatory power), what is the relevance of this fact for our understanding of the rationality of the Copernican Revolution? Let us define the Copernican Revolution as the transition from the belief that the essential structure of the Ptolemaic system is true to the belief that the essential structure of the Copernican system is true. Defined in this way, the rationality of the Copernican Revolution depends on the rationality of the following two decisions: (A) The rejection of the Ptolemaic theory as false; (B) The acceptance of the Copernican theory as true. By itself, the superiority of the Copernican theory over the Ptolemaic with respect to the explanation and systematic integration of the positional data fails to determine the rationality of either decision. This is because of two additional factors that are epistemologically significant in the historical context: the availability of a third alternative, the Tychonic theory, and the presence of important problems confronting the Copernican theory.

The Tychonic theory, published in outline by Brahe in 1588, proposes a geoheliocentric arrangement in which the Earth is central and stationary.[4] The superior and inferior planets all revolve about the Sun which in turn revolves about the Earth. Thus, in the Tychonic system all the planets have the Sun's apparent orbit as their deferent and their Copernican, heliocentric orbits as their major epicycle. If one ignores the fixed stars, the Tychonic system (unlike the Ptolemaic) is geometrically and kinematically equivalent to the Copernican. Thus despite the inelegance of having two main centers of revolution (the Earth and the Sun), the Tychonic system enjoys the same methodological superiority with respect to the positional data as does the Copernican system. In particular, it allows the derivation of Equation (1) and the determination of planetary distances from the Sun, and provides a simple explanation of the restricted elongation of the inferior planets. Thus any rational decision between the two theories must be made on other, that is, on physical and dynamical grounds.

The two most important problems for the Copernican theory were (1) the apparent dynamical evidence that the Earth is stationary, and (2) the absence of detectable stellar parallax. Neither of these two difficulties afflicted the Ptolemaic and Tychonic systems since they are both geostatic. The pro-Copernicans were forced to adopt the defensive strategy of arguing that neither (1) nor (2) are actually inconsistent with the two principal terrestrial motions required by their theory. When this strategy became successful is a difficult question to answer, but in regard to problem (1) I do not think we can place it any earlier than Galileo's mature writings on dynamics; and by this time (the 1630's) I think there were good grounds for rejecting the Ptolemaic theory as false. Even when this strategy did succeed, it still left the issue between the Copernican and Tychonic theories unresolved. In the light of these considerations I propose the following answers to the question of the rationality of the Copernican Revolution. Decision (A), to reject the Ptolemaic theory as false, first clearly became rational early in the 17th century as a result of Galileo's observations of the precise configurations of the phases of Venus. These observations (which also cleared up a long-standing problem for all three theories concerning the relative constancy in the observed brightness of Venus) are consistent with both the Copernican and Tychonic theories (in which Venus and Mercury both go round the Sun), but they cannot be reconciled with the essential geometric structure of the Ptolemaic system (in which Venus and Mercury do not go round the Sun). Decision (B), to accept the Copernican theory as true, only became rational when Newtonian gravitational theory provided reasonable physical grounds for rejecting the Tychonic theory as false towards the end of the 17th century.

The choice between the Copernican and the Tychonic theories is complicated somewhat by the appearance of semi-Tychonic systems in the late 16th and early 17th centuries. These systems, proposed by Reymers, Gilbert and Longomontanus among others, were identical with the Tychonic arrangement except that they attributed a diurnal rotation to the Earth while at the same time denying the Earth an annual revolution about the Sun.[5] The attraction of such semi-Tychonic systems were two-fold. First, they enjoyed all the advantages which the Tychonic system shared with the Copernican. Second, they avoided the conspicuously inelegant postulation of a separate diurnal motion to each celestial body, and in particular (ignoring precession) allowed the fixed stars to be stationary, as in the Copernican system. This was an important factor once it became reasonable to accept the principle of rectilinear inertia as a fundamental law of motion. The principle of rectilinear inertia (Newton's First Law of Motion) requires centripetal forces to sustain any circular motion of discrete bodies about an axis of revolution. As Newton explained in The System of the World, it is physically inconceivable that the centripetal forces required for diurnal stellar motion could be generated by the North-South celestial axis which is an imaginary line in space. (Newton 1687, pp. 553-4). Thus, given the principle of inertia, the diurnal rotation of the Earth becomes physically necessary. But until the overwhelming success of Newton's gravitational theory, it was not similarly plausible to maintain that the annual motion of the Earth about the Sun is also physically necessary.

What conclusions follow from this brief analysis of the Copernican Revolution? First, if we regard a theory as a set of statements from which other statements are logically deducible as in the hypothetico-deductive model of scientific theories (rather than regarding theories as fundamentally non-linguistic entities such as Kuhnian paradigms or as 'inference tickets') and if we regard the belief that a theory is true and the belief that it is false as appropriate kinds of epistemic attitude to hold towards it, then justification for these beliefs is rather difficult to come by. In the Copernican episode, the issue crucially turned not merely on observation but also on the rational acceptance of a new fundamental theory of dynamics. Observation alone was impotent to decide between the Copernican theory and its Tychonic alternatives. One begins to have some sympathy for those Jesuit astronomers, many of whom were proponents of the Tychonic theory, who challenged Galileo's claim to be able to conclusively demonstrate the truth of the Copernican theory.[6] (This is in no way intended as an endorsement of the way in which the Church dealt with Galileo.)

Second, we have seen that excellent objective grounds could be given and were given for taking the Copernican theory seriously and for trying to develop a theory of dynamics which would resolve some of its pressing problems. The fact that participants in the debate over the heliocentric theory also appealed to other factors such as compatibility with Holy Writ (Brahe), the perfection of the number six (the number of planets in the Copernican system) (Rheticus) and the metaphysical and theological appropriateness of placing the Sun at the center of the universe (Kepler), does not detract from this point. Of course the Copernicans were rationally justified in pursuing their new theory but then, if I am right, so too were the Tychonists.[7] Perhaps it is psychologically inevitable that most scientists who propose new theories and devote their time and energies to developing and defending them, experience a strong temptation to believe that they are true. Perhaps it is also sociologically more effective to have new theories championed by ardent believers than by disinterested spectators. But whether or not these beliefs are justified ones is a question that only an epistemological analysis of science can answer.

Notes

[1]Westman lists Digges, Hariot, Bruno, Galileo, Diego de Zuniga, Stevin, Rheticus, Maestlin, Rothmann and Kepler. As Westman notes, many astronomers in this period (1543-1600) conservatively adopted only those parts of the Copernican theory that were independent of the physical claim that the Earth is in motion, but were strongly reluctant to endorse the theory as a whole. Even after 1600 there were still prominent thinkers who argued against accepting the Copernican theory as true, notably Mersenne and Riccioli. See Stimson (1917) and Hine (1973).

[2]Aristotle's reasoning is based on what I call 'the friction argument'. Since the planetary spheres move in the opposite direction to the swifter diurnal motion of the sphere of the fixed stars, the planets move more slowly, the closer they are to the stellar sphere: "the nearest one is most strongly counteracted by the primary motion, and the farthest least, owing to its distance." (Aristotle 1939, p. 199, 291^{b}7-10.) Postulate $\underline{P}$ is not directly applicable to the Tychonic system in which there are two major centers of revolution, the Earth and the Sun. In the Copernican system, with the stellar sphere stationary, $\underline{P}$ is an intuitively plausible ordering principle but without an independent means of determining planetary distances and velocities, it remains untestable. As Kepler realised, the reason why the more distant planets have longer heliocentric periods is not simply because their orbits are larger but also because they move more slowly. Kepler proposed an (erroneous) physical explanation of this decrease of velocity with distance from the Sun. $\underline{P}$ follows as a deductive consequence of Newton's gravitational theory and applies not only to the solar system but also to the gravitational system of Jupiter's moons.

[3]Chalmers (1981) has recently criticized those who claim that the Copernican theory is superior to the Ptolemaic because in the former, but not in the latter, one can calculate the ratios of the average sizes of planetary orbits. Chalmers objects that the difference between the two theories in this respect is due solely to the position of the observer in the planetary system and thus should not be accorded any fundamental significance. For a critical reply to Chalmers on this point see Curd (1982).

[4]For details of the Tychonic system see Boas and Hall (1959), Moesgaard (1972) and Gingerich (1973). As Schofield (1965) notes, Brahe was not the first to suggest a geoheliocentric arrangement in the 16th century since such systems had already been suggested by Reinhold, Lonicerus, Rothmann and Viete. What distinguished Brahe was his conviction that the Tychonic system was true.

[5]I am indebted to Professor Westman for these examples. Gilbert in fact also assigned a precessional motion to the Earth's axis as well as a diurnal rotation and attributed both to magnetic causes. (Gilbert 1600, pp. 347-352).

10

^{6}Galileo (1632) presents two main physical or dynamical arguments for the Earth's motion around the Sun, the argument from sunspots (pp. 347-355) and the argument from the tides (pp. 416-465). As Galileo himself concedes, the argument from the <u>annual</u> variations in the paths of the sunspots is inconclusive since the same trajectories of solar spots can be produced in a geostatic model, as long as the Sun's axis of monthly rotation is inclined at a <u>constant</u> angle to the axis of the ecliptic plane. But as Drake (1970, pp. 177-199) convincingly argues, Galileo is also concerned with the <u>absence</u> of any <u>daily</u> variation in the positions of the spots. In a geostatic model this requires that the Sun always turn nearly the same face toward the Earth. The dynamical implausibility of the resulting complex of solar motions makes a powerful case for attributing the diurnal motion to the Earth and not to the Sun, but it leaves the question of the Sun's annual motion unresolved. Galileo's tidal argument is more difficult to evaluate because of the complexity of the phenomenon, the implausibility of Galileo's assumptions and the weakness of alternative theories. The classic criticism is Mach (1933), pp. 262-264. For more recent analyses which attempt to make a more favorable case for Galileo's theory, see Burstyn (1962), Shea (1970), Machamer (1973) and Brown (1976). The first really adequate explanation of the large-scale behavior of the tides was provided by Newton's gravitational theory.

^{7}Laudan (1977) distinguishes between and offers (different) criteria for acceptance and pursuit. In his reply to Westman, Laudan (1981) agrees that on his model it might have been rational to pursue but certainly was not rational to accept the Copernican theory in the 16th century. Laudan, however, means by "accepting a theory", <u>not</u> "believing that it is true" (as assumed in this paper) but "treating it as if it were true." (Laudan 1977, p. 108). Moreover, the judgment expressed in this paper that it was rational to pursue both the Copernican and the Tychonic theories in the 16th century is not based on Laudan's dictum that "it is always rational to pursue any research tradition which has a higher rate of progress than its rivals" (Laudan 1977, p. 111), though it is not obviously inconsistent with it. Rather, my judgment is based on a more accomodating notion of the rationality of pursuit, roughly the concept of promise that Laudan is trying to explicate, whether or not this is reflected in maximal rates of progress at a given time.

11

<u>References</u>

Aristotle. <u>De Caelo.</u> (As reprinted as <u>On the Heavens.</u> (trans.)
W.K.C. Guthrie. Cambridge: Harvard University Press, 1939.)

Babb, S.E., Jr. (1977). "Accuracy of Planetary Theories, Particularly
for Mars." <u>Isis</u> 68: 426-434.

Boas, M. and Hall, A.R. (1959). "Tycho Brahe's System of the World."
<u>Occasional Notes of the Royal Astronomical Society</u> 3: 252-263.

Brown, H.I. (1976). "Galileo, the Elements, and the Tides." <u>Studies
in the History and Philosophy of Science</u> 7: 337-351.

Burstyn, H.L. (1962). "Galileo's Attempt to Prove that the Earth
Moves." <u>Isis</u> 53: 161-185.

Chalmers, A. (1981). "Planetary Distances in Copernican Theory."
<u>British Journal for the Philosophy of Science</u> 32: 374-375.

Curd, M.V. (1982). "The Superiority of the Copernican Theory: A
Reply to Chalmers." Unpublished manuscript.

Drake, S. (1970). <u>Galileo Studies.</u> Ann Arbor: University of Michigan
Press.

Galilei, G. (1632). <u>Dialogo di massimi Sistemi del Mondo.</u> Florence:
G.B. Landini. (As reprinted as <u>Dialogue Concerning the Two Chief
World Systems.</u> (trans.) S. Drake. Berkeley: University of
California Press, 1953.)

Gilbert, W. (1600). <u>De Magnete.</u> London: Peter Short. (As reprinted
(trans.) P.F. Mottelay. New York: Dover Publications Inc.,
1958.)

Gingerich, O. (1973). "Copernicus and Tycho." <u>Scientific American</u>
229(6): 87-101.

Glymour, C. (1980a). <u>Theory and Evidence.</u> Princeton: Princeton
University Press.

----------. (1980b). "Explanations, Tests, Unity and Necessity." <u>Noûs</u>
14: 31-50.

Grafton, A. (1973). Michael Maestlin's Account of Copernican Planetary
Theory." <u>Proceedings of the American Philosophical Society</u> 117:
523-550.

Hall, R.J. (1970). "Kuhn and the Copernican Revolution." <u>British
Journal for the Philosophy of Science</u> 21: 196-197.

Heidelberger, M. (1976). "Some Intertheoretic Relations Between
Ptolemean and Copernican Astronomy." <u>Erkenntnis</u> 10: 323-336.

Hine, W.L. (1973). "Mersenne and Copernicanism." _Isis_ 64: 18-32.

Kuhn, T.S. (1957). _The Copernican Revolution._ Cambridge: Harvard University Press.

Lakatos, I. and Zahar, E. (1975). "Why Did Copernicus's Research Programme Supersede Ptolemy's?" In _The Copernican Achievement._ Edited by R. Westman. Berkeley: University of California Press. Pages 354-383. (This paper is reprinted with a hitherto unpublished postscript by Lakatos in Lakatos (1978). Pages 168-192.)

----------. (1978). _The Methodology of Scientific Research Programmes, Philosophical Papers._ Volume 1. (eds.) J. Worrall and G. Currie. Cambridge: Cambridge University Press.

Laudan, L. (1977). _Progress and Its Problems._ Berkeley: University of California Press.

----------. (1981). "The Philosophy of _Progress... ._" In _PSA 1978,_ Volume 2. Edited by P.D. Asquith and R.N. Giere. East Lansing, Michigan: Philosophy of Science Association. Pages 530-547.

Mach, E. (1933). _Die Mechanik in Ihrer Entwickelung historisch-kritisch dargestellt._ 9th ed. Leipzig: F.A. Brockhaus. (As reprinted as _The Science of Mechanics: A Critical and Historical Account of Its Development._ (trans.) T.J. McCormack. La Salle, Illinois: The Open Court Publishing Company, 1960.)

Machamer, P.K. (1973). "Feyerabend and Galileo: The Interpretation of Theories, and the Reinterpretation of Experience." _Studies in the History and Philosophy of Science_ 4: 1-46.

Millman, A.B. (1976). "The Plausibility of Research Programs." In _PSA 1976,_ Volume 1. Edited by F. Suppe and P.D. Asquith. East Lansing, Michigan: Philosophy of Science Association. Pages 140-148.

Moesgaard, K.P. (1972). "Copernican Influence on Tycho Brahe." In _The Reception of Copernicus' Heliocentric Theory._ Edited by J. Dobrzycki. Boston: Reidel. Pages 31-55.

Newton, I. (1687). _Philosophiae Naturalis Principia Mathematica._ London: Royal Society. (As reprinted as _Sir Isaac Newton's Mathematical Principles of Natural Philosophy and his System of the World._ 2 vols. (trans.) A. Motte, revised by F. Cajori. Berkeley: University of California Press, 1973.)

Palter, R. (1970). "An Approach to the History of Early Astronomy." _Studies in History and Philosophy of Science_ 1: 93-133.

Schofield, C. (1965). "The Geoheliocentric Mathematical Hypothesis in Sixteenth Century Planetary Theory." _British Journal for the History of Science_ 2: 291-296.

Shea, W. (1970). "Galileo's Claim to Fame." _British Journal for the History of Science_ 5: 111-127.

Stimson, D. (1917). _The Gradual Acceptance of the Copernican Theory of the Universe._ Ph.D. Dissertation, Columbia University. (As reprinted Glouster, MA: Peter Smith, 1972.)

Westman, R.S. (1980). "The Astronomer's Role in the Sixteenth Century: A Preliminary Study." _History of Science_ 18: 105-147.

Explaining Scientific Discovery

Noretta Koertge

Indiana University

The history of science provides two _prima facie_ reasons for believing that there is a "logic" to scientific discovery. The first is the phenomenon of multiple or simultaneous discoveries. The second is the fact that internalist historians of science (who eschew non-cognitive explanations) routinely explain _how_ a discovery was made and _why_ scientist X took an important step which scientist Y did not!

However, some philosophers have argued that purported historical explanations of discovery are all sleight-of-hand tricks--what we really have is a confused mixture of rational reconstructions of the justification steps in the process combined with a description, not an explanation, of the creative leaps. Others would claim that historians do indeed help us to _understand_ scientific discoveries, but the mode of understanding involved is not that provided by the Hempelian covering law model. Rather we empathize with the creative leaps of past geniuses.

In this paper I look first at various negative arguments which suggest that historians cannot explain discovery. I then propose an explication of the concept of discovery and analyze what is required for an adequate historical explanation of a scientific discovery.

1. Negative Argument #1: Historians Don't Explain--They Merely Fragment the Inexplicable

Nickles (1979) states a version of this argument (he does not endorse it) while describing continuist accounts of the history of science: "These historians do not _explain_ major discoveries so much as explain them _away_. Once a major innovation is placed in historical perspective, it will be seen not to be so major: its author has only modified or extended the views of his predecessors, which they in turn had done to the views of _their_ predecessors. Thus a major theoretical innovation, seemingly intractable to rational explanation, is pulverized into a zillion

---

PSA 1982, Volume 1, pp. 14-28

innovations small enough to give some account of--as stimu-
lated by factual data, external factors, accidental recom-
binations of ideas, etc." (pp. 94-95).

Without debating the merits of the continuity view of
the history of ideas, it should be noted that breaking up
a large, complex, improbable process into a series of
smaller, more probable events is generally an important
step towards providing an explanation.

Consider, for example, the simple case of a ball bounc-
ing down a peg board:

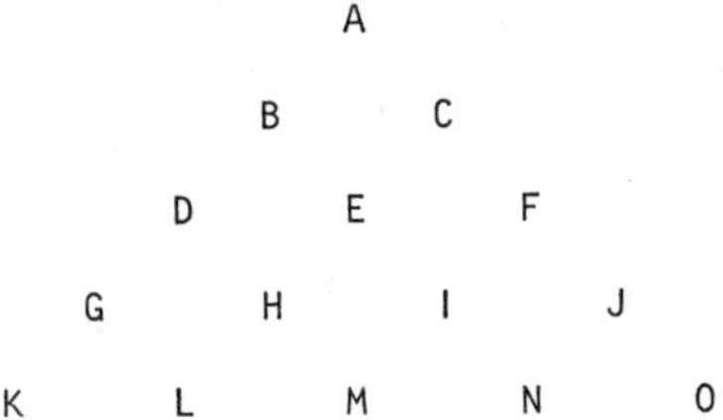

The odds that a ball dropped on A will come out between
K and L are very small indeed; but if we describe the
series of individual bounces taken by one ball, each step
in the process has a probability of 1/2, not enough for a
Hempelian I-S explanation to be sure, but much better than
the original probability. (And on Salmon's S-R account,
adding information about intermediate bounces increases the
homogeneity of the reference class.)

Partisans of the fragmentation argument might grant the
above formal point but nevertheless claim that to fragment
a discovery is to trivialize it. For example, if one wants
an explanation of why the English were the first to scale
the Matterhorn, it hardly seems satisfactory to be told
about how they ascended one little step at a time!

However, I would reply, if the route the English took was
significantly different from that of the Italian team, then
that fact in itself _would_ be interesting, regardless of
whether the path was chosen by chance or by design.

All I want to conclude at this point is that there's
nothing wrong _per se_ with fragmentation--the quality of the
proposed explanation will depend entirely on how good the
many mini-explanations are.

2. Negative Argument #2: Historians Just Describe the Creative Process, Not Explain It.

This argument which is hinted at in both Nickles (1979) and Salmon (1970a) goes as follows.

All the historian can do is to narrate the sequence of dreams, spontaneous hunches, aesthetically pleasing experiences, etc., which provide a story of the discovery. However, he can neither explain why that scientist had that dream, hunch, etc., nor why that stimulus led to the discovery in question.

According to this line of argument, Darwin's reading of Malthus or Kekulé's vision of snakes biting their own tails form a charming part of the stories of their respective discoveries but do not really serve any explanatory function. We still wonder <u>why</u> Kekulé had that particular vision, and why did he take it seriously? How exactly does one move from snakes to benzene rings?

There are two flaws with this argument. I will discuss them in turn. First, no explanations are ever ultimate, especially with respect to initial conditions. If a geologist explains the characteristics of a particular crater in terms of a meteor collision (plus laws of impact, etc.), one may, of course, go on to ask why a meteor came down at that spot. Often the answers to queries about initial conditions lie in other fields. (An astronomer might be able to cite patterns in meteor shower activity.) Often there may be no known explanation of the initial condition at all. But this circumstance does not vitiate the original explanation. It is a pragmatic question whether we find the initial condition more puzzling than the original explanandum.

Secondly, sometimes the historian does in fact provide at least a weak explanation of the stimulus event. (It is not surprising that Darwin read Malthus--it was a popular book at the time.) Sometimes the explanation, if there is one, lies in an entirely different field. (Perhaps there is a psycho-pharmacological explanation of Kekulé's vision.)

I conclude that the existence of unexplained (or even unexplainable) initial conditions in a narrative does not mean that no explanations are being proffered.

But, our critics might reply, does the recital of such felicitous interventions really explain the discovery? It is the nomic connection which is missing! This is a complicated issue which will be dealt with more fully below. A sketch of my answer goes as follows. I believe that there are patterns of analogical reasoning, problem-solving, etc., which can be given a formal characterization, which computers can be programmed to follow, etc. I also believe that humans have a propensity to think according to these patterns.

Thus in the Kekulé case, there is a formal analogy at work. Snakes, normally open curves, can change their topology and become a closed curve. Likewise for chains of atoms. See the discussion in Boden (1977). Furthermore, I conjecture that if two matched sets of innocent subjects were given Kekulé's problem and one group were shown movies of snakes biting their tails and another shown pictures of snakes which were merely wiggling around, the first group would exhibit a higher success rate in solving Kekulé's problem. Whether or not my conjecture is correct, I think it is the <u>belief</u> that it is correct which makes the Kekulé vision story <u>feel</u> like an explanation.

3. Negative Argument #3: "We Cannot Anticipate Today What We Shall Know Only Tomorrow."

The most intuitively appealing argument against the possibility of a strong theory of discovery is sketched by Popper in the <u>Poverty of Historianism</u>: "...we cannot anticipate today what we shall know only tomorrow... . No scientific predictor--whether a human scientist or a calculating machine--can possibly predict, by scientific methods, its own future results. Attempts to do so can attain this result only after the event, when it is too late for a prediction; they can attain their result only after the prediction has turned into a retrodiction." (1957, pp. vii-viii).

Finocchiaro extends the argument to undermine the possibility of providing an Hempelian explanation of discovery: "At the most general level, a scientific discovery is a historical event or development whereby new knowledge comes into being. If, to explain satisfactorily why this new knowledge came into being, we have to provide information showing that the new knowledge did in fact come into being, then clearly the explanatory information could not have been both available and taken into account before the discovery. Otherwise the new knowledge would have come into being before it actually did or it would have come into being differently from the way it did;... ." (1973, pp. 35-36).

Nickles (1979) correctly points out that Finocchiaro's argument assumes that all of the information in the explanans was acutally employed by someone prior to the time of the discovery. I would go even further. Since historians are, by definition, giving retrodictive accounts of discovery, why should they not avail themselves of information

which only became available after the explanandum events?
(Certainly geologists use laws and initial conditions discovered in the 20th century to explain events which happened before there were even any scientists around!) However, I think many philosophers and Popper himself on other
occasions (1972, p. 198) would deny the possibility of
giving even a retrodictive explanation of scientific discovery. We might formulate their argument as follows: In
at least some (and probably most) cases of major scientific
advance the new proposal N is either logically <u>inconsistent</u>
with the old system O or perhaps even conceptually <u>incom-
mensurable</u> with it (there are concepts in N which cannot be
defined within the language of O). No logic, whether deductive or inductive, can lead to conclusions which are in-
consistent/incommensurable with the premises. Therefore,
N cannot be explained, even after the fact, by O.

The above argument is clearly correct--Non-Euclidean
geometry cannot be derived from Euclid's axioms. But as my
example from mathematics quickly suggests, the argument is
also quite irrelevant. If we wish to explain the discovery
of non-Euclidean geometry, it is not Euclid's axioms <u>per se</u>
which we appeal to, but the problem situation within mathematics at the time. (People wondered if the 5th postulate
could be derived from the first four; they tried a <u>reductio</u>
proof but found that axioms #1-4 plus the negation of #5
seemed to form a consistent system; later Lobachevski...,
etc.)

Without pursuing the example further, several points
emerge immediately: the explanation of a <u>discovery</u> is not
at all like theoretical reduction or any other of the intertheoretic relations discussed by philosophers (such as correspondence relationsor comparative verisimilitude). In the
case of discovery, the explanans includes statements about
what scientists found problematic, what they were trying to
do, etc. Much of our understanding of the discovery comes
from an analysis of the general <u>form</u> of the problem situation, not its specific content. (Any mathematician trying
to investigate the independence of a postulate within a
system is likely to write down its negation and explore the
resulting new system.)

But this reference to <u>general</u> features of the problem
and its solution reveals one way in which we can understand
discovery--by subsuming particular cases under general
types of rational scientific activity. (Of course, a fuller
account would mention the content of the 5th postulate and
why people were nervous about talking about what went on at
infinity, etc.) Students of Bayesian confirmation theory
will recall that Salmon (1966) also refers to the general
<u>type</u> of a scientific hypothesis, not its content, in order

to assign values to prior probabilities.

Note that in the story of the discovery of non-Euclidean geometry sketched above there is no attempt to explain why negating the 5th postulate did <u>in fact</u> produce a consistent system. All that we have explained is why it was reasonable to try this ploy. Nature's actual response to our efforts must always be fed into the explanandum. Popper and Finocchiaro are quite right in saying the content of the new knowledge can neither be predicted in advance nor explained after the fact by the historian.

But let us turn from these negative arguments to the problem of presenting a positive account of the explanation of scientific discovery.

4. The Form of the Explanandum--What Is a Discovery that We Should Explain It?

We often speak informally as if the explanandum were the name of a more or less complex event, as when one discusses explanatory accounts of the Great Depression, the Scientific Revolution, or Halley's Comet. However, Hempel has made it quite clear that a proper explanandum consists of a proposition, not a name.

Likewise, for the contents of discoveries. Although we often casually talk about the discovery of radioactivity or the discovery of the Periodic Table, I propose that for explanation purposes we always describe the contents of discoveries with propositions--e.g., the discovery that uranium minerals emit radiation or the discovery that when the elements are arranged according to increasing atomic weight such and such a pattern emerges.

As soon as we employ sentences instead of nouns, it is easy to unravel old trick questions on history of science exams, such as "Who discovered oxygen?" One simply points out that Priestley discovered that heating mercury calx <u>produces</u> a gas which supports combustion, while Lavoisier discovered that the calx decomposes and <u>releases</u> such a gas.

Do the sentences describing the contents of discoveries have to be true? Here ordinary usage is not conclusive. It sounds very strange to say that Priestley discovered that dephlogisticated air supports combustion. Yet it would be equally odd to deny that Newton discovered Newton's Laws because Einstein later discovered that they are only approximately true!

20

Here again I will follow Hempel's example (as does Gutting 1980) and stipulate the strong requirement of _truth_. So I will say that no one _discovered_ the phlogiston theory; however, we can explain why people found it to be an interesting proposal. For the moment I will be similarly tough on the priority issue--only the temporally first uncovering of a proposition will count as its discovery. (Later, it will be useful to relativize the discovery to a scientific community and thereby allow ourselves to speak of independent discoveries.)

So far my proposed explication of the concept of discovery goes as follows.

"Jones discovered that P" means:
(1) P is _true_.

(2) Jones _believed_ P or at least took P seriously (the exact requirements on Jones' propositional attitude will be discussed later).

(3) Jones had fairly _good reasons_ for this attitude towards P (again, the detailed epistemic requirement will be discussed below).

(4) Jones was the _first_ to adopt the designated propositional attitude towards P.

Thus a complete explanatory account of Jones' discovery would explain all four conditions.

Before clarifying clauses (2) and (3), let me briefly point out some ramifications of conditions (1) and (4) for our problem of explaining discovery. Let us return to the oxygen example. What would a Hempelian explanation of the fact that it is true that mercury calx releases a gas which supports combustion be like? Clearly what would be required is a deep scientific theory about chemical reactions and chemical bonding. Several points emerge immediately.

(a) Historians, _qua_ historians, cannot explain the first element of discovery--that is the job of scientists. What historians typically do is just to assume the truth of P and spend their energy on Jones' activities.

(b) We now see one way in which modern science is relevant to an understanding of the history of science. If we've studied modern chemistry, there is a sense in which we have a more complete explanation of Lavoisier's discovery than he did.

(c) It is also clear that unless P is a trivial deductive consequence of a well established theory, there is no way that the truth of P could have been predicted before its discovery. Clause (4), the priority requirement, also poses severe problems for those who would try to predict discovery--one who has grounds for predicting P would in most cases thereby become the discoverer of P.

Clause (4) dictates that historians use comparative methods. If we are to explain the discovery of Snell's law, we must not only explain the success of a Harriott, Snell and Descartes, but also the failure of a Kepler, Wittelo and Ptolemy. Finocchiaro, following Agassi, has pointed out that frequently the most puzzling aspect of a discovery is how late it came or the difficulty which people had in arriving at it. For example, since Semmel-weis was using Mill's Method of Difference and since it was well known to him that the incidence of puerperal fever was higher in teaching hospitals, why didn't he concentrate immediately on the differences between medical students and midwives? In this case the main job of the historian is to describe how Semmelweis' perception of the problem was different from ours. (It turns out that the miasm theory of disease is the crucial element.)

Once we become immersed in a historical context, how-ever, we often become so impressed with an older conceptual scheme that we then become surprised that anyone should ever have abandoned it! As Galileo pointed out, at first it seems amazing that Copernicus, in the absence of infor-mation about the phases of Venus or the rudiments of a new physics, was able to question the Ptolemaic system! Here again the job of the historian is to fill in more details about Copernicus' problem situation so that we understand why he was uniquely qualified or situated to propose the new theory.

Let us now turn to clause (2) which describes the propo-sitional attitude of the discoverer. How positive must Jones' conviction be? Clearly, Kepler is not the discover-er of the sine law of refraction, although he toyed with its mathematical equivalent, for the simple reason that he re-jected it. (He was misled by Wittelo's erroneous data.)

Here I think our requirement shifts according to the cognitive context. Many important scientific theories begin life in a cognitive environment in which they are con-sidered to be logically impossible, but eventually end up being thought of as conceptual truths! Consider, for ex-ample, the proposition that two bodies can attract each other in the absence of any intervening medium. Descartes argued that this was logically absurd--if nothing lay be-tween two bodies, then they were contiguous! A little over a century later, Kant was convinced that this same proposition was _a priori_ true.

In a climate in which the motion of the earth is thought to be at least physically impossible, Copernicus' decision to treat the idea seriously and to calculate its consequences is an enormous achievement. In such a case, even if the Osiander preface had been a true description of Copernicus' attitude, I think we might still consider him to have made a crucial discovery.

I will be similarly wishy-washy regarding clause (3). How good must the discoverer's reasons be? Clearly, P can't be just a wild guess, a sentence buried in the protocol of an opium dream or randomly composed while seated in front of a word processor. Neither, however, can we require a high degree of inductive support for P. Before Galileo's telescopic discoveries and the new mechanics there was no strong empirical support for the Copernican theory. (Bohr was in an even more precarious epistemological situation.)

What Copernicus did have was what one might call abductive reasons for favoring his theory. If it were true, it would give nice astronomical predictions (though no better than did Ptolemy's--that's one reason why it didn't have much inductive support). Also its basic design was conceptually simpler because it accounted for retrograde motion without any epicycles. However we formulate requirement (3), it must not be narrowly positivistic. (One could, of course, ascend to the meta-level and try to argue that experience shows that non-_ad hoc_ theories have a better record in the history of science.)

This concludes my rather informal explication of the concept of discovery or perhaps I should say family of concepts. As many recent writers on discovery have pointed out, there is a multi-dimensional array of epistemological stances towards the successive modifications of a new theory. My main claim so far is that when historians explain the success of the discoverer of P, they typically do _not_ try to explain why P is true. Rather they take that for granted and try to explain how and why the scientist in question arrived at that true belief. I now turn to a closer examination of clause (2) because here is where the most puzzling issues arise.

5. Explaining Propositional Attitudes

How do we explain Jones' attitude towards P? Here it is useful to distinguish two stages: First, Jones must actually formulate and comprehend P. (I can't have a cognitive attitude toward a proposition which I've never thought of, nor towards one I can't understand.) Secondly, Jones must form a positive opinion regarding the epistemic status of P.

On the face of it, the second step should be the easier to explain. Although philosophers have generally focused on the rational explanation of actions, we can extend these discussions to the problem of explaining beliefs (Koertge 1975). Thus according to Popper's Rationality Principle (1967), Jones' attitude towards P will always be the one which is epistemically appropriate to his knowledge situation. To reinforce the approach, Popper also puts forward a Principle of Transference which says "...what is true in logic is true in psychology." (1972, p. 6).

But this clearly won't do. As those who try to program computers to derive theorems very well know, human beings have a wonderful ability to focus on interesting deductive consequences of their beliefs and ignore boring ones, so not every logical move is spontaneously made by human minds. And as every logic teacher knows, even when the content of an argument is nonthreatening and even when students are paying attention, human minds make fallacious moves. For example, a very common mistake is to confuse necessary and sufficient conditions. (Even great philosophers, such as Bacon and Hume, did this.) I think there are two "reasons" why this error is common. In mathematical proofs, many steps are reversible. (The plausibility of the ancient method of analysis and synthesis rests on this fact.) And in much of empirical inquiry, it is very rare to find that all A's are B's. More common is to find that A is positively relevant to or positively correlated with B--and both of these relations are symmetric. (If atheists are more likely than other folks to be college graduates, then it's true that going to college means that one is more apt to become an atheist--other things being equal.)

The upshot of all of this is that considerations of logic, rationality, or World-3 problem situations are very rough and incomplete guides to a cognitive science. (Adding in man's evolutionary history improves matters somewhat.) So, contrary to Popper, I think it is the "psychologic" of the agent's situation, not its logic, which allows us to explain propositional attitudes.

Having abandoned philosophy for the esoteric domain of speculative cognitive psychology, let me boldly move on to a discussion of the first stage in attitude formation. How could we ever hope to explain the fact that Jones happened to think of P in the first place, especially in the case where no human had ever previously formulated P? It seems

obvious to me that we could rarely satisfy Hempel's requirements for explanation--it would almost always be impossible to provide a strong inductive argument leading to
the conclusion "and so, not surprisingly, P popped into
Jones' mind." However, Salmon's relevance model of explanation (1970b) looks much more promising.

Let us now compile a list of typical factors which may
be relevant and hence of explanatory value.

(a) The first, and by far the most important factor is
the scientist's problem situation, including constraints
on admissible solutions. If P is a solution candidate to
a problem which intrigues Jones, surely Jones is more likely to entertain P than someone who is asking an entirely
different sort of question.

This point may seem trivial or obvious given the work
of Popper (1972) and Agassi (1963) as well as more recent
analyses by Laudan (1977) and Nickles (1981), yet the example of non-Euclidean geometry reminds us how important it
is to start with the scientist's <u>problem</u> situation not the
knowledge situation which was always stressed in traditional inductivist accounts of discovery.

(b) The scientist's scientific metaphysics and other
background beliefs also play a role in determining which
ideas will present themselves forcibly to a would-be problem solver. Thus I think it is no accident that Semmelweis
first conjectured that cadaveric matter was a cause of
childbed fever immediately after he started reading up on
contagion theory in anticipation of taking a job in England.
(Of course, other factors were also relevant.)

(c) In some cases the move to P follows rather straightforward inductive principles. Consider the following problem situation (which is a slightly simplified version of
what really happened): A team of scientists has already
decided to pursue the idea that DNA has a helical structure.
They have just found out that whereas a one-strand model
has a density which is too low, a three-strand model has
too high a density. So next (surprise, surprise) they think
of trying... .

The reader can easily fill in the blank because s/he believes scientists think in accordance with some sort of
Principle of Interpolation (unless there is some special
reason to expect discontinuities or inflection points).

(d) Another common element in the explanation of discovery involves analogical inferences which may or may not
carry any epistemological warrant, but nevertheless seem to

be characteristic of the way people think.

Consider first a question taken from the Miller Analogies Test:

Taurus : Toros :: (a. ethnology, b. morphology, c. astrology, d. psychology) : Architecture

Now consider the following poetic image: "Like the ancient tree who dies at the top, our father's graying hair and fading brain... ."

I submit that in both examples, we can grasp the connection. In neither case, however, does the stated analogy provide any <u>grounds</u> for an inference. (At least not strong grounds. I suppose one might argue that there could be a causal similarity between the processes of aging in trees and humans. But certainly the analogy between architecture and astrology is completely devoid of epistemic significance.)

The point I wish to emphasize is the following: For justification purposes only a small sub-set of analogical arguments are worthwhile. Hesse (1966), following Mill and Jevons, has tried to characterize these. However, for the purposes of understanding discovery many more analogical inferences can be useful. (Compare the Kekulé case above.)

6. A New, Temporary Role For <u>Verstehen</u>.

Critics of Hempel's approach to explanation have often claimed that in the social sciences and history, a special empathetic mode of understanding is required. The standard analytic response is to say that although <u>verstehen</u> may be a useful heuristic device for constructing explanations, the methods of justifying and determining the cognitive status of the explanations thus obtained are the same as in the natural sciences.

However, I believe that the laws of thought which figure in explanations of scientific discovery are not only discovered through empathetic means, but also <u>at present</u> the major source of justification comes through their intuitive appeal. A possible analogy: Before the invention of the thermometer, the most reliable way of detecting temperatures (in the middle ranges) was to use a human being as instrument. Now that is not necessary.

<u>At present</u>, the best way to <u>test</u> (not just discover) the strength of an analogical inference or the plausibility of a step in the story of a discovery is to ask a human being to place himself in the scientist's problem situation. This

is <u>not</u> to say that historical explanations of scientific discoveries cannot be critically discussed. Neither is it to imply that we shouldn't try to articulate these tacit laws and test them more rigorously. It is important to realize, however, that the laws of thought required are not just psychological correlates of logic and methodology.

7. Conclusion

Historians cannot hope to explain the content of scientific advances (e.g., why cadaveric matter causes puerperal fever)--that is a job for scientists. What historians can do is explain why a certain scientist was the first to conceive of and take seriously a given idea (which later turned out to be successful).

Since most new ideas are constructed piecemeal (although in retrospect certain moves may appear to be by far the most important), the typical historical explanation will follow Hempel's genetic model (1965, p. 449). Each step in the sequence will describe the current problem faced by the agent or what Popper would call the "logic" of his/her situation (1972, pp. 170-75).

Since the initial conditions thus described deal with belief systems (the agent's perceptions of his/her problem situation), the laws which generate the explanandum state must be what used to be called "laws of thought", the domain now studied by cognitive scientists. The regularities which govern such phenomena will often involve probabilities less than one-half. Thus Salmon's statistical relevance approach to explanation will have to be used.

Sometimes the cognitive laws invoked will apply to a very narrow set of circumstances and hence will be of low content. ("Anyone faced with Kekulé's problem who then contemplates snakes... .") As our understanding of the formation of new ideas grows, the laws we utilize will deepen and increase in content. ("Anyone faced with a problem whose solution requires a change in topology... .")

As to the age-old question of whether there is a secret to success, this would require a different sort of inquiry, not just a psychological study of which ideas are likely to pop into people's heads and find favor, but also a follow-up on what sort of conjectures turn out in the long run to constitute an advance. In the old days such speculations were the delight of metaphysicians. Today, as the history of science becomes more sophisticated and more complete (including the stories of ideas which didn't work as well as those of successes), we might hope to use it as a base for a quasi-empirical study of the problem of heuristics.

References

Agassi, Joseph. (1963). _Towards an Historiography of Science._
    The Hague: Mouton and Co.

Boden, Margaret. (1977). _Artificial Intelligence and Natural Man._
    New York: Basic Books.

Finocchiaro, Maurice A. (1973). _History of Science as Explanation._
    Detroit: Wayne State University Press.

Gutting, Gary. (1980). "Science as Discovery." _Revue Internationale
    de Philosophie_ 34: 26-48.

Hempel, Carl G. (1965). _Aspects of Scientific Explanation._ New York:
    The Free Press.

Hesse, Mary B. (1966). _Models and Analogies in Science._ Notre Dame:
    University of Notre Dame Press.

Koertge, Noretta. (1975). "Popper's Metaphysical Research Program for
    the Human Sciences." _Inquiry_ 18: 437-462.

Laudan, Larry. (1977). _Progress and Its Problems._ Berkeley:
    University of California Press.

Nickles, Thomas. (1979). "Review of _History of Science as Explanation_
    by M.A. Finocchiaro." _Erkenntnis_ 14: 93-102.

----------------. (1981). "What is A Problem That We May Solve It?"
    Synthese 47: 85-118.

Popper, Karl R. (1957). _The Poverty of Historicism._ London:
    Routledge and Kegan Paul.

----------------. (1967). "La Rationalité et le Statut du Principe de
    Rationalité." In _Les Fondements Philosophiques des Systèmes
    Économiques._ Edited by E.M. Classen. Paris: Payot. Pages
    142-150.

----------------. (1972). _Objective Knowledge._ London: Oxford
    University Press.

Salmon, Wesley E. (1966). _The Foundations of Scientific Inference._
    Pittsburgh: University of Pittsburgh Press.

----------------. (1970a). "Bayes's Theorem and the History of
    Science." In _Historical and Philosophical Perspectives of
    Science. (Minnesota Studies in the Philosophy of Science._ Volume
    V.) Edited by Roger Stuewer. Minneapolis: University of
    Minnesota Press.

----------------. (1970b). <u>Statistical Explanation and Statistical Relevance.</u> Pittsburgh: University of Pittsburgh Press.

Curried Lakatos
or, How Not to Spice Up the Norm-Ladenness Thesis

Stephen J. Wykstra

University of Tulsa

Though Imre Lakatos's ideas have been of first importance to recent
History and Philosophy of Science, one of his central theses has been
ill-received by philosophers and historians alike. I refer to his
thesis that historiography of science is (and should be) so laden with
normative philosophical doctrines about scientific rationality that
"history of science without philosophy of science is blind." (1971,
p.91).   In defending this 'norm-ladenness thesis' (as I shall call
it), Lakatos employs a notion of "rational reconstruction" and
"internal history" that has been indicted by Thomas Kuhn as "not
history at all but philosophy fabricating examples." (1971, p. 143).
Kuhn's charge, echoed in varying degrees by Holton (1974, p. 68),
Laudan (1977, p. 170), McMullin (1970, p.33), Suppe (1977, p. 669),
and others, has such wide currency that I shall call it 'the Received
Critique' of Lakatos's norm-ladenness thesis.

A rather different critique of this thesis has recently been given
by Gregory Currie. One of my aims here will be to criticize Currie's
arguments, but motivating this is a deeper aim. The Received Critique
of Lakatos, it seems to me, rests on a misinterpretation so severe
that the real issues we need to raise about Lakatos's thesis have
scarcely been raised. Currie, I shall argue, avoids the standard
misinterpretation only to fall into others; but nevertheless, his
arguments do point us toward the issues that need addressing. In
criticizing his arguments, my deeper aim is to reconstruct Lakatos's
position in a way that will facilitate the further discussion it
deserves.

1. The Received Critique

Lakatos's main case for the norm-ladenness thesis is in "History of
Science and its Rational Reconstructions" (hereafter 'HSSR'), and most
of his critics' fire is drawn by what Lakatos there says about
"internal history" and "rational reconstruction". In the first half

---

PSA 1982, Volume 1, pp. 29-39

of HSRR, Lakatos applies the predicates "internal" and "rational" mainly to explanations of scientific episodes: an "internal" historical account seems to be one which explains the theory choice of a past scientist or group of scientists by displaying it as scientifically rational, i.e., as motivated by reasons "internal" to science. By contrast, an account is "external history" when it portrays an episode as transpiring for other than scientific considerations. "Rational reconstructions"--again, until about midway through the essay--seem to be synonymous with "internal history" in the above sense: as a detective reconstructs a crime, so a historian reconstructs past episodes; and when a reconstruction portrays the episode as rational, it is a "rational reconstruction". I shall call these the primary senses of the above terms.

But midway through HSRR some odd mutations seem to occur. The first portent of a change is Lakatos's statement that "when history differs from its rational reconstruction, it (external history) provides an explanation of why it differs." (1971, p. 105). This is puzzling: for if a rational reconstruction is a type of historical narrative, a difference between it and history would tell us simply that it is false narrative. That this is not just a slip of Lakatos' pen becomes clear on the next page, where he writes:

> Thus, in constructing internal history the historian will
> be highly selective: he will omit everything that is irrational
> in the light of his rationality theory. But this normative
> selection still does not add up to a fully fledged rational
> reconstruction. . . . Internal history is not just a selection
> of methodologically interpreted facts: it may be, on occasions,
> their radically improved versions. One may illustrate this
> using the Bohrian programme. Bohr, in 1913, may not even have
> thought of the possibility of electron spin. He may have had
> more than enough on his hands without the spin. Nevertheless,
> the historian, describing with hindsight the Bohrian programme,
> should include electron spin in it, since electron spin fits
> naturally in the original outline of the programme. Bohr might
> have referred to it in 1913. Why Bohr did not do so, is an
> interesting problem which deserves to be indicated in a
> footnote. (1971, p. 104.)

Recognizing that his "reconstructions" of "internal history" will often portray a fabrication from which the real past "deviates". Lakatos here seems to relegate such deviations to the footnotes of a text of reconstructed history. So construed, his proposal is enough to make even his followers blanch, and it tends to make his critics apoplectic. Responding to an earlier version of this proposal, Kuhn promptly charged that Lakatos has forfeited his right to address historians, since ". . . a historian would not include in his narrative a factual report which he knew to be false. If he had done so, he would be so sensitive to the offence that he could not conceivably compose a footnote drawing attention to it." (1970, p. 256). As noted in my introductory paragraph, many critics have followed Kuhn's lead; and at

a conference five years later, Kuhn, finding no reason to change his
verdict, still felt able to charge "that what Lakatos conceives as
history is not history at all but philosophy fabricating examples."
(1971, p. 143). In the view of many, Lakatos's case that history of
science needs normative methodology (and hence philosophy of science)
is purchased at the price of defining "internal history" as something
no real historian of science would want to do.

In the passages cited, Lakatos clearly does give "internal history"
a new and remarkable sense: it no longer refers to reconstructions
portraying the actual past in so far as it was rational; instead, it
seems to reconstruct the past as it should have been had it been
rational. But the crucial exegetical question is whether Lakatos intends these prescriptivist reconstructions to be a surrogate for what
we have always thought historical narrative should provide. Does he
intend them to replace the old-fashioned concern with what really
happened? Critics like Kuhn assume he does, for only so can Lakatos
be accused of perversely abandoning one of the historian's first
duties, viz., to portray the past faithfully as well as intelligibly,
not to concoct a "fabrication" of it.

2. Why the Received Critique is Wrong

To read Lakatos aright, and to acquit him of Kuhn's indictment, I
shall now argue that we need to see his norm-ladenness thesis as
comprising two claims: for as Lakatos's attention turns from one to
the other of these claims, the meaning of "internal history" and
"rational reconstruction" shifts in an intelligible way.

The first claim, which I shall call his "constitutive claim", is
that normative statements enter as premises--constitutively--into
"internal explanations" of the theory-choices of past scientists. To
explain a scientist's preference of some theory T over its rivals, the
historian must do more than recite the reasons adduced by a scientist for his choice; he must also--as Lakatos sees it--judge whether
the reasons adduced by (or available to) the scientist are good
reasons (making the theory-choice intelligible as a piece of
scientifically rational behavior), or whether the adduced reasons are
mere epiphenomena to a theory-choice caused by factors external to
science. This requires the historian to judge whether the theory
preferred by the scientist was, in the scientist's evidential situation, superior to its rival; and to judge this the historian must rely
upon a normative methodology. In the first half of HSRR, Lakatos
argues that historiographies of science have been shaped in just this
way by the tacit normative suppositions of historians. His claim is
that in so far as these suppositions are normatively inadequate , the
historian will overlook the rationality of episodes that are
internally explicable, and will instead scratch about for an external account where none is needed. In building his case for his
constitutive claim that norms enter as premises into historical
accounts, Lakatos uses "internal history" in what I have called its
"primary sense".

32

But as we saw above, Lakatos also uses "internal history" in a
sense which seems not at all concerned (except perhaps in footnotes)
with the actual behavior of scientists: rather it reconstructs the
past-as-it-should-have-been. Is there any way to make sense of this?

I believe there is: we need to see that about midway through HSRR
Lakatos shifts from his primary constitutive claim to a second (and
secondary) heuristic claim, in the context of which "internal history"
and "rational reconstruction" receive secondary meanings. The second
claim is that in seeking to generate explanations of past episodes, it
is heuristically useful for the historian first to give a
preconstruction that prescribes the theory-judgments that ought to
have been made in the historical situation. In developing this idea,
Lakatos uses "internal history" and "rational reconstruction" to refer
to these heuristically-intended preconstructions. The key point is
that such preconstructions no longer even purport to be explanatory
historical narrative about scientists' theory-choice behavior: they
are instead a kind of normative discourse about the theories at issue
in an episode, telling which theories were superior at various times,
and why. Further, such preconstructions are not meant as surrogates
for historical narrative, but rather as a preliminary to it: the idea
is that by first providing an explicit heuristic preconstruction of
this sort, historians can more reflectively and self-critically ask
whether scientists' actual behaviour is internally intelligible, or
whether an external account is needed.

For the clearest evidence for this interpretation, we must turn to
Lakatos's essay on "Falsification and the Methodology of Research
Programmes", which contains Lakatos's "rational reconstructions" of
the Proutian and Bohrian programmes. Critics have taken these, in
particular, as illustrating the cavalier attitude toward historical
fact that Lakatos allegedly goes on to advocate in HSRR. But the
critics do not usually attend to the fact that Lakatos introduces
these quasi-historical sketches with an explicit and italicized
caveat: as rational reconstructions, they are, he says, intended to
illustrate the sort of "heuristic study" that any "historical study"
should be "preceded by". (1970, p. 138).

Though space does not permit documentation here, I believe that
close study will support the claim that in the second half of HSRR,
Lakatos uses "internal history" and "rational reconstruction" in this
second sense: he means by these terms what I shall call "rational
preconstructions". His presentation is not entirely felicitous, and
it is no doubt somewhat odd for him to urge that a special class of
"internal historians" be so concerned with giving these preliminary
preconstructions that they relegate to footnotes the historical study
the preconstructions are meant heuristically to precede. But this
Lakatosian idiosyncrasy is easily understandable as a reaction against
the usual practice of doing history of science in the dimness of
unarticulated and unacknowledged preconstructions whose deeply buried
normative suppositions can seldom endure the light of explicitness.

Lakatos's stylistic alternative (preconstructions in the text and actual history in the footnotes) is no doubt an overreaction: but it will not do to impugn him for advocating "a historical narrative that demands footnotes to point out its fabrications." (Kuhn, 1971, p. 143). It will not do because the text to such footnotes is intended to be, not historical narrative, but a heuristic preliminary to it. Kuhn is therefore mistaken; and so are many others.

3. Currie's Interpretation and Critique of Lakatos's Heuristic Claim

To clarify Lakatos's view further, I now turn to Currie's critique. Unlike most critics, Currie sees that when Lakatos urges the historian to begin by writing an "internal history", this term refers to a normative preconstruction about the theories at issue in an episode, and is a preliminary to explaining the behavior of scientists in relation to those theories. But precisely how is the historian supposed to utilize these preconstructions?

Currie (1980, pp. 456-63) construes Lakatos's view as follows. Suppose we (as historians) want to explain the choice of some past scientist X who preferred theory T over its rival T'. Following Lakatos's advice, we have written an "internal history" which, by specifying "the intrinsic scientific merits of the theories" (p. 459) prescribes which theory was superior in the situation, and why. This prescription is of course relative to the norms of some methodology M we have selected. Now if the behavior of scientist X "fits our picture of rational action" (p. 458), as given by our internal history, Lakatos would have us (says Currie) "assume that the behavior of X was motivated by an awareness on the agent's part of those very reasons which make it (the behavior) seem rational to us." (p. 458.) That is, we are to explain the behavior "simply by invoking the claim that scientists make correct decisions based on rational assessment." (p. 457.) On Currie's account, Lakatos would have us engage in empirical investigation of the scientist's motives only if the scientist's behavior does not "fit our picture of rational action." (p. 457.) The "fit" between our internal preconstruction and the choice of the scientist is thus to licence an assumption which functions as a surrogate for empirical investigation of the scientist's motives. This claim, which Currie attributes to Lakatos, I shall therefore call "the investigation-surrogate (or I-S) claim".

We can avoid certain ambiguities to which Currie is prone here by distinguishing two uses of the term "rational". Sometimes this term is predicated of a specific person's act of doing or believing something, as for example one were to say "Madalyn O'Hare is rational in believing that God does not exist." And in this act-specific sense, to call a belief (qua act of believing) rational is to say something about the reasons motivating just that person to so believe. But there is also a different use, as for example if one says "Today, atheism is the rational view." Here one would mean, I take it, that in the evidential situation today, there are good reasons available such that on balance, atheism is what ought to be believed. To mark

34

this distinction I shall use the term "rationable" for the second use,
reserving the term "rational" for the first. The two are clearly
different: for to say "atheism is rationable" does not at all mean
(or entail) that any specific person (say, Mrs. O'Hare) is rational
in believing atheism. (Cf., Wykstra 1980, p. 212.)

The I-S claim Currie attributes to Lakatos can now be put this way.
When a historian "armed with methodology M" (p. 459) judges that the
theory-choice of scientist X of T over T' "fits" an M-based
preconstruction, the historian is not, so far forth, judging that
"scientist X was rational to prefer T to T'" (as Currie misstates it
on p. 459). The historian is judging only that a preference of T over
T' was rationable in the scientist's situation—relative of course to
the norms of M. Now according to Currie, once the historian has made
this judgment, Lakatos would have him go on to "immediately assume"
that the scientist chose T because of those very reasons which, by our
preconstruction, made it a rationable choice. And this, Currie
objects, is surely wrongheaded. His main argument (p. 459) is that it
is generally possible for two different and incompatible methodologies (say, M and M') to issue in the same verdict about T being
rationable, though for very different reasons. From the M-ration-
ability of T over T', it is therefore a mistake for the historian to
"immediately assume that X's decision was governed by the appraisal
criteria implicit in M." Now by "immediately assume" Currie here
means "conclude straightaway": he clearly means to say that Lakatos
endorses this way of drawing conclusions. Currie adds that merely
because X's choice is "prima facie rational" (that is, rationable), we
"cannot afford to ignore psychological facts about the agent's
decision"; we still need to investigate his reasons. So Lakatos
evidently is supposed to regard such investigation as superfluous.
Finally, while owning that Lakatos's alleged procedure might be used
in desperation in uncommon cases where historical records are lacking,
Currie says that normally—where such records do exist—it "seems
strange to suggest that we should rest content with the assumptions
that the scientist's beliefs and motives were those of a well-informed
methodologist of the late twentieth century." Lakatos, we are thus to
understand, would have us "rest content" with this assumption.

To summarize: according to Currie, Lakatos advocates a heuristic
procedure for arriving at historical explanations which is objectionable in two main respects. First (I-S1), the procedure authorizes a
historian armed with methodology M to conclude that a scientist's
theory-preference (_qua_ act of preferring) was M-motivated whenever,
and on the sole grounds that, the preference (_qua_ theory preferred)
was M-rationable. Second (I-S2), in such cases the procedure
dispenses entirely with investigation of the aims and beliefs of the
scientist. The I-S claim for which Currie faults Lakatos comprises
both I-S1 and I-S2. The two are distinct in that I-S2 could be held
even by one who rejects I-S1. But I shall now argue that neither is
part of Lakatos's position: Currie misconstrues Lakatos's heuristic
claim.

4. Why Currie is Wrong about Lakatos's Heuristic Claim

   Begin with I-S1. Stated vaguely, Lakatos's heuristic claim is that
rational preconstructions (as I have defined the term), based upon a
selected normative methodology, ought to play a vital function in
arriving at explanations of the theory-choices of past scientists.
But Currie is mistaken, I think, about how Lakatos thinks they should
function. To see why, we must take seriously Lakatos's proposal that
normative methodologies should function as "hard cores" of "historiographical research programmes". (1971, pp. 116 ff.) The crux can
best be broached by considering a point about scientific research
programmes, for this is where the function of a hard core is clearest.

   Consider the following situation. A scientist, armed with a
vortex-aether hard core, wants to explain the behavior of compass
needles. His hard core, given certain theoretically plausible statements about the earth's vortex, suggests that a compass needle should
behave in a certain way near the earth's poles; and precisely this
behavior is found to occur. Let us describe this situation by saying
that the hard core has "generated a solution" to the _explanandum_
(which may or may not be a novelly-predicted one): the principles and
conceptual machinery of the scientist's hard core, plus various
auxiliary conditions, entail a description of the _explanandum_. Now
the crucial question is: in just this situation, how would Lakatos
have the scientist view the generated solution? Specifically: would
Lakatos have him view it as in any sense an acceptance-worthy causal
account of the phenomenon?

   I submit that Lakatos would not: for he is (and would have the
scientist be) acutely aware that some rival programme may well generate a different solution, which renders the phenomenon explicable in
terms of some incompatible causal mechanism. One is not entitled, for
Lakatos, to regard one generated solution as preferable to others
merely because it was generated by the hard core one has oneself opted
to use: hard cores do not function as justifying solutions in this
simple-minded and egocentric way. To be sure, the significance of a
hard core for solutions lies not merely in "the context of discovery", but in "the context of appraisal" as well. But the order of
appraisal is for Lakatos a subtle one, roughly coming to this:
through its hard core, a programme heuristically generates problem-
shifts and problem solutions; these go on the "track record" of a
programme as posits; the programme is then assessed by comparing its
diachronic track-record with those of its rivals; and a superior
track-record rating for the programme then accrues to the prefer-
ability of specific solutions generated under its auspices. To assess
the solution he has generated, the aether-theorist in my example must
thus assess the overall progressiveness of his programme relative to
rival programmes.

   Just so, I submit, for historiographical hard cores. Here, an
explanatory solution is "generated" when a hard core, applied to a
historical theory-choice situation, leads one to expect (as ration-

able) certain theory-choice behavior. Now Currie imputes to Lakatos the proposal that once a historian has so generated a solution, the historian should conclude straightaway ("immediately assume") that this solution is a correct causal account--that is, that the theory-choice behavior was caused by those factors which make it rationable. This caricatures the justificatory relation Lakatos generally envisions between a hard core and the solutions it generates. Indeed, Lakatos is explicitly aware (though Currie touts it as news) that rival hard cores in historiography (as in science) may generate rival solutions for a given phenomenon. (1971, pp. 117 ff.) He would regard as anathema a historian who regarded a solution as correct merely because it was generated in this way by the normative methodology he has himself opted to employ. To evaluate the credentials of a solution, once it has been generated by one's hard core, one must assess the programmatic power of this hard core by comparing its overall progressiveness with rival programmes. In imputing I-S1 to Lakatos Currie is therefore mistaken.

Currie might concede that Lakatos does not hold I-S1, but object that if we want to assess a proffered explanation of why some given scientist, say, Newton, chose some theory T, what is needed is more specific study of Newton's motives, and that Lakatos's call instead for assessment of the hard core that generated the explanation fails to see this. Indeed, it might be pressed, does not Lakatos dismiss as irrelevant "false consciousness" the evidence of what a scientist adduces as his reasons whenever this conflicts with a rational reconstruction? Currie might, in short, make his stand on the wrong-headedness of I-S2.

But it seems to me equally unclear that Lakatos holds I-S2. Indeed when carefully examined, Lakatos's references to false consciousness indicate he does not hold it.

The key to interpreting Lakatos here is, I think, that he regards work on false-consciousness explanations as one band in the "protective belt" around any historiographical hard core, along with work on other "externalist" solutions to problems. That is, Lakatos sees a historiographical research programme as generating new problems (often involving interesting problem-shifts) both when scientists chose what was (by ones preconstruction) the "wrong" theory, and when they chose the "right" theory but adduced the wrong reasons for their choice. (1971, p. 102,106.) And Lakatos holds that one way for a historiographical programme to progress is by showing that there are good "external explanations" for what it must view as "deviations". (1971, p. 119: cf. p. 94, 96, 98.) Now two points need to be brought out here. First, for Lakatos's work in the protective belt--in historiography as in science--is in response to problems precisely because they are relevant problems. So if one needs protective-belt work given what a scientist says about his reasons, this is precisely because what he says is relevant to an explanation of his choice. Second, if what a scientist says about his reasons conflicts with what one's preconstruction says his reasons ought to have been, this

problem is not, by anything Lakatos says, to be exorcized merely by
waving one's hands and incanting "false consciousness". What are
wanted are contentful psychological theories of false consciousness,
which have genuine explanatory power. If one fails to produce such
theories in the protective belt, one has an anomaly that counts again
against one's programme. So long as rival programmes suffer similar
anomalies this failure may not weigh relatively against one's
programme; but it is a source of anomalies nonetheless, and like all
anomalies these "will eventually have to be explained either by some
better rational reconstruction or by some 'external' empirical
theory." (1971, p. 118.)

Summarizing, my claim is this. In giving a role to protective belt
work on "false consciousness" explanations, Lakatos is recognizing
that the historian must come to terms with the evidence of what a
scientist says about his beliefs, aims, and reasonings. (The
constitutive claim of the norm-ladenness thesis does not hold that
psychological attributions about these are irrelevant to explanations,
but that they are insufficient: one needs normative premises as well.)
It therefore seems to me dubious to attribute I-S2 to Lakatos. It is
perhaps worth noting, too, that Currie does not cite texts
substantiating any such attribution.

5. Conclusion

Lakatos's norm-ladenness thesis, I have argued, comprises both a
heuristic claim and an constitutive claim. The Received Critique of
Lakatos fails to see that "internal history" and "rational
reconstruction" receive a special meaning (by which they designate
"rational preconstructions") when used in the context of the heuristic
claim. Currie avoids this mistake, but attributes to Lakatos an
"investigation-surrogate claim" which misrepresents the heuristic
claim, oversimplifying the relation Lakatos envisions between hard
cores and the solutions they generate.

Space does not here permit an examination of Currie's arguments
against what I have called Lakatos's "constitutive claim", which is
Currie's real target. But it is worth noting that Currie may not
grant that there is a real distinction here. Immediately following,
and summarizing, his case against the I-S claim, Currie writes:

> Thus we ought, I believe, to be suspicious of any attempt to
> generate an historical explanation by first stating premises
> about the intrinsic scientific merits of theories and then
> linking these appraisals with the actor's psychological state
> by means of an assumption that it was exactly these merits which
> weighed with the actor himself. Historical explanations in
> science (as elsewhere) must contain only descriptive-
> psychological assumptions about the agent's beliefs and aims;
> there is really no place for 'normative-internal' premises such
> as are described by Lakatos. (p. 459.)

The first sentence here summarizes Currie's misgivings about Lakatos's alleged I-S claim, which is a heuristic claim about how best to generate historical explanations. But in the second sentence, Currie seems to take this as establishing the wrongness of Lakatos's claim that normative statements enter into such explanations as premises--that is, constitutively. If this is Currie's intention, it seems to me he is mistaken. For even if Lakatos's heuristic claim is false, this does not entail that his constitutive claim is false. Currie does, however, go on to give additional arguments against Lakatos's constitutive claim; but space does not permit an analysis of them here. And a discussion of the logical and conceptual relations between Lakatos's heuristic claim and his constitutive claim must, likewise, await another occasion.

Notes

I should like here to thank Larry Laudan for his comments on an earlier version of this paper, and the University of Tulsa for a Summer Research Grant supporting the research for it.

References

Buck, Roger and Cohen, R.S. (eds.). (1971). _PSA 1970. (Boston Studies in the Philosophy of Science,_ Volume 8.) Dordrecht: Reidel.

Currie, Gregory. (1980). "The Role of Normative Assumptions in Historical Explanations." _Philosophy of Science_ 47: 456-473.

Holton, Gerald. (1974). "On Being Caught Between Dionysians and Appollonians." _Daedalus_ 103: 65-81.

Kuhn, Thomas. (1970). "Reflections on My Critics." In Lakatos and Musgrave (1970). Pages 231-278.

-------------. (1971). "Notes on Lakatos." In Buck and Cohen (1971). Pages 137-146.

Lakatos, Imre and Musgrave, Alan. (eds.). (1970). _Criticism and the Growth of Knowledge._ Cambridge: Cambridge University Press.

-------------. (1970). "Falsification and the Methodology of Scientific Research Programmes." In Lakatos and Musgrave (1970). Pages 91-195.

-------------. (1971). "History of Science and Its Rational Reconstructions." In Buck and Cohen (1971). Pages 91-135.

Laudan, Larry. (1977). _Progress and Its Problems._ Berkeley: University of California Press.

McMullin, Ernan. (1970). "The History and Philosophy of Science: A Taxonomy." In _Historical and Philosophical Perspectives of Science. (Minnesota Studies in Philosophy of Science,_ Volume V.) Edited by Roger Stuewer. Minneapolis: University of Minnesota Press. Pages 12-67.

Suppe, Frederick. (ed.). (1977). _The Structure of Scientific Theories._ 2nd ed. Urbana: University of Illinois Press.

Wykstra, Stephen. (1980). "Toward a Historical Meta-method for Assessing Normative Methodologies: Serendipity, Rationability, and the Robinson Crusoe Fallacy." In _PSA 1980,_ Volume 1. Edited by P.D. Asquith and R.N. Giere. East Lansing, Michigan: Philosophy of Science Association. Pages 211-223.

Part II

Causation

Causal and Explanatory Asymmetry[1]

Daniel M. Hausman

University of Maryland
College Park, MD 20742

Explanations sometimes involve a certain asymmetry. For example, given Ohm's Law and the voltage of a battery, one can calculate the current in an electric circuit if one knows the resistance, or one can calculate the resistance if one knows the current. Both of these calculations are (roughly) deductive-nomological arguments, but only the calculation of the current from the resistance is an explanation. Why should only one of these calculations be an explanation?

One plausible diagnosis of such explanatory asymmetries is that they arise from the asymmetry of causal relations (see, for example, Salmon 1978 or Beauchamp and Rosenberg 1981, p. 313). The direction of explanation is the direction of causation. The resistance of the wire is a causal condition influencing the amount of the current, while the amount of current does not (as a first approximation) influence the resistance. The resistance of the wire explains the quantity of current and not vice versa, because causes explain their effects, while effects do not explain their causes.

But not all explanatory asymmetries reflect causal asymmetries. As Peter Railton has argued at length (1980, pp. 191-224, 385-410), there are different kinds of explanatory asymmetries. For example, one explains why a substance has a certain dispositional property by identifying those features of underlying structure which are "responsible" for the disposition, while one does not explain why the substance has such an underlying structure by pointing out that it has a certain disposition. Even if there is a one-to-one relation between a particular underlying structure and a particular disposition, the explanatory asymmetry remains. Yet an explanation of why a substance has a disposition is not a causal explanation. The underlying structure constitutes, rather than causes the disposition. Not all explanatory asymmetries can be attributed to the asymmetry of causation.

PSA 1982, Volume 1, pp. 43-54

44

Some explanatory asymmetries do nevertheless depend on causal asymmetries. Examples are cases like the one above involving Ohm's Law or the often-cited flagpole example (where the height of a flagpole explains the length of its shadow, but not vice versa--see Bromberger 1966, p. 71 for a Manhattan version). Such asymmetries of causal explanation can be of practical importance. Monetarists maintain that the quantity of money explains the value of gross national product and not (to any appreciable extent) vice versa. They are right only if the relationship between the quantity of money and the value of GNP is an asymmetric causal relation.

But there is a further question concerning this diagnosis of explanatory asymmetries which has not been carefully considered. <u>Why should the asymmetry of causation give rise to an asymmetry in explanation</u>?

The question may seem an odd one. What could be more obvious than that causes explain their effects, while effects generally do not explain their causes? To give a cause is historically, etymologically and perhaps even by definition to explain. We probably know nothing about explanation which is more secure or informative than is the claim that to give a cause of a phenomenon is to explain it.

Yet it is still important to ask why the asymmetry of causation gives rise to an asymmetry in explanation. In attempting to clarify and to answer this question, one may learn a great deal about the relations between causation and explanation. Even if it is obvious that causes explain their effects but not vice versa, it is not obvious <u>how</u> causes explain or how causal explanations are related to other kinds of explanations. It is no more evident now that 1 + 1 = 2 than it was before the foundational studies at the turn of the century, but we now understand the relations between logic, sets and numbers better than we used to. In a similar way, in considering why causes explain their effects, while effects do not generally explain their causes, we may be able to understand the relations between causality and explanation better.

There is also a specific reason to consider why causal asymmetries should give rise to explanatory asymmetries. On the dominant Humean view of causation, it seems that the amount of current should explain the resistance in a circuit just as well as the amount of resistance explains the amount of current. On Hume's view of causation, the only difference between causes and effects is, of course, that causes come before their effects in time. The constant conjunction between cause and effect can be equally constant from either direction. The tendency of the mind to pass from the impression of the cause to the idea of the effect seems, in Hume's view at least, to be no stronger than the tendency of the mind to pass from the impression of the effect to the idea of the cause. All there is is a difference in time order. Why should position in time matter so much to explanation? Perhaps people are irrational and mistaken when they deny that the current in the wire explains the resistance or that the length of the shadow of a flagpole explains its height. But let us postpone discussing this implausible suggestion until we have considered a more moderate alternative.

One possible way to retain a Humean analysis of causation without denying that there are genuine asymmetries of causal explanation is to insist that explanation is largely subjective or pragmatic. Bas van Fraassen offers such an account of explanation and of explanatory asymmetries (1980, ch. 5). He argues, "If that is correct, if the asymmetries of explanation result from a contextually determined relation of relevance,. . .it should then also be possible to account for specific asymmetries in terms of the interests of questioner and audience that determine this relevance." (1980, p. 130). In van Fraassen's treatment, explanations are answers to context-dependent why questions.

One might then attempt to connect causal and explanatory asymmetries by arguing that why questions are often requests for the salient or relevant cause of a phenomenon. To identify an effect of the phenomenon will clearly not satisfy such a request. But this way of connecting explanatory and causal asymmetries is trivial. It reformulates rather than answers the basic question: Why, in asking "Why?" should one be asking for a cause rather than an effect? One needs some account of why causal order (time order, if one is a Humean) should be of so much significance when one's object is explanation or understanding.

One cannot make sense of the asymmetry of causal explanation without noticing that there is more to causal priority than merely temporal priority. Either one needs some other analysis of causal asymmetry than Hume's, or one must show how a difference in time order gives rise to other significant differences between causes and effects. In attempting to give an alternative analysis of causal asymmetry, philosophers like Collingwood (1940), Gasking (1955) and von Wright (1975) have pointed out that causes are, as it were, levers for moving effects. If one jacks up or bends a flagpole, its shadow will grow longer or shorter, while if one alters the shadow (perhaps by sloping the ground), the flagpole's height will be unaffected.

As an independent _analysis_ of causal asymmetry, manipulability runs into intractable difficulties (Beauchamp and Rosenberg 1981, pp. 203-8; Ehring 1981, ch. 3). But the brute fact remains that agents manipulate effects via their causes. One would not normally attempt to change the resistance in a circuit by changing the amount of current flowing in it. Moreover, one can, to a considerable extent, account for this asymmetry of manipulability by the difference in time order Hume relies on. Agents are, after all, able to affect the future, but not the past. Manipulability then forms one leg for Humean-pragmatic accounts of explanatory asymmetries to stand on.

The fact that causes are often "levers" for moving effects, coupled with a pragmatic theory of explanation like van Fraassen's, makes it possible to explain why causal asymmetries should give rise to explanatory asymmetries. Agents are usually much more interested in the causes than in the effects of a certain phenomenon, because they want to know how to bring about or to prevent the phenomenon. Given this practical interest, we ask "Why?" to find out about causes, not effects. Sometimes, of course, causal explanations are of no practical help, but one might argue that agents extend their general course of questioning to such cases, too. Causes thus explain their effects, while effects do not generally explain their causes.

One can summarize this Humean-pragmatic connection between causal and explanatory asymmetry as follows: Explanations are answers to context-dependent why questions. Such questions are frequently requests for causes and rarely requests for effects, because in the pervasive context of potential actions, knowledge of causes is especially important. In the context of potential action, knowledge of causes is crucial, because one can manipulate events via their causes, but not via their effects. This asymmetry of manipulability results (perhaps) from the two facts that causes precede their effects and that present action can alter the future, but not the past. So the temporal priority of causes should result in asymmetries in causal explanations.

The above account of why requests for explanation are requests for causes and not for effects is overly simple. It misleadingly suggests that one is particularly interested in causes mainly when one is about to act and when one is concerned with phenomena within one's control. The context within which, in asking for an explanation, one wants to know the causes is in fact much more general. As actors our construal of counterfactuals leads to a more general perspective within which knowledge of causes is especially important.

As actors or interveners, agents often ask, "What if. . .?" Consider two such questions: (1) What if the resistance in this circuit were larger? (2) What if there were more current flowing in this circuit? Given Ohm's Law, there are as many possibilities in the first case as in the second. But the usual answer to these questions reveals an asymmetry. In answer to (1), one would say that if there were more resistance in the circuit, less current would be flowing. In answer to (2), in contrast, one would assert only that if there had been more current flowing, there would have been less resistance or a higher voltage (see Simon and Rescher 1966, pp. 125-30).

The different way in which we respond to the two "What if?" questions reflects our perspective as actors and the fact that when one intervenes and changes something, one generally influences only its effects. This asymmetry of manipulability leads us imaginatively to hold fixed the voltage when considering a circuit with a larger resistance and to believe that there would be less current. The supposition that there is more current, on the other hand, leads to no determinate results. From the perspective of action in this context, the latter supposition is sterile. All we know is that something must have been different.

Our interests as actors thus lead us to regard effects as counterfactually dependent on their causes and not vice versa. Moreover, as actors, agents are interested not only in what does happen, but in what can and must happen. Given this general interest in counterfactuals and this general interpretation of counterfactuals, agents ask "Why?" to find out why phenomena <u>had</u> <u>to</u> (or were made more likely to) occur (where these modal notions are to be understood as suggested in the paragraph above). An agent's desire to know the cause does not follow <u>immediately</u> from any practical interest. It derives instead from an interest in understanding not only what does, but what might or must happen. This general interest in turn follows from practical interests in action. When one asks why something occurred, one wants to know the cause, not, typically, because one has any immediate interest in doing anything, but because, knowing the cause, one can construct an account of what might and must occur, which in turn is vital for action. In this more indirect and sophisticated way, our interests lead us to regard causes as explaining their effects, but not vice versa. (This view is, I think, implicit in ch. 5 of van Fraassen's <u>The Scientific Image</u> (1980).)

Notice that there is nothing in the above discussion to which Hume need object. Although the more sophisticated account relies on counterfactuals, the truth of these is determined by the way in which agents think about the world, not by any fact about necessities in nature. As van Fraassen puts it, ". . .there is nothing in science itself--nothing in the objective description of nature that science purports to give us--that corresponds to these counterfactual conditionals." (1980, pp. 115-16). What we have in effect noticed in the above discussion of the relations between causal and explanatory asymmetries is that the "customary transition of the imagination" (Hume 1748, p. 86) from the idea of a cause to the idea of its effect differs in important ways from the propensity of the mind to pass from the idea of an effect to the idea of its cause.

Although subtle and ingenious, I believe that such pragmatic interpretations of explanatory asymmetries are mistaken. What bothers me is that there is little difference between, on the one hand, showing how (given a pragmatic account of explanation) causal asymmetries should give rise to explanatory asymmetries and, on the other, showing how (given an objective account of explanation) our perspective as actors causes us mistakenly to believe that there are such asymmetries. On a pragmatic view of explanatory asymmetries, only our practical interests provide us with grounds to prefer a purported explanation that cites a cause to one that cites an effect. It is only because actors seek to trace events to their causes that people regard causes as explaining their effects and deny that effects explain their causes.

The above observation is hardly a decisive criticism. One way to challenge such a pragmatic interpretation of explanatory asymmetries is to concede and indeed to emphasize the importance of counterfactuals, but then to argue that counterfactuals state objective truths. One can then offer a non-Humean counterfactual analysis of causal asymmetry (see Lewis 1973, p. 190 or Swain 1978, p. 11) and relate causal and explanatory asymmetries by examining the connection between explanation and counterfactuals. I shall not, however, follow this path. Nor shall I directly attack Humean-pragmatic interpretations of explanatory asymmetries. I shall instead attempt to show that, apart from our practical interests and any discussion of counterfactuals, explanations in terms of causes are far better explanations than ones in terms of effects. To establish this objective basis for the asymmetries of causal explanations, I shall rely on a non-Humean analysis of causal asymmetry (see Mackie 1974, esp. ch. 7 for the classic modern formulation of the problem of causal asymmetry). This analysis of causal asymmetry will incidentally clarify the fact that the manipulability theorists rely on and will purge it of its anthropomorphism. I shall also venture some comments on the nature of scientific explanation.

As a first step, let us consider briefly the account of causal asymmetry presented and defended in chapter 8 of Bernard Berofsky's book, Determinism (1971). I shall later argue against Berofsky's account, but it vividly reveals an important aspect of the asymmetry of causal explanation.

Stripped to its bare essentials, Berofsky's account maintains that event X is causally prior to event Y if and only if the law which links X to Y is part of the most comprehensive adequate systematic theory of phenomena like Y, while there are more comprehensive adequate systematic theories of phenomena like X than any which include the law which links X to Y (pp. 265-67). Theories of phenomena like shadows include the law that light travels in straight lines. Theories of phenomena like the size of poles or towers do not include laws of optics. Berofsky's view thus correctly implies, for example, that the height of a flagpole is a cause of the length of its shadow.

From the perspective of this paper, the most interesting feature of Berofsky's account of causal asymmetry is the way in which it enables one to connect causal and explanatory asymmetries. When one links Y to one of its causes, X, through some generalization, one is ultimately bringing Y under a general systematic comprehensive theory of phenomena like Y, while the generalization cannot be embedded in any comprehensive theory of phenomena like X. The isolated link between X and Y does little to explain any aspect of X. Isolated linkages provide worse explanations than do relationships which may be embedded in comprehensive theories.

One feature of explanations that has been recognized by many authors is that they reduce the extent to which the phenomena they explain appear to be merely contingent happenings. Some such intuition is involved in the otherwise unacceptable view that explanations involve a reduction of the unfamiliar to the familiar. Hempel attempts to capture this intuition with his notion of "explanatory relevance" (1966, p. 48)--his view that an explanation should show that the phenomenon to be explained was to be expected. "Clearly, then, the phenomenon. . .is here explained, and thus understood, by showing that its occurrence was to be expected under the specified circumstances in view of certain general laws. . ." (Hempel 1965, p. 364). The most satisfactory way of capturing this intuition is, however, Michael Friedman's (1974). Friedman argues that "unification or reduction in the number of independent phenomena is the essence of explanation in science. . ." (1974, p. 15). In explaining a happening or a fact, one shows that it is not just some disconnected occurrence. A good explanation reveals the extent to which the happening to be explained is linked to other, more general phenomena. The more extensive and systematic the linkages (and thus the more general the phenomena to which the happening is linked), the better the explanation. If one crucial virtue of scientific explanations is the extent to which they unify phenomena and reduce the number of independent phenomena, then "locating" the fact or happening to be explained in a general systematic comprehensive theory will explain it better than will citing an isolated regularity. (Although I believe that Friedman's technical apparatus can be adapted to my needs, I am only relying on his basic intuition. Notice, by the way, that his explication of the notion of "independence" is quite different than mine.)

If one grants that an event is better explained by a generalization when the generalization can be embedded in a comprehensive and systematic theory of phenomena like the particular event, then Berofsky's account enables one to show why causes explain their effects, while effects do not explain their causes very well. Only when a phenomenon is linked to its cause is it exhibited as an instance of some general systematic feature of the world. (From this perspective one can, incidentally, appreciate more fully the explanatory importance of counterfactual determination.)

Notice that this difference in explanatory adequacy is not subjective or pragmatic. On Berofsky's view, it is not which theories we happen to have which matters, but which theories ideally exist. A generalization which links an event to one of its causes forges in principle links to a whole range of phenomena. Such a linkage enables one, as it were, to recognize the event's definite theoretical location. From a generalization connecting an event to one of its effects, on the other hand, one can construct no such system of interconnections. The cause remains, as it were, unlocated, unconnected, "contingent".

Berofsky's account thus enables one to show how causal asymmetries can give rise to explanatory asymmetries. It also permits one to demonstrate that the asymmetry of explanation is not only a result of our perspective as actors. But Berofsky's account of causal asymmetry is inadequate. Will it account for the Ohm's Law example above? It clearly fails, as Douglas Ehring has pointed out (1981, pp. 137-39), when the cause is a random phenomenon. In such a case, the most systematic thing one may ever be able to say of phenomena like the particular cause is that they are linked by various laws to certain effects. In such a case, on Berofsky's view, there would be no causal asymmetry.

Underlying both Berofsky's account and the facts concerning manipulability is, I believe, a single basic objective asymmetry which constitutes the asymmetry of causation. There are different ways to get at this basic asymmetry. David Sanford (1976) and Douglas Ehring (1981, ch. 8) have, in their own ways, focused on the same fundamental feature. But I shall stick with my own account (Hausman forthcoming). The basic asymmetry that accounts for both the asymmetry of manipulability and the asymmetry of inclusion in comprehensive systematic theories is an asymmetry of "connectedness".

If an event X is causally connected to an event Y, then (and only then) there will typically be some sort of an association or correlation or stochastic dependence between events like X and events like Y. This fact provides a "quasi-operational definition" of the notion of a causal connection. Notice that I am not saying that the existence of a correlation or a stochastic dependence is either necessary or sufficient for the existence of a causal connection. Correlations or dependencies give us empirical "purchase"^{2} on the notion of a causal connection--that is, they reliably indicate when two events are causally connected and provide some of the meaning of the notion of a causal connection.

Reflection and experience reveal that two events X and Y are causally connected when and only when X causes Y, Y causes X, X and Y have a common cause or X and Y are mutually dependent (if there is such a thing as mutual dependence). This last claim is <u>not</u> a definition of "causal connection". We know independently when causal connections obtain from the "quasi-operational definition" above. They are the sorts of relations reliably indicated by correlations or dependencies. We discover from experience that causal connections come in only the three or four varieties mentioned.

Given the notion of a causal connection (which is, of course, in need of further analysis), one can give the following truth conditions for "X causes (or is a causal condition of) Y":

> X causes Y if and only if
> (1) X and Y are causally connected,
> (2) everything causally connected to X is causally
>   connected to Y, and
> (3) something is causally connected to Y but not to X.

For example, everything causally connected to the flagpole having a particular height (certain bungling on the part of those who built it, perhaps) is causally connected to its having a shadow of a particular length at a particular time, while something is causally connected to the length of the shadow (such as the angle of elevation of the sun), but not to the height of the flagpole.

On this account the difference between causes and effects lies in the connectedness of effects of a common cause and the relative independence of the various causes of a single effect. When X causes Y there will be some other cause or causal condition of Y which is not causally connected to X, while everything that X causes will be causally connected to Y.

The full elaboration and defense of this view of causal asymmetry is too large a task for this paper and has been undertaken elsewhere. My goal here, instead, is to show how this view of causal asymmetry (and, in similar ways, Ehring's and Sanford's views) relates causal and explanatory asymmetries.

When one identifies a cause or a causal condition of some event, one links that event to everything causally connected to that cause or causal condition. On the other hand, when one identifies an effect of some event, one brings about no such comprehensive linkage to that event. In establishing such comprehensive linkage, one is giving a better explanation.

On a view of scientific explanations as reducing "mere contingency" through unifying and establishing systematic interconnections among phenomena and through reducing the number of independent phenomena, causes will explain their effects better than effects will explain their causes. Through its relationship with its cause, an event is comprehensively linked to a whole network of events. The relation of an event to its cause should thus be embedded in a systematic comprehensive theory of happenings like the given event. What is right about Berofsky's asymmetry of inclusion in systematic theories thus follows from the asymmetry of connectedness. Given the asymmetry of connectedness, knowledge of causes leads to theoretical unification of our knowledge of the event phenomena and not vice versa. Causal asymmetry should give rise to explanatory asymmetry.

The objective asymmetry of connectedness upon which my account of causal asymmetry is based also explains why causal explanations are so important from the perspective of human action. For the manipulability theorists, human ability to act lies at the core of causal asymmetry. The manipulability theorist claims that X causes Y if and only if, acting and affecting X, one can affect Y, but acting and affecting Y (if we can), except via X, one does not affect X. There is a considerable measure of truth in these claims. How on my account of causal asymmetry are the facts of manipulability to be understood?

Manipulability is not some independent fact about causality, but is instead a direct corollary of my analysis of causality and the fact (which von Wright implausibly contests, 1975, pp. 48-62) that actions are events. Doing something which is causally connected to the cause will register in the effect, because everything causally connected to the cause is causally connected to the effect. Doing something which is causally connected to the effect need not register in the cause, because not all of the various causes of any given event are causally connected to one another. The practical difference between causes and effects follows from the objective difference in connectedness.

In this paper I have offered an answer to the question of why causal asymmetries should give rise to explanatory asymmetries--of why causes should explain their effects so much better than effects explain their causes. The answer is that effects are linked by their causes to a local causal network. From knowledge of causes of certain phenomena, one obtains more systematic and comprehensive "placement" of the phenomena than one obtains from knowledge of effects of the phenomena. The alternative "Humean" way of making sense of the asymmetry of causal explanation stresses manipulability and the pragmatics of explanation. On this alternative--as van Fraassen has argued--explanatory asymmetries are based on pragmatic considerations. They are a function of our interests as actors and of our consequent construals of counterfactuals. The account of the asymmetry of causal explanation defended here--in terms of an objective asymmetry of connectedness-- need not repudiate the pragmatic aspects of causal and explanatory asymmetries. Instead it reveals their basis in the central objective difference between causes and effects.

Notes

[1] I am deeply indebted to many more people than can be named here. Particularly to be thanked are Douglas Ehring, without whose help I could never have made progress with the problems of causal asymmetry, and Eugene Pidzarko, David Sanford and Paul Thagard, who offered helpful criticisms of the paper. My research was generously supported by the National Science Foundation (Grant #SES 8007385).

[2] This terminology was suggested by Sidney Morgenbesser.

<u>References</u>

Beauchamp, T. and Rosenberg, A. (1981). <u>Hume and the Problem of Causation.</u> Oxford: Oxford University Press.

Berofsky, B. (1971). <u>Determinism.</u> Princeton: Princeton University Press.

Brand, M. (ed.). (1976). <u>The Nature of Causation.</u> Urbana: University of Illinois Press.

Bromberger, S. (1966). "Why Questions." In <u>Mind and Cosmos: Explorations in the Philosophy of Science. (University of Pittsburgh Series in the Philosophy of Science,</u> Volume 3.) Edited by R. Colodny. Pittsburgh: University of Pittsburgh Press. Pages 86-111. (As reprinted in Brody, B. (ed.). <u>Readings in the Philosophy of Science.</u> Englewood Cliffs: Prentice Hall, 1970. Pages 66-87.)

Collingwood, R. (1940). "Causation." In <u>An Essay on Metaphysics.</u> Oxford: Clarendon Press. Pages 285-327. (As reprinted in Brand (1976). Pages 167-211.)

Ehring, D. (1981). <u>Causal Asymmetry.</u> Unpublished Ph.D. Dissertation, Columbia University. Xerox University Microfilms Publication Number 81-25278.

Friedman, M. (1974). "Explanation and Scientific Understanding." <u>Journal of Philosophy</u> 71: 5-19.

Gasking, D. (1955). "Causation and Recipes." <u>Mind</u> 64: 479-487. (As reprinted in Brand (1976). Pages 215-223.)

Hausman, D. (forthcoming). "Causal Priority." <u>Noûs.</u>

Hempel, C. (1965). <u>Aspects of Scientific Explanation and Other Essays in the Philosophy of Science.</u> New York: Free Press.

----------. (1966). <u>Philosophy of Natural Science.</u> Englewood Cliffs: Prentice Hall.

Hume, D. (1748). <u>Philosophical Essays Concerning Human Understanding.</u> London: Printed for A. Millar. (As reprinted as <u>An Inquiry Concerning Human Understanding.</u> (ed.) Charles W. Hendel. Indianapolis, Indiana: Bobbs-Merrill, 1955.)

Lewis, D. (1973). "Causation." <u>Journal of Philosophy</u> 70: 556-567. (As reprinted in Sosa (1975). Pages 180-191.)

Mackie, J. (1974). <u>The Cement of the Universe.</u> Oxford: Oxford University Press.

Railton, P. (1980). <u>Explaining Explanation: A Realist Account of Scientific Explanation.</u> Unpublished Ph.D. Dissertation, Princeton University. Xerox University Microfilms Publication Number 80-16165.

Salmon, W. (1978). "Why Ask, 'Why?'? An Inquiry Concerning Scientific Explanation." <u>Proceedings and Addresses of the American Philo-sophical Association</u> 51: 683-705.

Sanford, D. (1976). "The Direction of Causation and the Direction of Conditionship." <u>Journal of Philosophy</u> 73: 193-207.

Simon, H. and Rescher, N. (1966). "Cause and Counterfactual." <u>Philosophy of Science</u> 33: 323-340. (As reprinted in Simon, H. <u>Models of Discovery.</u> Dordrecht: Reidel, 1977. Pages 107-131.)

Sosa, E. (ed.). (1975). <u>Causation and Conditionals.</u> Oxford: Oxford University Press.

Swain, M. (1978). "A Counterfactual Analysis of Event Causation." <u>Philosophical Studies</u> 34: 1-19.

van Fraassen, B. (1980). <u>The Scientific Image.</u> Oxford: Oxford University Press.

von Wright, G. (1975). <u>Causality and Determinism.</u> New York: Columbia University Press.

<u>Mackie's Singular Causality and Linked Overdetermination</u>[1]

Robert H. Ennis

University of Illinois at Urbana-Champaign

In <u>The Cement of the Universe</u>[2] J. L. Mackie offered--with supple-
mentation--a necessary-condition analysis of singular causal claims:
"It seems that if any <u>X</u> is both necessary in the circumstances for and
causally prior to <u>Y</u>, we shall say that <u>X</u> caused <u>Y</u>; also, wherever we
are prepared to say that <u>X</u> caused <u>Y</u> we are prepared to say that <u>X</u> was
necessary in the circumstances for and causally prior to <u>Y</u>." (p. 51).
As Mackie interprets 'necessary', this is a counterfactual analysis.

Although there are difficulties with treating necessity (supplemented with the indicated qualifications) as sufficient for our being
justified in saying that <u>X</u> caused <u>Y</u> (argued by Berofsky 1977 and
Earman 1976, among others), I want to challenge the second half of the
analysis, because refuting it seems to shake more deeply my sympathies
for necessary-condition analyses. To show that causes are not necessarily necessary conditions would seem to strike at the heart of the
basic idea that a singular cause is a necessary condition for its
effect.

Many philosophers have offered a necessary-condition, or counterfactual, analysis of singular causal claims. I pick Mackie's approach
because it seems to me to be the strongest and one of the best elaborated. David Lewis' approach, the only other well-elaborated approach
of which I know, has been shown by Wayne Davis (1980) and Martin Bunzl
(1980) to be vulnerable to the same problem I offer against Mackie:
linked overdetermination.

1. Linked Overdetermination

One commonly finds cases in which <u>X</u> caused <u>Y</u>,but in which if <u>X</u> had
not occurred, something else would have stepped in (fail-safe cases),
or would have proceeded (preempted-cause cases), to cause <u>Y</u>. Since in
these cases the non-operation of the potential cause is <u>linked</u> to the
operation of the actual cause, the label 'linked overdetermination'

PSA 1982, Volume 1, ppl 55-64

(suggested by Michael Scriven 1966), seems appropriate. Some linked overdetermination cases, I shall argue, show that causes do not have to be necessary conditions. In such cases the effect would have occurred even if the cause had not.

In challenging Lewis' counterfactual analysis Davis offered a simple switch-operated-light example involving one light, one switch, one power source, and one set of wires: "Suppose that in fact Chuck flipped the switch at t, causing the light to come on. However, David was backing Chuck up, and would have flipped the switch at t, if Chuck had not. It is not true that if Chuck had not flipped the switch, the light would not have come on." (1980, p. 174). This simple case appears to show that, contrary to Mackie's claim, we are sometimes prepared to say that X caused Y even though we are not also prepared to say that X was necessary in the circumstances for Y.

In a review of Cement, Earman (1976, p. 391) claimed that he could construct an example that does what the Davis example does; but he did not offer the detailed discussion of the Mackie responses that is needed. Mackie's position does not collapse with the mere presentation of the Davis example or one like it.

In Cement Mackie offered three kinds of response that apply to linked overdetermination, a facts-and-events response, a different-effect response, and an in-the-circumstances response. I shall elaborate each and try to show that none succeeds.

2. Facts and Events

In dealing with a punctured-poisoned-canteen linked-overdetermina-tion case discussed by Hart & Honoré (1959, pp. 219-220), Mackie employed a distinction between facts and events. In this case a desert traveler's canteen water was poisoned by an enemy. Another enemy then punctured a hole in the canteen, not knowing that the water was poisoned. The traveler died. This preempted cause linked-overdetermination case might be thought to count against a necessary-condition analysis on the ground that the puncturing of the can caused the death although the traveler would have died anyway. Intuitions vary however.

Mackie's facts-events distinction (p. 46), he believed, avoided the refuting force of the example, if any, and explains our conflicting intuitions about the case. Events for Mackie apparently incorporate the full details of the result, while facts incorporate only what the words require. In particular Mackie distinguished between the fact that the traveler died, and the event of the traveler's death: The fact-result is described by the words, "The traveler died", which words could have been satisfied by a wide variety of occurrences, including a thirst-type death and a poison-type death. The event-result is described by the words, "the traveler's death", which words refer to that particular death in its full detail, including all of its thirst-type characteristics. Since the fact-result-words could have been satisfied by a poison-type death, Mackie agreed that the puncturing of the can was not a necessary condition for the fact that the traveler died. But he contended that the puncturing of the can was a necessary condition for the traveler's death, that particular death being as it was a thirst-type death. If the can had not been punctured, the death would have been a different kind of death. From this Mackie concluded that the puncturing of the can caused the event of the traveler's death, but <u>did not bring about the fact that he died</u>.

In so concluding, Mackie, if he is right, would also have explained the puzzle we feel about the case: we are puzzled about whether to pick the puncturing as the cause of the result because we regard the result in different ways, so the explanation goes. If we regard the result as an event, it was caused by the puncturing (which was a necessary condition for the death). If we regard the result as the fact that he died, then the puncturing was not a necessary condition for it and, according to Mackie, did not cause it.

Let us first concentrate on his handling of "fact-results". He did not eschew fact-results, as Lewis and Donald Davidson (1967) appear to have done. Instead he appeared to believe that his counterfactual analysis holds for fact-results. However, in linked overdetermination cases in which the fact would have been brought about by some other means than the putative cause, the putative cause would not then, according to him, be the cause of the fact result.

Interestingly, some kind of extension of his stance on fact-results seems to be required by what I shall call 'infinitive-results', which seem to have the same relevant characteristics as fact-results. The phrase 'him to die' indicates an infinitive-result. In the same spirit as Mackie's reasoning about fact-results, we must conclude that the puncturing of the can was not a necessary condition for <u>him to die</u>, since he would have died anyway (though not in the same way). The satisfaction of the phrase, 'him to die', just like 'that he died', in that context does not require that the death be a thirst-type death. So the phrase would have been satisfied if the can had not been punctured. Thus the Mackie doctrine must be that linked-overdetermined fact-results and infinitive-results, when described broadly enough for the descriptions to be satisfied by what would have been the result of the preempted or fail-safe causal factor, were not caused by the putative cause. Applied to the punctured-poisoned-canteen case, Mackie's conclusion not only was that the puncturing of the can did not bring about the fact that the traveler died, but also should be that it did not cause him to die.

The Mackie approach for fact-results and infinitive-results might do for a preemptive case like the canteen case, but it does not do for fail-safe cases like the switch-light case. Chuck's flipping the switch brought it about that the light went on (fact result) and caused the light to come on (infinitive result), even though the light would have gone on if he had not flipped the switch. Mackie's approach to

58

fact results requires us to say that Chuck's flipping the switch did
not cause what we know it caused.

So the switch-light example calls for Mackie to retreat to limiting his analysis to event-causation, as Lewis and Davidson have done.
This limitation is required by a combination of what Mackie said about
the canteen case together with what we can see to be true about the
switch-light case. It would however, be a significant and, I think,
unfortunate, limitation. What are we then to say about causal claims
like 'Chuck's flipping the switch brought it about that the light came
on' and 'Chuck's flipping the switch caused the light to come on'?

But would the analysis thus limited even then meet the challenge of
the switch-light example? I think not. Although Davis described the
result with infinitive and fact expressions in the material I quoted,
the example can be reformulated with a clear event-type description in
the crucial position. I have reenacted the case and describe it as
follows:

> <u>R</u>: Suppose that in fact Chuck's flipping the switch at <u>t</u>
> caused <u>the coming on of the light</u>. However, David was
> backing Chuck up, and would have flipped the switch at <u>t</u>,
> if Chuck had not.

In this reenacted case (called '<u>R</u>' for 'revised') it seems appropriate to
say, even given the event-type description of the result, that <u>the
coming on of the light</u>, was caused by Chuck's flipping the switch at <u>t</u>,
but would have occurred even if Chuck had not flipped the switch at <u>t</u>.
(Even the electricity going through the bulb would have been the same.)
Hence singular causal claims containing event-type descriptions of the
result do not appear to require that the cause have been necessary in
the circumstances for the effect.

I have put the conclusion in tentative terms because there are still
two further questions that, as I read Mackie, would apply to this
linked-overdetermination case: Is the effect that occurred actually
the <u>same</u> effect that would have occurred if the cause had not occurred?
And even if it is, would that effect actually have occurred, <u>in those
circumstances</u>? I shall argue for an affirmative answer to both questions.

3. Different Effects?

Assuming that our concern is now limited to the analysis of singular
causal claims containing event-type result-descriptions, a possible
response to the <u>R</u> case is that the event that occurred is actually
different from the one that would have occurred if Chuck had not flipped
the switch, the difference being that they would have had different
causes. This response would be in line with Davidson's criterion for
event identity: "Events are identical if and only if they have exactly
the same causes and effects." (1969, p. 231). (I do not here mean to
imply that Davidson would, or would not, employ such a response.) By

this criterion the coming on of the light that was caused by Chuck's flipping the switch would have to be different from the coming on of the light that would have been caused by David's flipping the switch. Mackie suggested such a response when he said, "For a concrete event effect, we require a cause, or causal chain, that leads to it, and it is the chain, puncturing-lack-of-water-thirst-death, that was realized, whereas the rival chain that starts with poison-in-can was not completed." (p. 46).

My reply has some similarity to Brand's (1977) view that event identity has to do with sameness of spatio-temporal region, but my emphasis instead is on speaker reference. In describing my reenactment of the light-switch case, I was the speaker. In an act of speaker reference I referred to an event that occurred in the vicinity of the light bulb. That is all that I referred to. I have a right to refer to whatever I choose, and that is what I chose to refer to: the coming on of the light.

Suppose that Chuck had not flipped the switch at $\underline{t}$, but David had done so at $\underline{t}$. Would there have been any difference in what occurred in the vicinity of the ceiling, which is where the event to which I referred took place? No. I referred to something that occurred up there in the vicinity of the ceiling, not something that occurred down here in the vicinity of the switch. The same effect (the one to which I referred) would have occurred, if Chuck had not flipped the switch at $\underline{t}$.

So, when the details of this act of speaker reference are made clear, we can see that the different-effect response to the $\underline{R}$ case does not succeed. Consider now whether there is an interpretation of 'in the circumstances' that will save the Mackie analysis.

4. 'In the Circumstances'

To my knowledge Mackie did not in _Cement_ explicitly resort to a special interpretation of 'in the circumstances' to handle linked over-determination counterexamples. He relied instead upon the different event-result approach that I earlier considered, perhaps because the examples that he considered were ones in which the concrete event-result would have been different, if the cause had not occurred.

However in handling a collateral-effects counterexample, he offers an interpretation of 'in the circumstances' that implies that <u>in the circumstances</u>, it is false that if Chuck's switch flipping had not occurred, then the light's coming on would have occurred anyway--since the actual circumstances included David's not flipping the switch. I shall consider this and three other possible interpretations of 'in the circumstances'.

A collateral-effects counterexample is one such that two different things were each caused by the same thing, giving at least the appearance that each result, according to counterfactual analyses, caused

the other, even though they clearly did not so do: As Mackie describes
a case, "Labor's defeat at the election pleases James but saddens John,
who, as it happens, are quite unknown to each other. Then James' being
pleased does not cause John's being sad, and yet we might well say that
in the circumstances John would not have been sad if James had not been
pleased. [Because if it had not been the case that James was pleased,
then it would not have been the case that Labor was defeated, and thus
John would not have been sad.]" (p. 33). Thus the analysis appears to
imply that James' being pleased caused John's being sad, though we
know full well that this is not the case.

Mackie's defense against the Labor collateral-effects counterexample employs a let-a-possible-world-run-on approach: "There is a
way of handling the words 'if' and 'in the circumstances' that will
defeat this supposed counterexample. Construct a 'possible world' in
the following way: take something that is just like the actual world
up to the point where, in the actual world, James is pleased; keep out
from your possible world, James' being pleased but otherwise let your
possible world run on in the same manner as the actual world; only if
John does not become sad in your possible world can you say 'in the
circumstances John would not have been sad if James had not been
pleased.'" (p. 33).

Mackie describes as follows his strategy for interpreting 'in the
circumstances' so that collateral effects do not refute his analysis:
"We thought of constructing a possible world which was just like the
actual world up to the point where, in the actual world, the proposed
antecedent occurs, of excluding this antecedent from the possible
world and the letting that world run on: if and only if the result
did not occur in that possible world were we prepared to say that the
antecedent was in the special sense necessary in the circumstances for
the result." (p. 52). Mackie then goes on to connect this approach
with his notion of causal priority, and to make an adjustment to
accommodate what he thinks to be a possibility, backwards causation.
This adjustment, which is couched in terms of "fixity", produces a
nest of problems pointed out by Earman (1976), Berofsky (1977), and
Beauchamp & Rosenberg (1977), among others. I shall consider the above
interpretation of 'in the circumstances' as it stands.

In the general account Mackie used 'if and only if' (after the
colon) rather than the 'only if' of the original Labor-case account.
Since he needs the 'if' in order that he have a way of telling that
something is necessary in the circumstances, I shall assume he means
to have it there. For my purposes, however, the important phrase is
the 'only if' since its implication is the one challenged by the
counterexample.

Applying this general approach to the _R_ case we _are_ entitled to say
that in the circumstances if Chuck had not flipped the switch, then
the light would not have come on. This is so because by the time that
the possible world branches out from the actual world it is too late
for David to conclude that Chuck is not going to do it, and to do it

himself. In order that David flip the switch at t, he had to conclude prior to t that Chuck was not going to do it, so that he could
set about doing it himself. But the way Mackie interprets 'in the
circumstances', nothing about the possible world can differ from the
actual world prior to t. For David to have acted at t, he would
have had to do something prior to t that did not happen in the actual
world--that is, conclude that Chuck was not going to do it. But by
the assumption that the possible world is like the actual world up to
t, David probably could not do this concluding in time.

A minor adjustment in the case will show that the given Mackie
interpretation of 'in the circumstances' yields a linked overdetermination counterexample to his analysis. The idea is to enable David's
reaction to come after t, but still be in time to set off the effect
at the same time that it would have been set off by Chuck's switch
flipping. I introduce a (roughly) two-second time-delay relay
between the switch and the light, creating a slightly more complicated
chain between cause and effect. Chuck's switch flipping at t activated
the relay (located near the switch), which at t plus two closed the
circuit to the light (on the ceiling). The workings of the relay were
out in the open and the circuit could also be closed by a touch of a
screwdriver. David had a screwdriver and knew how to do this.

In this case ('R2') Chuck's flipping the switch at t caused the
light to come on, but David was backing Chuck up, and would have
closed the light circuit at the relay at t plus two, if Chuck had not
flipped the switch at t. So, even though Chuck's flipping the switch
caused the coming on of the light, it is false that[3] in the circumstances (as Mackie interprets this) if Chuck had not flipped the switch
at t, the coming on of the light would not have occurred. David's
touch of the screwdriver would have caused it.

(For those who are finicky about exact timing, I can introduce a
cyclical time-delay relay switch that is set to go off at the end of
any five-second period, if it has been activated at any time during
that period. David's timing then need not be absolutely precise.)

So, according to Mackie's interpretation of 'in the circumstances'
Chuck's switch flipping, contrary to the assumed facts in the case,
would not have caused the coming on of the light: we take a possible
world that is just like the actual world up to t, the time at which
Chuck flipped the switch; we keep Chuck's switch flipping out from the
possible world, but otherwise let it run as the actual world. Now at
about t David notices that Chuck's switch flipping does not occur, so
he, following his back-up plan, uses a screwdriver and sets off the
relay, which fires at t plus two. The effect occurs in this possible
world so, by Mackie's formula, it is false that in the circumstances
the coming on of the light would not have occurred if Chuck's switch
flipping had not occurred. So, even though Chuck's act was the cause
of the coming on of the light, Mackie's analysis combined with his
interpretation of 'in the circumstances' implies otherwise.

62

Suppose instead that in order to avoid David's intervention in the
possible world, the phrase 'in the circumstances' is interpreted to
imply that everything in the possible world, except for the removal
of the cause, is like the actual world up to the occurrence of the
effect, and the world is allowed to run on from there. The trouble
with this is that it preserves all the effects of the cause, including
the activating of the relay (as well as David's not intervening), so
the effect occurs, and the Mackie counterfactual again is false.

Next, as a third possibility, suppose that Mackie interprets 'in
the circumstances' to include all the circumstances surrounding the
actual causal chain. This also would not do, for although it would
include David's not setting off the relay, it would also include his
firm realizable intention to set it to go off at the same time, if
Chuck did not flip the switch at t. The combination of all the circumstances surrounding the actual causal chain together with the
assumption that Chuck did not flip the switch leads us to an inconsistent commitment both to David's doing it and his not doing it.[4]

A fourth possibility: Suppose that 'in the circumstances' is interpreted to mean that there was at least one set of circumstances such
that if the cause had not occurred, the effect would not have occurred.[5] This could be a set that includes David's not intervening but
does not include his realizable intention to intervene. The crucial
part of the analysans then would be: 'there existed a set of circumstances such that if X had not occurred, then Y would not have
occurred'.

According to this approach, we can select any of the existing
circumstances, thus leaving out any that we choose. This unfortunately opens the door too wide. It would make a cause out of a
back-up factor, for example. In the R2 case let us select as the
circumstances a set including all the mechanisms, but not including
Chuck's flipping the switch. Then if David's intention to intervene
(X) had not occurred, the coming on of the light (Y) would not have
occurred. So the back-up factor is established as the cause under
this approach.

There no doubt is a limitless array of possible interpretations of
'in the circumstances', and one cannot attempt to meet them all in
advance. But given what Mackie said, it appears that his counterfactual analysis is defective. As Davis said in concluding his examination of the Lewis approach, "The prospects for a counterfactual
analysis of causation look dim." (1980, p. 175).

5. Summary

J. L. Mackie's counterfactual analysis of singular causal claims
is vulnerable to linked overdetermination counterexamples. When the
effect is treated as a fact-result or an infinitive result, Mackie
would in some cases agree that the putative cause was not necessary

in the circumstances, but he then mistakenly would hold it not to be the cause. When the effect is treated as an event-result, he mistakenly holds the cause to have been necessary in the circumstances. This seems so in spite of his suggestion of the effect's changing with a change in the cause, and his and other interpretations of 'in the circumstances'.

Notes

[1] For suggestions and comments I am indebted to David Bantz, John Barker, John Canfield, Nancy Cartwright, Jaegwon Kim, Adele Laslie, John Mackie, Robert Monk, Stephen Norris, Shirley Pendlebury, Denis Phillips, Shekhar Pradhan, and Stephen Wagner. I regret that Mr. Mackie did not have the opportunity to see this version.

[2] Henceforth '_Cement_'; undated numbers refers to pages in _Cement_.

[3] Mackie used the locution 'We cannot say that . . .' rather than 'It is false that . . .' because he did not believe that counterfactuals (and hence singular causal claims) can be true or false. Someone who agrees with Mackie on this point can substitute 'We cannot say that . . .' in appropriate places, if so desired.

[4] Berofsky (1977, pp. 107-8) has made a point similar to this one and the one two paragraphs above with respect to an example in which the effect and the would-be effect were different: Scriven's stroke-heart-attack example. (1964, p. 411).

[5] Mackie has suggested this in a letter to me.

<u>References</u>

Berofsky, B. (1977). "J.L. Mackie: <u>The Cement of the Universe.</u>" <u>The Journal of Philosophy</u> 74: 103-118.

Beauchamp, T.L. and Rosenberg, A. (1977). "J.L. Mackie, <u>The Cement of the Universe.</u>" <u>Canadian Journal of Philosophy</u> 7: 371-404.

Brand, M. (1977). "Identity Conditions for Events." <u>American Philosophical Quarterly</u> 14: 329-337.

Bunzl, M. (1980). "Causal Preemption and Counterfactuals." <u>Philosophical Studies</u> 37: 115-124.

Davidson, D. (1967). "Causal Relations." <u>The Journal of Philosophy</u> 64: 691-703.

-----------. (1969). "The Individuation of Events." In <u>Essays in Honor of Carl G. Hempel.</u> Edited by N. Rescher, <u>et al.</u> Dordrecht: Reidel. Pages 216-234.

Davis, W. (1980). "Swain's Counterfactual Analysis of Causation." <u>Philosophical Studies</u> 38: 169-176.

Earman, J. (1976). "J.L. Mackie, <u>The Cement of the Universe.</u>" <u>The Philosophical Review</u> 85: 390-394.

Hart, H.L.A. and Honoré, A.M. (1959). <u>Causation in the Law.</u> Oxford: Clarendon Press.

Lewis, D. (1973). "Causation." <u>The Journal of Philosophy</u> 70: 556-567.

---------. (1979). "Counterfactual Dependence and Time's Arrow." <u>Noûs</u> 13: 455-476.

Mackie, J.L. (1974). <u>The Cement of the Universe.</u> Oxford: Clarendon Press.

Scriven, M. (1964). "Critical Study: <u>The Structure of Science.</u>" <u>The Review of Metaphysics</u> 17: 403-424.

-----------. (1966). "Causes, Connections and Conditions in History." In <u>Philosophical Analysis and History.</u> Edited by W.H. Dray. New York: Harper & Row. Pages 238-264.

The "Established Maxim" and Causal Chains[1]

A. David Kline

Iowa State University

1. Introduction

There is an ancient principle for which a number of writers on time
and causation feel a certain allegiance. Bertrand Russell puts the
idea with his characteristic flare for style: "... it seems strange--
too strange to be accepted, in spite of bare logical possibility--that
the cause, after existing placidly for some time, should suddenly
explode into the effect, when it might just as well have done so at an
earlier time, or have gone unchanged without producing its effect."
(Russell 1912, pp. 178-179). The maxim has a habit in the history of
thought of lying unnoticed but once spotted leading its viewer to less
than cautious theses. With maxim in hand, David Hume (1739, p. 76)
concludes that simultaneous causation is impossible, Myles Brand
(1980) concludes that every causal relation is one of simultaneity,
Russell$_2$ (1912) declares that the very notion of causation is inco-
herent.[2]

To deny the maxim would be strange; let us assume it is true and
see where it leads. The above conclusions are excesses, nevertheless
some surprising results do follow. Given the maxim and that there are
causal chains, then it must be the case that time has a discrete order
or that at least some causally related events form a dense (in the
mathematical sense) sequence.

2. Prima Facie Conflict Between the Maxim and Causal Chains

The live options with respect to the temporal dimensions of an
event are that events have an infinitesimal duration or that they
exist unchanged for a finite period of time. For certain practical or
ordinary purposes it might be convenient to think of an event as
existing for a finite period of time during which it changes, the
Battle of the Marne, for example. It will be assumed that this possibility is to be analyzed in terms of the more fundamental possibili-

---

PSA 1982, Volume 1, pp. 65-74

ties.

Consider the temporal relations between a causally related pair of events, e and f. The following possibilities suggest themselves:[3]

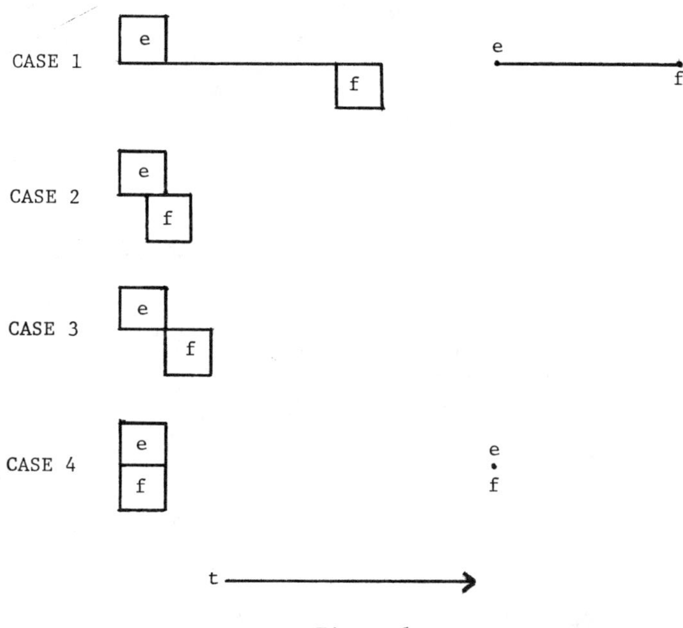

Figure 1

It will be instructive to consider the four cases in light of the maxim but first a sharp positive statement of the maxim is needed.

> (K) For events e and f, if e occurs at T and e is the
> cause of f, then there is no time $T+n$ (where $n>0$)
> such that e has occurred and f has not occurred.
> (Substitutions for f are understood to range over
> effects of e which are not effects in virtue of the
> transitivity of causation.)

In case 1, for both event ontologies, clearly there is some time between the occurrence of e and the occurrence of f. In cases 2 and 3, infinitesimal events have no counterpart of the static conception. Infinitesimal events can neither partially overlap nor be contiguous with one another. For static events the maxim prohibits both case 2 and 3, since in each case there is some time after the occurrence of e and before the occurrence of f. Case 4, simultaneous causation, appears to be the only possibility consistent with the maxim.

But if all causal relations must be understood as in case 4, the problem of causal chains awaits. Hume gives the classical statement of the problem. He believed that if all causal relations were relations of simultaneity there would be no causal chains. As if that were not absurd enough, he went on to argue that there would not even be time.

> 'Tis an establish'd maxim both in natural and moral
> philosophy, that an object, which exists for any time
> in its full perfection without producing another, is not
> its sole cause; but is assisted by some other principle,
> which pushes it from its state of inactivity, and makes
> it exert that energy, of which it was secretly possest.
> Now if any cause may be perfectly co-temporary with its
> effect, 'tis certain, according to this maxim, that they
> must all of them be so; since any one of them, which
> retards its operation for a single moment, exerts not
> itself at that very individual time, in which it might
> have operated; and therefore is no proper cause. The
> consequence of this wou'd be no less than the destruc-
> tion of that succession of causes, which we observe in
> the world; and indeed, the utter annihilation of time.
> For if one cause were co-temporary with its effect, and
> this with _its_ effect, and so on, 'tis plain there you'd
> be no such thing as succession, and all objects must be
> co-existent. (Hume 1739, p. 76).

Hume claims in this curious argument that if there is one case of simultaneous causation then all causal relations will be simultaneous. I am not concerned here with this feature of the argument but rather the absurdity that is alleged to follow from _all_ causal relations being simultaneous.[4] Richard Taylor gives a modern version of the idea.

> ...that _all_ causes are contemporaneous with their effect...
> is probably false. For if this were true, then there
> would be no such thing as a causal chain. Indeed, it
> would be impossible for any two events whatever to have
> any causal connection with each other in case there were
> any lapse of time at all between their occurrences. If
> some event A, for example, causes B, which in turn causes
> C, which in turn causes D, then in case every cause is
> simultaneous with its effect, it follows that when A
> occurs, then the others, and indeed every event in the
> universe that is in any way causally connected with A,
> must occur _at the same time_. This, however, is false.
> There are causal chains, and sometimes temporally sepa-
> rated events are causally related in one way or another.
> (Taylor 1966, p. 38).

In typical philosophical fashion, a few simple and apparently innocent moves when combined have perplexing results. Collecting the

68

skeleton of the argument we have: (1) The established maxim. (2)
If the maxim then all causal relations are simultaneous. (3) If all
causal relations are simultaneous then there are no causal chains.
(4) Therefore, there are no causal chains.

3. Brand's Compatibilism

Myles Brand (1980) in his paper "Simultaneous Causation" has provided what he regards as the proper way to avoid the above conclusion.
His work will be considered in some detail, for in discovering its
inadequacies a correct understanding of the maxim and its relation to
causal chains will emerge.

According to Brand, the maxim does require simultaneous causation;
nevertheless causal chains are possible. The real problem is with
premise (3). Central to the defense of (3) in Hume's and Taylor's
arguments is what can be called the causal chain requirement or (CCR):
There are causal chains only if there are causally connected but
temporally separated events. Brand argues against (CCR). He does so
by providing an alternative to the standard picture of causal chains.
Consider an example from pool, the mother of all causal insight. The
cue ball strikes the 4 ball which strikes the 13 which unfortunately
knocks the 8 ball into the side pocket. To simplify matters suppose
that in each collision all the momentum is transferred--lots of
reverse English. This sequence fits what Brand suggests (1980, p. 141)
as a model of a causal chain.

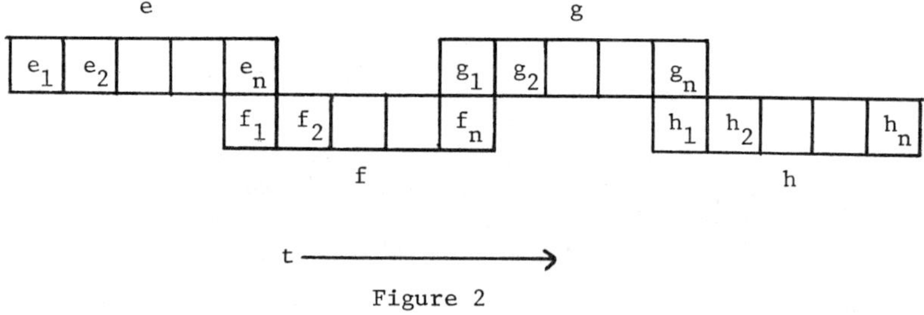

Figure 2

In Figure 2 only the pairs of events, e_n/f_1, g_1/f_n, g_n/h_1, (the
collisions in the example) are causally connected. e, f, g, and h are
sequences of events, the movements of the balls. These events
supposedly are not causally connected, but are nomically connected.
Hence we have a _prima_ _facie_ situation that violates (CCR)--all the
causal relations are relations of simultaneity yet there is a causal
chain.

There are two serious problems with this alternative picture of
causal chains. (a) The events, e_1--e_n, f_1--f_n, and so on (the

movements of the balls in the example) appear to be causally related.
To deny this seems little more than stipulating a new way of talking.
Brand clearly appreciates the problem but his solution is unacceptably
weak. He asserts that the events which compose a "link" in the chain
are related not causally but by a kind of physical necessity. As
evidence for this he points out that they support the appropriate
counterfactuals; for example, in the pool case the following is true:
(i) If the cue ball had not been at the e_2 position it would not
have been at the e_4 position.

This, of course, is of little help since the counterfactual
supporting proclivity is equally accounted for by the supposition
that the events are causally related. Brand goes on to try to
strengthen his case by giving examples of counterfactuals that are
not supported by causal relations. His examples (1980, p. 149) are:
(ii) If the coming of this night had not occurred, then the coming of
this day would not have occurred; and (iii) If Jones had not raised
his arm in order to attract the chairman's attention, he would not
have raised his arm.

If Brand is correct, which he is, in urging that there are non-
causal relations which support counterfactuals, what does that show?
At best, only that events in a link <u>could be</u> related by a non-causal
physical necessity. We want a reason to believe that they are so
related.

Perhaps Brand wants us to understand physically necessary relations as one or the other relation behind his sample counterfactuals.
(iii) does not help for as Brand himself remarks (iii) rests on a
connection of events in some "complicated part whole or conventional
way." It is most implausible to think that the events composing the
cue ball's path are related by a part whole or conventional relation.

(ii) is more interesting. Similar cases are familiar from discussions of explanation. The barometer/storm case is well known. Brand
points out that what lies behind the counterfactuals in these cases is
not a causal relation but a correlation. The falling of the barometer
does not cause the storm and similarly for night and day. The crucial
point here is that we do not regard correlations as supporting counterfactuals unless we are armed with a deeper <u>causal</u> story about why
the correlation holds. For example, the drop in pressure causes the
barometer to drop and also causes the storm to occur. Since there is
no causal story which explains the correlation of the ball at the e_2
and e_4 positions other than that e_2 causes e_4, it is unpromising to
understand Brand's physical necessity as correlation.

I suggest that Brand has done little more than stipulate that
causal talk be reserved for the interaction of bodies, such as the cue
ball and the 4 ball in the example, and that it not be applied to

sequences involving a single material object. He has given no theoretically compelling reason for doing so nor has he spelled out the
alternative kind of nomic relation involved in sequences.

(b) Brand reads the maxim as necessitating simultaneous causation.
It is hard to see why the insight expressed in the maxim should be
restricted to _causally_ related events. Put directly, even if one
grants that the events in sequences, such as, e, f, g, and h, are
connected by some special non-causal necessity, a version of the
maxim should still apply to them--rendering the elements simultaneous.
The lack of parsimony in Brand's view foretells its _ad hoc_ nature.

4. How to Read the Maxim

Brand was forced into his rather unpersuasive "solution" by a misinterpretation of the maxim. In his paper he concentrates on Hume's
version of the maxim. Collecting the relevant parts from Hume's
argument quoted earlier we have:

(H) ...an object, which exists for any time in its full
     perfection without producing another, is not its sole
     cause; but is assisted by some other principle, which
     pushes it from its state of inactivity, and makes it
     exert energy, of which it was secretly possest.

. . .

     any one of them [purported cause], which retards its
     operation for a single moment, exerts not itself at
     that very individual time, in which it might have
     operated; therefore is no proper cause. (Hume 1739,
     p. 76).

Brand's gloss of (H) is (B).

(B) For any events _e_ and _f_ and time interval _t_, if _e_ occurs
     during _t_ and _e_ is the cause of _f_, then _f_ occurs during
     _t_. (Brand 1980, p. 147).

(B) comes down to a straightforward claim that all causal relations
are relations of simultaneity. Given (B) one must argue that somehow
simultaneous causation is compatible with causal chains, i.e., argue
against premise (3). We have already seen Brand's efforts along this
line.

Reading (H) as (B) is natural enough. After all, the earlier conclusion arrived at by examining cases 1-4 was that only simultaneous
causation was compatible with the maxim. Nevertheless that was a
mistake. The real problem is with premise (2), if the maxim then all
causal relations are simultaneous. The earlier rigorous statement of
the maxim, (K), is not, even though it may appear to be, logically
equivalent to Brand's (B). Actually (B) is a consequence of (K) and

certain implicit assumptions. Given the naturalness of the assumptions it is easy to slip into not distinguishing (K) and (B). The crucial implicit assumptions are: A_1--Time is dense; and A_2--For any cause, e, there is a first member in the chain of e's effects.

In what follows two possible causal worlds will be described. The worlds are constructed in such a way that (K) is true and (B) is false. They will show three important points: (i) (B) and (K) are not logically equivalent. (ii) Premise (2) is false. The maxim can be true, understood as (K), without necessitating simultaneous causation. (iii) The truth of (ii) rests crucially on denying A_1 or A_2. To avoid universal simultaneous causation and the concomitant problem of causal chains either A_1 or A_2 must be denied. The situations are constructed in such a way that along with the previous discussion it should be obvious that if A_1 and A_2 are true, then (B) will follow from (K).

Situation I: Assume that time is discrete and that events have an infinitesimal duration. For simplicity let time be modeled by the positive integers. Furthermore, suppose that effects always occur at the moment immediately following their cause. For example if e is the cause of f and e occurs at t_1, then f occurs at t_2. In this world (K) will be true since there is no time, t+n, between e and f. But (B) is false since e and f do not occur at the same time. We have a situation where (K) is true and simultaneous causation is not necessitated. This was accomplished by denying A_1, i.e., asserting that time is discrete.

Situation II: Assume again that events are of infinitesimal duration but now that time is continuous. Furthermore, assume that causally related events form a dense series. Now consider e, a cause occurring at t. Since by hypothesis causally related events form a dense series, there is no immediate effect of e (that event that occurs later than t and is the first effect of e). Since e has no immediate effect, (K) will be true in this world. There is no specific time, t+n, that occurs after e and before e's immediate effect. Furthermore it does not follow that e's effects must be simultaneous with e. It can be shown mathematically that a dense sequence of causally related events can "compose" a causal chain of finite temporal duration. Again we have a situation where (K) is true and simultaneous causation is not necessitated. This was accomplished by denying A_2, i.e., asserting that causally related events form a dense sequence.[5]

Before summarizing the argument a few final points are in order. Though the goal in this paper has not been historical scholarship it should be indicated that Brand's paraphrase of (H) fails to capture Hume's text or intentions. First, Brand incorporates into the maxim a specific view of events, viz., as occurrences that endure, unchanging, for a finite interval of time. Hume's maxim is silent on the issue of event-time ontology. Secondly, Hume uses the maxim as part of an argument, as we have seen, against the possibility of simultaneous causation. If Hume understood the maxim as (B) it would lead, and in a very obvious way, to the contradiction of the conclusion he wished to establish. Lastly, Hume's point in (H) is that <u>a cause can not retard its operation at a single moment</u>--the idea captured by (K).

(K) fits nicely with the kind of rationale expressed by Hume that inclines us to accept (H). The established maxim is actually a specific version of the principle of sufficient reason. Suppose e and f are events and that e occurs at T and f occurs at a later time $T+m$. Further suppose, contrary to (K), that there is a time $T+n>T$ and $T+m>T+n$. Now we can ask why f came into existence at $T+m$ rather than $T+n$. Since there is no sufficient reason forthcoming we are forced to deny that e is the real or whole cause of f.

5. Conclusion

It has been argued that if one accepts the established maxim then one must be prepared to accept the view that time is discrete or that some causally related events form a dense sequence. The argument began by assuming the truth of the maxim and noticing that it appeared to lead to universal simultaneous causation and in turn to the problem of causal chains. Brand's way out was rejected. It was suggested that there is a way to read the maxim that does not lead to universal simultaneous causation and hence renders the problem of causal chains moot. That reading, in order to avoid simultaneous causation, requires either discrete time or a dense causal order.

<u>Notes</u>

[1]I would like to thank my colleague, William Robinson, for helping me think about this topic. A version of this paper was read at Marquette University.

[2]Though it may not be obvious the maxim has consequences for a standard issue in the philosophy of science. In Kline (1980) I showed that the uncritical acceptance of cases of simultaneous causation has lead to premature counterexamples to traditional theories of explanation and premature alternative theories. I went on to argue on empirical grounds that there are no genuine cases of simultaneous causation. The maxim is relevant to metaphysical discussions of the possibility or necessity of simultaneous causation.

[3]The style of the figures follows that used by Brand (1980).

[4]Beauchamp and Rosenberg, (1981) Fogelin, (1976) and Munsat (1971) discuss the argument as a whole. Unfortunately none of these authors appreciates the dependency of the argument on Hume's discussion of the nature of time which precedes the argument.

[5]Situation II does not depend crucially on the assumption of infinitesimal events. The possibility can be formulated just as well with the static conception of events.

References

Beauchamp, T.L. and Rosenberg, A. (1981). _Hume and the Problem of Causation._ Oxford: Oxford University Press.

Brand, M. (1980). "Simultaneous Causation." In _Time and Cause. (Philosophical Studies Series in Philosophy,_ No. 19.) Edited by Peter van Inwagen. Dordrecht: D. Reidel Publishing Company. Pages 137-153.

Ducasse, C.J. (1969). _Causation and the Types of Necessity._ New York: Dover Publications Inc.

Fogelin, R.J. (1976). "Kant and Hume on Simultaneity of Causes and Effects." _Kantstudien_ 67: 51-59.

Hume, D. (1739). _A Treatise of Human Nature._ London: John Noone. (As reprinted (ed.) L.A. Selby-Bigge. Oxford: Oxford University Press, 1888.)

Kline, A.D. (1980). "Are There Cases of Simultaneous Causation?" In _PSA 1980,_ Volume One. Edited by P.D. Asquith and R.N. Giere. East Lansing, Michigan: Philosophy of Science Association. Pages 292-301.

Munsat, S. (1971). "Hume's Argument That Causes Must Precede Their Effects." _Philosophical Studies_ 22: 24-26.

Russell, B. (1912). "On the Notion of Cause." _Proceedings of the Aristotelian Society_ 13: 1-26. (As reprinted in Russell (1957). Pages 174-201.)

----------. (1957). _Mysticism and Logic._ New York: Anchor Books.

Taylor, R. (1966). _Action and Purpose._ Englewood Cliffs, NJ: Prentice Hall.

Part III

Scientific Realism and Observation

Hermeneutical Realism and Scientific Observation

Patrick A. Heelan

State University of New York at Stony Brook

This paper will summarize positions explained and defended more
fully in my book, Space-Perception and The Philosophy of Science
(Heelan 1982).[1] The philosophical genre of this paper is that of a
hermeneutical phenomenology, it addresses questions from the point of
view of such writers as, E. Husserl, M. Heidegger, M. Merleau-Ponty,
P. Ricoeur, and H-G. Gadamer (see references).[2] Those who are not
familiar with this kind of writing may well find the exposition too
brief to be persuasive; those who are, may wonder at the audacity of
applying such methods of analysis to the natural sciences. It is
unlikely that either group will be entirely pleased with the content
of what I have to say.

My principal theses are the following: (1) reality is the content
of World,[3] and this comprises whatever is or can be given directly in
a public way within perception; what is given in perception--a perceptual object--is not an "internal representation", but a state of
the World. This thesis states the ontological primacy of perception.
(2) Acts of perception are epistemic, contextual, and hermeneutical.
(3) The objects of scientific observations are perceptual objects. I
shall not, for lack of time, dwell on the first two theses, though
what they imply should become clear in the course of the paper. I
shall then plunge directly into the third thesis.

1. Scientific objects as perceptual objects

A theoretical scientific entity is a part of a substructure detectable only by instruments, and hidden to unaided perception. By this
fact, theoretical scientific entities are, I claim, no more than
candidates for inclusion in the real order; to be accepted for inclusion in that order, they must present themselves with the proper
credentials which, according to the principle stated above, means to
be capable of being exhibited in genuine horizons of scientific observation; in these such quantities are or could be manifestly and

---

PSA 1982, Volume 1, pp. 77-87

directly given as perceptual objects to a public of experienced scientific observers. I claim that there are such horizons of scientific
observation.

2. Scientific Observation as 'Reading' of 'Text'

The argument for this thesis is based (1) on an analogy between
reading a text and "reading" an instrument--both are hermeneutic, and
contextual acts--and (2) on the conditions for a perceptual act, which
involve the "picking up" of "information" about the present and actual
state of the World from the environment directly _via_ the Body.

Theoretical entities or quantities become known by attending to the
response of appropriate empirical procedures, such as the use of
measuring instruments. There are, however, two ways of attending to
the response of an instrument: the first way is where the instrumental
response is treated as a physical event associated on the basis of a
theoretical inference with a certain theoretical scientific quantity--
to do this, one must know the relevant scientific theory; the second
way is where the value of that theoretical quantity is 'read' directly
from this response--to do this, one must be experienced and skillful
in the use of instruments. I want to focus on the latter, and shall
call it 'reading' (with single quotes to mark a special sense).

To the extent one 'reads' the thermometer, the thermodynamic argument remains in the background, being merely the historical reason
why thermometers came to be constructed in the first place. One can
'read' a thermometer. however, whether or not one knows anything
formal about thermodynamical theory.   Provided the instrument is
standardized, and so can function as (what I call) a "readable
technology", _the_ _instrument_ _itself_ _can_ _define_ _the_ _perceptual_ _pro-_
files _and_ _essence_ _of_ _temperature_ _to_ _this_ _culture_.

The process of 'reading' is something like this: a 'text' is
'written' causally on the thermometer by the environment under
standard circumstances (_ceteris_ _paribus_ conditions), this 'text' is
'read' as being 'about' a presented state of some scientific system,
in this example, the state of temperature; the piece of empirical
knowledge so acquired--the current temperature--is expressed in a
language that uses scientific terms, like "temperature", in a
descriptive way about the World. Such a process, as I shall argue
below, is essentially both _hermeneutical_ and _perceptual_.

3. 'Reading' as hermeneutical

The process of reading, understanding or interpreting a literary
text is guided by what Heidegger calls a "fore-structure of understanding", that is, an anticipation about the kinds of things or
objects about which the text speaks (Heidegger 1953, pp. 98-114, 188-
195). This fore-structure is also called a "hermeneutical circle", in
it is "hidden a positive possibility of the most primordial kind of
knowing. To be sure, we genuinely take hold of this possibility only

when, in our interpretation, we have understood that our first, last
and constant task is never to allow our fore-having (<u>Vorhabe</u>), foresight (<u>Vorsicht</u>), and fore-conception (<u>Vorgriff</u>) to be presented to us
by fancies and popular conceptions, but rather to make the scientific
theme secure by working out these fore-structures in terms of the
things themselves."(p. 195). These fore-structures may be part of
antecedently developed, already entrenched, cognitive systems, or
merely heuristic structures awaiting testing and deployment.

The task of hermeneutical interpretation does not have as its goal
just some or any understanding consistent with the text, but a reading
that attains to "the things themselves" (Husserl's term originally)
about which the text speaks. Such a task is not, as Heidegger says, a
work of arbitrary fancy, but controlled, on the one hand, by the
totality of the text and its parts, and on the other, by the fore-
structure of understanding that permits us to read the text as referring to specific kinds of things and objects. This fore-structure of
understanding, according to Heidegger, has three parts: (1) <u>Vorhabe</u>, a
set of praxes, embodiments, skills, etc. that mediate applications of
the descriptive categories or terms to that to which they refer, (2)
<u>Vorsicht</u>, or a set of common descriptive categories, a common descrip-
tive language, as it were, and (3) <u>Vorgriff</u>, a particular hypothesis
about the subject matter in hand (p. 191).

The hermeneutical task is <u>circular</u> in a peculiar but "virtuous" or
"non-vicious" way, because it involves the simultaneous and mutual
determination of the (meaning of the) whole by the (meaning of the)
parts, and vice versa; on the one hand, the fore-structure of under-
standing--the "hermeneutical circle"--provides a conjectured meaning
for the text as a whole and for its parts, but, on the other hand,
what kinds of conjectures one entertains about the (meaning of the)
whole depends on clues scattered in the text itself. One moves from a
partial disjointed set of insights (let me say, clues) to an understanding of the whole and back to the not-yet-understood portions of
the text, the process guided by the attempt to discover the outlines
of "the things themselves".

A satisfactory solution to a hermeneutical inquiry would fulfil the
following conditions, (1) all the clues lie on (or <u>sufficiently</u> near)
the proposed solution, and (2) one is persuaded that none of the as
yet undiscovered clues lies <u>too far off</u> the proposed solution; the
terms "sufficiently", and "too far off" imply reference to the goals
and purposes of the inquiry. <u>Every hermeneutical inquiry then is
fraught with a certain ambiguity and uncertainty. It is also laden
with values, and directed by a teleology.</u>

When the text is imperfect or corrupt, or when the interpreter is
removed historically or culturally from the subject matter, the hermeneutical task is in addition accompanied by a special kind of effort, it is clouded by obscurity, and possesses an essential incompleteness (Gadamer 1965, pp. 235-240 and passim). The variety and
tentativeness of scientific traditions, that were the concern of Duhem

and Poincaré, and are currently topics of lively discussion in the
Conventionalist debate, are also connected with the possibility of
alternative scientific 'readings' of experimental data (Fleck 1935;
Hesse 1980; Holton 1973; Kuhn 1977, and Rorty 1979). I am, however, in
this paper less concerned with the indeterminacy aspect, than with
displaying the existence of a hermeneutical component in scientific
observations, the presence of which is easily obscured both by the
practised familiarity with which these are often performed, and by the
ideology that represents them as unproblematic, because "observational".

All hermeneutical processes possess the <u>dual</u> <u>structure</u> associated
with the acquisition or expression of <u>information</u>. The term "information" refers both to the signs that are read and to the meaning given
to those signs. I shall distinguish between these by calling the text
or the 'text', "information$_1$", and its meaning or content, "informa-
tion$_2$". Note that, in the literary analogue, syllables, phonemes, or
other linguistic signs--information$_1$--once read for their meaning--
information$_2$--cease to be objects in the World, like houses or trees,
and become more like <u>windows</u> <u>to</u> <u>a</u> <u>room</u> that by their (more or less)
transparent quality give direct access to the contents of the room
beyond. One does <u>not</u> <u>perceive</u> the syllables of a text: one <u>reads</u> them.
Polanyi has already pointed this out (Polanyi 1964, chap. 4). In a
reading, the physical character of the text disappears from direct
view leaving no objective trace whatsoever, it becomes to that extent
"non-objective", and I take that to mean "belonging to the conditions
of the knowing subject"; it becomes in fact physically part of the
cognitive subject <u>as</u> <u>embodied</u>; it becomes a modulation of the particular somatic information channel of the embodied cognitive subject used
in this particular reading.

Although, on the whole, one is oblivious to the syllables and marks
as things on paper, one is not totally unaware of them: it is through
them that one is guided through the meaning of the text. There is
connected with this a subliminal awareness of pleasure, or perhaps of
frustration and discomfort, that arises from the activity itself, from
the clarity or unclarity of its typographic and syntactical form, from
its natural or forced rhythm, from the musical quality of the sounds,
and from the resonances of heard or imagined speech. In all of this,
lie the esthetic qualities of a text (sometimes, called the "experience" of reading).

The kind of transformation I want to describe is like that in which
the syllables or phonemes of a strange language at first engage our
attention as curious objects for possible theoretical study, and end
up by being dropped from awareness when we have become familiar with
the language and have learned to read the syllables as text or to
listen to the phonemes as spoken words. When one knows the language,
one obtains direct access to the meaning of a text or spoken word.
Such meanings are expressed in judgments about the subject matter--the
things themselves--referred to by the written text or spoken word,
whether this be a historical event, or a fictional narrative, or an

abstract logical construction. The transformation just described, wherein intermediaries (information$_1$) in the acquisition or expression of empirical data (information$_2$) "drop out of consciousness", has been noted, for example, by Schrödinger (Schrödinger 1958, p. 8) who ascribed such transformations to processes perfected during the long course of evolution, however, it is a commonplace that many processes perfected through the painful process of learning share this characteristic that intermediaries drop out of consciousness. The paradigm example is reading, but there are other processes that are similar, like playing a musical instrument, sight reading from music, driving a car, and reading an instrument: to a suitably experienced person, processes like these have "the same subjective ease and immediacy as the simplest perceptions." (Sloman 1980, p. 403).

4. 'Reading' as perceptual

The instrumental response as a 'text' shares in the information theoretic aspect of literary texts. In its genesis, a literary text is written in a standard typography, using the vocabulary and background root metaphors paradigmatic for a particular domain, etc., by a writer, following out the rules of language (langue, in the structuralist sense). By comparison, a 'text' in its genesis is 'written' by the ambient environment on a standard instrument under paradigmatic circumstances, and the production follows scientific laws and theories (in something like a model-theoretic or structuralist sense). In both cases, the process of interpretation--as opposed to genesis--involves (1) the physical causality of the text or 'text' stimulating "resonance" in some somatic information channel of the reading or 'reading' subject, and (2) the use of some hermeneutical circle to get the meaning of the text or 'text'. In both cases, when the hermeneutic task is accomplished, one is in direct possession of that meaning. The meanings, however, in the two cases are of different kinds.

While literary texts may speak about the World, they do not manifest, show, or exhibit states of the World. A 'text', however, can and usually does exhibit some state of the World actually present and manifesting itself. (Note, however, that a scientific observation may also be about some past state of the World, but this is something I shall not deal with here.) Returning, for the purpose of illustration, to the thermometer; the position of mercury on the scale functions as a 'text'. Through a 'reading' of this 'text', one gains knowledge in the form of judgment of the current thermodynamic temperature, "The present ambient temperature is [say] 70°"; this judgment is empirical, direct, and uses scientific terms descriptively of the World. I now claim that this 'reading' fulfills all the characteristics of perceptual knowledge. These are (1) that it is direct knowledge (that is, it is not mediated by inferences, nor does it terminate at an "internal representation" or a "model" of the known object, constructed, say, out of sensations or using logical or mathematical structures); (2) that it "picks up" information (information$_2$) from experience, this connotes the existence of a somatic information channel constituted in part, for the (potentially) skilled or experienced scientist, by some standard instrument (or "readable" technology), the "resonant" states (or states of information$_1$) shape what it is about the object that is revealed, and are coded (by the instrument) for the profiles of the object known (information$_2$); (3) that it is hermeneutical, that is, it acquires its meaning through the employment of a hermeneutical circle using the terms of a scientific theory descriptively; (4) that it's object, a state of the World, is experienced as given directly in a public way to the knower by the World, and (5) that it terminates in a judgment which purports to describe what is actually here and now existent in the public World, present and manifest to the knower appropriately embodied, and of which the expression is a descriptive statement using the terms of a scientific theory.

Like all perceptual knowledge, a scientific observation is not apodictic in the natural attitude, that is, though usually perceivers suitably trained perceive _unproblematically_ in the natural attitude, they do not perceive _apodictically_, for to perceive apodictically, is to be able to sample at will the perceptual profiles of the object, and presupposes the reflective attitude. However, like all perceptual knowledge, to the extent that profiles and invariants can be clearly articulated, a scientific observation too is capable of aspiring to apodicticity in the reflective attitude. It is this capacity to become apodictic that establishes the possibility of genuine scientific horizons in the World. This capability distinguishes the hermeneutics of literary texts, from the hermeneutics of perception and scientific observation: while both are underdetermined, the former is about meanings as unexhibited, possibly unexhibitable, possibilities (as states of the World), and the latter is about exhibited possibilities.

Returning to the question of the ontological status of scientific entities: if 'reading' a thermometer is a perceptual process, then, because of the ontological primacy of perception, I argue that thermodynamic temperature and other scientific entities like it, enter the World as recognizable objects of definite kinds described by scientific theories, and so belong to the furniture of the earth. The perceptual hermeneutical process of scientific observation is then a horizon- or World-building process, it is reality in the process of constitution.

A striking example of the use of instrumentation to generate a new field of perceptual knowledge is the Tactile Visual Substitution System (TVSS) of Collins and Bach-y-Rita (Morgan 1977, pp. 197-200). This is a device for use by the blind: it consists in a video camera attached to the temples, from which the output is fed to a ten-inch square array of electically driven vibrators in contact with the skin of the back or abdomen. The blind person who uses this device initially experiences only a tickling sensation, which then gives way to the experience of solid objects external to the subject, and located in a three dimensional space. Morgan writes: "In general, there is little doubt that the TVSS allows the blind to see. Or to 'see,' as Bach-y-Rita prefers to say. . . as experience with the TVSS proceeds, the judgements become more and more automatic, until they are ac-

complished in much the same way as a sighted person perceives objects."
(p. 203).

5. Observation as Theory-Laden

The claim that theoretical entities or states are observable runs
contrary to a basic principle both of traditional empiricism and
traditional phenomenology; for both, something is <u>observable</u>, exactly
if it could under appropriate circumstances be observed with the
<u>unaided</u> senses. This is also van Fraassen's position in his recent
work (van Fraassen 1980).

Now the position I have been defending is exactly the contrary: I
claim that theoretical states and entities are or become directly
<u>perceivable</u> (alternatively, "observable", in the new stipulated sense)
because the measuring process can be or become a "readable technology", a new form of embodiment for the scientific observer. In this
view, the term "observation" no longer means <u>unaided</u> perception. It
implies that theoretical states and entities are real and belong to
the furniture of the earth, <u>because (and to the extent that) they are
perceivable in the perceiver's new embodiment</u>. It also implies that
the nature and aim of scientific explanation is to make manifest the
processes and structures of the real, the real now being taken as what
is or can be given publicly in some World.

6. Some conclusions

(1) Hermeneutical activity is never unique, final and definitive;
consequently, the contents of perception, and a fortiori of scientific
observation, are never unique, definitive, final, absolute, apart from
history and particular social and cultural milieux (see Heelan 1982,
chaps. 10, 11, 14, and 15, and Ihde 1979).

(2) Acts of perception, and consequently, acts of scientific
observation, are always contextual, and involve an antecedent <u>realized
embodiment</u> and a set of learned skills for the subject (<u>Vorhabe</u>)--in
the case of scientific observation, this includes the use of standard
instruments. Different perceptual contexts may involve different sets
of contextual conditions; it should not then be a matter for surprise
that different contexts whether of perception or of scientific observation sometimes turn out to interfere with one another <u>performa-
tively</u> in a way that suggests an explanation of the phenomenon
referred to by Bohr, as <u>complementarity</u> (see Heelan 1970).

(3) If scientific inquiry is essentially hermeneutical, then the
same categories must be used <u>from the start</u> in both the explanandum
and in the explanans: this is contrary to the assumptions of the
<u>deductive-nomological</u> model of scientific explanation. Moreover, a
hermeneutical account would also rule out a purely <u>inductive</u> account
of scientific explanation: in this latter the descriptive scientific
categories are <u>first</u> exemplified in the explanandum and then used in
the construction of explanatory premises; in the hermeneutic account,

84

however, they arise simultaneously for the explanans and explanandum, and transform the observational field from the start of the inquiry.

(4) Turning to the current debates in epistemology and the philosophy of science: in the first place, my position is _neither_ that of _Scientific Realism_, _nor_ of _Instrumentalism_. It could be called "_Hermeneutical Realism_" (in Heelan 1982, I called it "Horizonal Realism " to stress the perceptual character of this kind of realism): science has the intent of describing the elements and structures of reality, hidden to (theoretically and instrumentally) unaided perception, but manifested as genuine perceptual essences-- _horizons_--with the aid of theoretically structured instruments used as "readable" technologies.

(5) In the second place, contrary to many theories of scientific reference, Hermeneutical Realism holds that there is no _identity_ _of_ _reference_ between individual objects given to unaided perception (such as this patch of sensed color, or other object of a manifest image) and individual objects of the relevant scientific image (such as this spectral mix of wave lengths, or other "theoretical" object); there are only many-to-one and one-to-many mappings of perceptual objects contextually defined within mutually incompatible but complementary contexts.

(6) In the third place, Hermeneutical Realism is _neither_ that of _Conventionalism_, nor of _Cultural Relativism_. Like them, however, it admits of plural incompatible empirically descriptive frameworks among which, I would hold, some are complementary, but it differs from both by insisting on the fact, necessity and limitations of _Vorhabe_ (in this case, trained bodily expertise and the ability to construct "readable" technologies) for linking descriptive categories to their appropriate empirical objects; the plasticity of _Vorhabe_ places limits on the descriptive frameworks that can be conventionally chosen or used in a culturally relative way, I surmise that these limits can only be known through empirical, historical, and cultural studies.

Notes

[1]In Part I of this book, I attempt to establish the thesis that under some very general conditions we can and do see a World of hyperbolic objects (shaped by a hyperbolic visual metric); once the appropriate theoretical model is set up, evidence for the thesis is provided by optical illusions, the history of pictorial art, and daily experience. The theses proposed in this paper are defended in Part II of this book.

[2]Phenomenology as a philosophical tradition has grown up around the critique of certain pervasive attitudes, such as _objectivism_ and _scientism_, that define the traditional status of experimental science

in our culture, and has attempted to counter these claims by appealing
to the intuition of directly experienced objects. Hermeneutics deals
with signs and their interpretation, particularly with the interpretation of literary texts (see Bleicher 1980). Other philosophers who
have used phenomenological methods about various aspects of the natural sciences are, to name a few, John J. Compton, Hubert Dreyfus, Gary
Gutting, Aaron Gurwitch, Don Ihde, Joseph Kockelmans, Theodore Kisiel,
Wolfe Mays, Hans Seigfried, Elizabeth Ströker.

[3]The terms "World", "horizon", and "Body", are technical terms. The
horizon of an object is taken in this paper to mean _inner horizon_,
this is the set of perceptual profiles of an object generated by its
essential law (perceptual essence). _World_ is the general background
reality context that is experienced as given to our perception together with the individual objects that we perceive. _Body_ (with capi-
tal letter) designates the perceiver, and connotes the specific embodiment or _somatic information channel_ used by the perceiver for a
specific class of perceptions; the somatic information channel may
comprise structures of the neurophysiological system, somatic processes such as hands, feet, etc., technological extensions ("readable
techologies"), as well as some stimulus field that is directly modu-
lated by the objects known.

References

Bleicher, J. (1980). _Contemporary Hermeneutics: Hermeneutics as Method, Philosophy, and Critique._ London and Boston: Routledge and Kegan Paul.

Fleck, L. (1935). _Entstehung und Entwicklung einer wissenschaftlichen Tatsache: Einführung in die Lehre vom Denkstil und Denkkollektiv._ Basel, Switzerland: Benno Schwabe & Co. (As reprinted as _Genesis and Development of a Scientific Fact._ (eds.) Thaddeus J. Trenn and Robert K. Merton. (trans.) Fred Bradley and Thaddeus J. Trenn. Chicago: University of Chicago Press, 1979.)

Gadamer, H.G. (1965). _Wahrheit und Methode._ Tübingen: Mohr. (As reprinted as _Truth and Method._ (trans.) G. Barden and J. Cumming. New York: Seabury Press, 1975.)

Heidegger, M. (1953). _Sein und Zeit._ 7th ed. Tübingen: M. Niemeyer. (As reprinted as _Being and Time._ (trans.) J. Macquarrie and E. Robinson. New York: Harper and Row, 1962.)

Heelan, P.A. (1970). "Complementarity, Context-dependence, and Quantum Logic." _Foundations of Physics_ 1: 95-110.

------------. (1982). _Space-Perception and the Philosophy of Science._ Berkeley and Los Angeles: University of California Press.

Hesse, M. (1980). _Revolutions and Reconstructions in the Philosophy of Science._ Bloomington: University of Indiana Press.

Holton, G. (1973). _Thematic Origins of Scientific Thought: Kepler to Einstein._ Cambridge, Mass.: Harvard University Press.

Husserl, E. (1954). _Die Krisis der europäischen Wissenschaften und die transzendentale Phänomenologie._ (_Husserliana_ Volume VI.) (ed.) W. Biemel. The Hague: M. Nijhoff. (As reprinted as _The Crisis of European Philosophy and Transcendental Phenomenology._ (trans.) D. Carr. Evanston, Ill.: Northwestern University Press, 1970.)

Ihde, D. (1979). _Technics and Praxis._ Dordrecht and Boston: Reidel.

Kuhn, T.S. (1977). _Essential Tension._ Chicago: University of Chicago Press.

Merleau-Ponty, M. (1945). _Phénoménologie de la Perception._ Paris: Librairie Gallimard. (As reprinted as _Phenomenology of Perception._ (trans.) Colin Smith. London: Routledge and Kegan Paul, 1962.)

Morgan, M. (1977). _Moleneux's Question: Vision, Touch, and the Philosophy of Perception._ Cambridge: Cambridge University Press.

Polanyi, M. (1964). _Personal Knowledge._ New York: Harper and Row.

Ricoeur, P. (1981). _Hermeneutics and the Human Sciences._ (ed. and trans.) J.B. Thompson. Cambridge and New York: Cambridge University Press.

Rorty, R. (1979). _Philosophy and the Mirror of Nature._ Princeton: Princeton University Press.

Schrödinger, E. (1958). _Mind and Matter._ Cambridge: Cambridge University Press.

Sloman, A. (1980). "What Kind of Indirect Process is Visual Perception?" _The Behavioral and Brain Sciences_ 3: 401-404.

van Fraassen, B. (1980). _The Scientific Image._ Oxford: Clarendon Press.

<u>The Historical Objection to Scientific Realism</u>

Jarrett Leplin

University of North Carolina, Greensboro

1. Realist Assumptions

    Scientific Realism, as I shall understand it, holds that (a) the purportedly referential terms of current theory are genuinely referential, and (b) the theoretical, lawlike statements of current theory are at least approximately or partly true. Realism recommends itself as an explanation of the success of current science at the level of explanations and predictions of experientially accessible phenomena. That is, the case for realism assumes:

    (1) Current theory in fact achieves such success, and achieves it on balance, <u>i.e.</u>, is free of serious disconfirmations.
    (2) Realism explains such success; <u>i.e.</u>, the truth of (a) and (b) would justify us in expecting current theory to be successful.
    (3) There is no viable nonrealist explanation of the success of current theory.
    (4) No features of current theory independent of predictive failure discredit (a) or (b).

I shall comment briefly on each of these assumptions.

    (1) That science is successful, and markedly so relative to other forms of intellectual inquiry at least in modern times, should be uncontroversial. Some explanation of scientific success may reasonably be required as a condition of adequacy for any philosophy of science. But it does not follow that success can be or ought to be attributed to any single aspect of science--that it has any single, comprehensive explanation. Success comes in many forms, and the fact that realism is unnecessary or insufficient as an explanation of some of its forms is no refutation of an explanationist defense of realism. Thus the realist may locate the reliability of scientific conclusions in the methodology of testing, the unifying character of theories in the

PSA 1982, Volume 1, pp. 88-97

methodology of theory construction, or the longevity of theories in the evolutionary competition among scientific ideas. It is incumbent on him only to identify some form(s) of success which the truthlikeness of theories is uniquely able to explain.

(2) From the fact that the truth of theories would explain (some forms of) their success it does not follow that their approximate or partial truth explains this. At least two objections may be brought against this inference: ·(i) there is available no analysis of approximate truth on which the approximate truth of a theory explains its success; (ii) because of the involvement of auxiliaries in the extraction of predictions from a theory, a theory could easily be unsuccessful even if approximately true.

Regarding (i), extant analyses of approximate truth typically presuppose success, identifying the degree of approximation to truth with the degree of approximation of actual empirical results to predicted results. Thus the margin for error in experimental results is often supposed to divide approximate truth from falsehood, although that division is thereby relativized to the state of technology. Even philosophers who accept a correspondence analysis of truth typically adopt a coherence analysis of approximate truth: approximately true theories do not lead us, in association with other beliefs, to expectations in conflict with independently decidable statements. (Giere (1979) is a recent example; Popper (1963), a classic one.) Thus theories are sometimes described as approximately true if they "cohere", within reasonable limits, with the established body of scientific knowledge at the empirical level. A statement could give us a description of reality radically different from that given by statements true in the correspondence sense and yet satisfy this condition for approximate truth. It should be clear that realism can make no use of such a notion of approximate truth. Realism requires a correspondence type analysis of approximate truth, if that is to be its vehicle of explanation, no less than of truth. For only on such an analysis does approximate truth inherit the explanatory status of truth with regard to scientific success. The evidence that such an analysis is possible is that we do seem able to distinguish radical difference from approximate similarity of descriptions of the world without essential reference to empirically accessible consequences. Maxwell's and Lorentz's theories are closer in this regard than is either to Newton's; the relativistic theory of gravitation is closer to Descartes's theory than to Newton's.

Regarding (ii), it may be said that the involvement of auxiliaries has the potential to impede the success of rigorously true theories no less than of approximately true ones. Yet rigorous truth is typically counted a sufficient explanation of success. This may be because just as defective auxiliaries can jeopardize the credibility of meritorious theories, clever auxiliaries can protect the credibility of undeserving theories. It is convenient to assume that on average these effects balance so that the role of auxiliaries need not be an issue.

(3) Of course realists are not in a position to deny the possibility of alternative explanations even of those forms of success identified as special explananda for realism, let alone of success generally (unless, of course, they interpret success in realist terms). Transcendental arguments do not establish their conclusions uniquely. But realism is no worse off on this account than other metaphysical theses. The difference between our epistemic stance with respect to realism and that with respect to the existence of enduring physical objects is, arguably, one of degree. And so long as there are no serious competitors to a realist explanation of any avowed form of scientific success, the inference to realism is as legitimate methodologically as that from evidence to hypothesis generally.

(4) Conceptual problems within or among current theories are _prima facie_ objections to realism as here formulated. They are not conclusive inasmuch as the retreat to partial truth leaves open the possibility of a semantics on which irreconcilable or internally inconsistent theories admit of realist interpretation. This point is of great importance to realism in view of the conflicts between relativity and quantum mechanics and the conflicts of each with deep intuitions. However, it cannot simply be assumed that weakening the truth claim of realism obviates problems posed by logical relations among theories. On the other hand, realism might be sustainable with respect to the conjecture that some new theory free of conceptual problems will become current. This approach is misleadingly reflected in formulations of realism in terms of the goals or aims of science, as though one could be a realist about science without cognitive endorsement of anything science has in fact achieved. (See van Fraassen 1980 .) Realism is not a thesis about the aims of science. It claims that there really are such entities as science, read literally, claims there to be, and that what science tells us about the nature of these entities is at least not completely false, whether science aims to discover such entities and their natures or not. Whether the aims of science could be served without such discoveries is beside the point; the realist need not deny this. What he can assert conjecturally, in the face of conceptual problems of current science, is that at least some of the references of current science will be preserved in future theories free of such problems.

2. Objection and Replies

The major source of difficulty for realism which these four assumptions, heavily qualified and inadequately supported as they are, leave outstanding is the alleged fact that theories successful in whatever respects the realist cites as explananda have turned out to be nonreferential and therefore false. The classic examples are the phlogiston theory and Maxwell's theory (although there is a tradition dating from Maxwell's time of divorcing his theory from his models of the ether). Evidently a realist interpretation of those theories, offered while they were current, would not have been the correct explanation of their success. What entitles us, then, to hold that realism is the correct explanation of the success of theories which are current? Although an interpretation of current theory, not of history, realism is plausible only insofar as it does not make of current theory too much of an exception. For our epistemic situation with respect to current theory is not all that exceptional; theories can be cited which once achieved successes comparable in their inducement to credence to those achieved now. Thus even granting (1) — (4), there is the difficulty that realism predicts that from the perspective of future science current theories will continue to be regarded as referential and partly or approximately true, whereas the prediction warranted historically, if any, is that current theories will fail altogether.

Of course this difficulty is recognized by realists. The classic response is to impose some retentionist requirement, beyond empirical success, on new theories. Realism no more requires that all successful theories be referential and verisimilar than a gravitational explanation of free fall requires that all free bodies fall. What realism does require is that in individual cases of successful, false theories there be some account of how success was possible despite the falsity, _i.e._, despite the absence of that which normally explains it. Retentionist requirements on successor theories are meant to guarantee the possibility of such an account.

It may be required, for example, that each new theory explain why its disconfirmed predecessor succeeded where it did. This move conveniently delegates to science itself the responsibility that would otherwise fall to philosophy of establishing violations of the _ceteris_ _paribus_ clause which realism is entitled to invoke in cases of successful, false theories. But we should note that such a requirement on new theories is not, strictly speaking, a condition of acceptability for _them_. For so taken, it is clearly gratuitous. There is no condition of acceptability for scientific theories beyond their own empirical success (although I wish there were). Rather the requirement is a condition of acceptability for a realist interpretation of theories. And the requirement is not that the particular theory which happens to replace a disconfirmed theory explain the success of that previous theory. The requirement at most is that science generally explain its own past successes. So the realist response to the difficulty posed by successful, false theories is not to claim that particular new theories nor science generally ought to be retentionist, but to claim that in fact science is retentionist and this fact exonerates realism.

What is retentionist about the requirement? The natural suggestion for implementing it is that successful, disconfirmed theories be retained as limiting cases of future theories. The truthlikeness of a theory T_2 explains the success of a disconfirmed theory T_1 by yielding laws which reduce to successful lawlike implications of T_1 "in the limit". That is, given (2) the successes of T_2 are explained, but these properly include successes of T_1 which are

thereby explained as well. Unfortunately this suggestion fails. It
amounts to claiming that the success of T_1 is explained by explaining
the success of certain empirical lawlike generalizations which T_1
yields. T_1 was successful because it yields successful laws. To be
sure, but then the problem is to explain how T_1, which has been found
to be nonreferential and false, manages to yield such laws.

It does no good to extend the limiting case requirement to the
"deep structure" of T_1, insisting that the theoretical mechanisms of
T_1 be retained as well and appealing to lawlike connections between
the deep structure of a theory and its empirical consequences. For
this amounts to claiming that T_1 is referential after all, whereas the
problem is to explain the successes of certain avowedly nonreferential
theories. A retentionist scheme must take care not to retain too much.
It becomes inappropriate to regard T_1 as a rejected theory; rather it
is an early stage of a single, developing and still viable theoretical
program. If T_1 is to relate to T_2 roughly as the Lorentz theory without contraction relates to the Lorentz theory with contraction, then
T_1 is simply not the sort of theory that the realist response must
address.

The whole limiting case approach simply fails to provide the
explanation realism needs of why a rejected theory succeeded where it
did. The deep reason for the failure is that the very question on
which this approach founders, _viz._ "how is it possible for a false
deep structure to yield laws which are limiting cases of correct
laws?", is ill posed. For what generally happens is not that T_1
starts with some deep structure, motivated independently of empirical
considerations by metaphysical speculations about the inner workings
of nature, which then generates empirical laws that turn out roughly
correct. Rather T_1 starts with the empirical laws themselves, induced
from observations or interpolated from the laws of successful background theories, and proceeds to develop a theoretical basis for them.
So Newton began with Kepler's and Galileo's laws and invented universal
gravitation; Maxwell began with Faraday's, Coulomb's, Biot's laws,
and others and invented the electromagnetic ether; Planck began with
Wien's law and Rayleigh's, or at least with results indicating a low
frequency proportionality of energy and temperature supportive of
Rayleigh's, and developed an empirical distribution law for which he
subsequently found a theoretical basis by quantizing the energy of
resonators interacting with enclosed radiation; Einstein began with
Maxwell's and Newton's laws and generalizations of facts about light
propagation and developed a theoretical structure based on the
relativity of motion and leading to the spacetime continuum. What
there is to be explained in such cases is not how false theoretical
structures manage to yield roughly correct empirical laws; they were
so constructed as to yield them. What is to be explained is how
false theoretical structures manage to go beyond the empirical laws
they were designed to yield in successfully predicting and explaining
novel phenomenon unanticipated in their construction—how they manage
to be better than "_ad hoc_" with respect to the phenomena they started
with.

At this point, however, the explanandum has been so weakened as to lose its grip on the historical authority of the initial challenge posed by discarded science. The realist will be tempted simply to deny it, to insist on the applicability of theses (a) and (b) to any theory which successfully anticipates novel experience. This is, in effect, to narrow the range of successes which realism purports to explain to cases in which theories make significant advances in empirical knowledge over those facts which they can reasonably be held responsible for explaining or those problems which they can reasonably be held responsible for solving.

What are the consequences of this response to the historical objection to realism?

3. Consequences

I shall comment on three areas in which realism must be developed if it is to rely on this line of response: the notion of novelty, the notion of approximate truth, and the theory of reference.

3.1 Novelty

First, the notion of novelty must be generally circumscribed so as to disqualify a wide range of apparently successful historical theories as counterexamples. Theories are not designed to explain particular empirical facts but general empirical laws speculatively induced from such facts. These are laws of potentially infinite application and no experimental results construable as instances of them will count as novel with respect to theories designed to explain them, however significantly they differ from results from which the laws were induced. Thus even a theory confirmed by experiments "different in kind" from those which motivated its development need not, on that account, be credited with novel success.

The Kennedy-Thorndike experiment of 1932 (KT) differed in many important respects from the Michelson-Morley experiment of 1887 (MM), which was instrumental in the development of ether-based electrodynamics. Philosophers have seized on these differences to argue that the contraction hypothesis used by Lorentz to account for the MM result was not _ad hoc_, since it was, after all, testable independently of MM by KT. (See Laymon 1980 .) Yet the predictions of ether-based electrodynamics for the KT result are not novel, because the explanatory task of that theory, induced from MM and other second order experiments of its day, was to show that no effect of the earth's motion on the velocity of light could be manifested in any terrestrial experiment. This task was fundamental to Lorentz's program as of 1904. The discovery of Neptune was not a novel success of Newtonian gravitation theory, because it was fundamental to the original Newtonian program to reduce all motions of celestial bodies to compositions of motions resulting from the attractions of neighboring bodies.

94

On the other hand, the excess precession of the perihelion of
Mercury was a novel prediction of General Relativity, although
documented half a century before the inception of that theory. It was
not an explanatory task of Einstein's program to account for empirical
regularities which violate Newton's theory of gravitation. The task,
rather, was to construct a theory of gravitation which, unlike
Newton's, would be consistent with the general requirements imposed by
Special Relativity on the forms of natural law. In this Einstein
could have been successful had his new theory yielded no predictions
at variance with Newton's. As General Relativity is still a viable
theory, its novel successes pose no problem for realism.

In short, there is a distinction to be drawn between experimental
results which confirm a theory and results which confirm a realist
interpretation of it. Confirmation in general is a wider notion which
includes indications of explanatory or problem solving adequacy for
certain antecedently circumscribable ranges of phenomena. Results
confirmatory in the sense of realism must qualify as novel in the
respect described.

However, the claim that novel status is sufficient for realist
confirmation requires additional argument. It is clearly insufficient
that a successful result exceed the scope of lawlike assumptions for
which a theory merely happens to have been held accountable, and which
one might imagine, counterfactually, to have been rejected or judged
irrelevant without effect on the development or final content of the
theory. It would be difficult to argue that a theory gains more or
different support from such a result than from ones which instantiate
laws for which it was thought accountable. The status of novelty
cannot be thus accidental if it is to make a difference as to confirmation. The point, however, is that the problems and explanatory tasks
for which theories are designed typically influence or direct the
course of theory construction to such an extent that one cannot consistently imagine them changed while holding the theory constant. A
theory developed in a context sufficiently different to affect the
novel status of an impending result would be a different theory. Thus
it cannot be maintained that a nonnovel result confirming a theory
would have been novel for the theory had the theory's original
conditions of adequacy been different. Confirmation is not realist by
chance. This result underlies the realist claim that a theory so
structured as to yield certain laws is more likely also to yield
results in other areas if its structure captures the actual physical
basis of known instances of the laws, than if it yields them by
happenstance.

3.2 Truth

Second, anticipating the objection that some rejected theories did
achieve successes which qualify as novel even in the restricted sense
just described, there must be at least a partial resurrection of them.
This is possible because thesis (b) does not commit the realist to
maintaining that any theory, however successful, is true. Mere

approximate truth is supposed capable of explaining success. But, as
already noted, from the fact that the truth of a theory would explain
its success it by no means follows that its approximate truth explains
its success. Whether this inference fails only because of the
problems already considered, the character of extant analyses of
approximate truth and the role of auxiliary assumptions, or whether
there are further reasons for its failure is debatable. But unless he
has a suitable correspondence type analysis of approximate truth to
offer, the realist would be well advised to take as his explanandum
not success itself, but increases in the successfulness of theories
as science changes. Increases of successfulness, are, in any case, a
more manifest feature of science than success as such, since it is
unclear by what standard to assess the latter, and at any stage of
science one may point to failures as well as successes. The realist
may then offer increases in the partial truth of theories as his
explanans. "If truth explains success then increases in truth explain
increases in success" is a more likely inference than that from the
success of truth to the success of approximate truth.

The view of truth to which this leads, however, is a relative one.
We must take 'false' to mean 'incompletely true', and maintain that
some rejected, false theories are truer than others which were less
successful.   If this program is not to founder on well known
objections to the comparability of the truth contents of false theories
(see, _e.g._, Oddie 1981), which may be deployed against the successfulness of approximate truth, bivalence will have to be abandoned. An
infinity of truth values will be necessary for which the semantics will
be degrees of accuracy of descriptions. We suppose that even if there
is such a thing as a unique, total condition of the universe serving
as the ultimate object of scientific knowledge, there is no such thing
as a unique, maximally accurate, linguistic description of the universe.
This accords with the intuitions that any description of a feature of
the world is in principle refinable, and refinability does not imply
inaccuracy.

3.3 Reference

Third, if we are to attribute some degree of truth to successful,
rejected theories, we will have to provide for referential stability
through theory change. For it is unclear how to speak intelligibly
of truth in the absence of reference. We therefore require a theory
of reference, or at least of coreference, with the historical consequence that rejected theories which successfully predicted and
explained novel phenomena instantiate theoretical programs through
which reference can be traced back from current theory. For each
referential term which figured crucially in a rejected theory's
successful predictions of novel facts, there must be a coreferential
term, though perhaps one of wider extension, in that theory's more
successful successor.

The idea underlying such backward projection of reference is that a novel, successful prediction of a rejected theory becomes part of the evidential basis of its successor. Coreference is then an extra-theoretic interpretation which explains how it was possible for the rejected theory to achieve novel success. It was successful because it identified correctly the phenomenon causally responsible for the predicted effect, even if it misdescribed that phenomenon. As referential success alone is insufficient as an explanation of empirical success, this idea must be supplemented by the attribution of some degree of truth to the rejected theory's misdescriptions.

References

Giere, R. (1979). _Understanding Scientific Reasoning._ New York: Holt, Rinehart and Winston.

Laymon, R. (1980). "Independent Testability: The Michelson-Morley and Kennedy-Thorndike Experiments." _Philosophy of Science_ 47: 1-38.

Oddie, G. (1981). "Verisimilitude Revisited." _The British Journal for the Philosophy of Science_ 32: 237-265.

Popper, K. (1963). _Conjectures and Refutations._ New York: Harper and Row.

van Fraassen, B.C. (1980). _The Scientific Image._ Oxford: Clarendon Press.

<u>Realism, Miracles, and the Common Cause</u>[1]

James Robert Brown

University of Toronto

Scientific realism is the doctrine that our theories are intended to be literal descriptions of the world. It is the claim that there is a realm of non-observable entities and processes, and that science legitimately appeals to this realm in its explanations of the observable world. This is a sketchy characterization, and with it go some sketchy arguments. These are the miracle arguments for scientific realism. Here is the way J.J.C. Smart puts his version:

> If the phenomenalist about theoretical entities is correct we must believe in a *cosmic coincidence*. That is, if this is so, statements about electrons, etc., are of only instrumental value: they simply enable us to predict phenomena on the level of galvanometers and cloud chambers. They do nothing to remove the *surprising character* of these phenomena....[But is] it not odd that the phenomena of the world should be such as to make a purely instrumental theory true? On the other hand, if we interpret a theory in a realist way, then we have no need for such a cosmic coincidence: it is not surprising that galvanometers and cloud chambers behave in the sort of way they do, for if there really are electrons, etc., this is just what we should expect. A lot of surprising facts no longer seem surprising. (Smart 1963, p. 39)

We can understand the miracle argument as going something like this:

1. Conclusion P can be deduced from theory T.
2. P is observed to be the case.
3. If T is true then the argument for P is *sound* and P had to be true.
4. If T is false then the argument for P is *merely valid* and the probability of P being true is very small (i.e., it would be very surprising if P were true, a miracle.)

PSA 1982, Volume 1, pp. 98-106

∴ The argument for P was probably sound.
∴ T is probably true   (i.e., all of T's statements, including
   ones about theoretical entities, are probably true).

The principle of the common cause was first introduced explicitly
by name by Hans Reichenbach (1956); but it has been a working rule of
thumb down through the ages. Most recently, it has been utilized by
Wesley Salmon in a work mainly on explanation in which he also uses
the principle to give a brief argument for scientific realism (Salmon
1978). The principle goes like this: *Every significant statistical
correlation must be explained through a common cause.* (Or, as Reichenbach (p. 157) puts it, "If an improbable coincidence has occurred, there
must exist a common cause.") This methodological rule means that we
are required to posit a single cause whenever there are events or kinds
of events which are statistically correlated. For instance, to use
Salmon's example, should all the lights in the house go out at once it
is unlikely that we would be tempted to say it was just a coincidence
that they all burned out at the same time. The principle of the common
cause says we should look for a single culprit, such as a blown fuse,
to account for things.

Here is how the argument may be used for scientific realism: *Often
when we have a significant correlation in the phenomena the common
cause can be found in the observable realm* (e.g., the fuse); *but this is
not always so. Sometimes no common cause can be found in the phenomena,
so appeal must be made to something non-observable. Consequently, the
imperative to explain is sometimes an imperative to posit theoretical
entities.*

The principle of the common cause as it stands is both naive and
imprecise, though it obviously has a great deal of intuitive appeal.
To make it *precise* is just to cast it in the formalism of the probability calculus. Making it more *sophisticated* is a complicated under-
taking, as we shall see below. But it will have to be made more sophisticated as the following example shows. Since the example comes from
quantum mechanics it plays, as one might guess, on the notion of *cause*.

Quantum mechanics is the skeleton in the closet of causality and
like every skeleton it should be brought out and made to dance. This
is exactly what van Fraassen (1980, 1981) does with the aim of subvert-
ing scientific realism. Let me begin first by describing a particular
situation which arises in quantum mechanics, then I will show how it
allegedly becomes a counter-example to the principle of the common
cause.

The following kinds of situations (called EPR situations) first arose
in a thought experiment designed by Einstein together with Podolsky and
Rosen (1935). Einstein's response to the Copenhagen interpretation of
quantum mechanics was not a favourable one; it is summed up in his
remark "God does not play dice with the Universe." The central feature
which he objected to was the claim that quantum mechanics is *both* a
complete theory (that is, it said everything there was to say) and that

100

it is an irreducibly non-deterministic theory. The original purpose
of the EPR thought experiment was to argue that the theory is not
complete; but more recently, other conclusions which are really quite
remarkable, have been drawn from EPR situations by J.S. Bell (1964,
1966; Wigner 1970; Clauser and Horne 1974; Clauser and Shimony 1978).
Before getting on to the Bell results, I will first review the original
EPR argument because it sets the stage so nicely.

EPR situations arise when we have something like the following: We
start out with a quantum mechanical system which is initially coupled
and then separated. For instance, we might have a disintegrating particle which becomes a pair of photons that move apart in opposite
directions. After the parts are sufficiently separated they are measured for the presence of some property or other. Spin is a handy one
since it is quite a remarkable quantum mechanical property; in the case
of photons, e.g., it can only come in one of two forms: up or down,
for each of the three spatial coordinates. The law states that the
quantity of spin must remain the same throughout the evolution of the
system. For instance, in the case of pair production, if we start out
with zero spin then we must have the same amount of total spin (for
each of the three spatial coordinates) after the production of the two
photons as we had before. The law is obeyed this way: If the photon
going to the left has spin down then the photon going to the right must
have spin up (i.e., spin up + spin down = spin 0). The quantum
theory can only make statistical predictions; the conservation of spin
law, on the other hand, *demands* that there be the following correlation: If one is up, then the other is down.

The Copenhagen interpretation maintains that properties such as the
spin components of the photon do not exist until the measurement takes
place; measurements *create*, in some sense or other, the property that
they measure and do not simply ascertain what is *already* there. It was
claims such as this that the EPR argument was designed to confute. We
are asked to imagine an EPR situation such as the following:

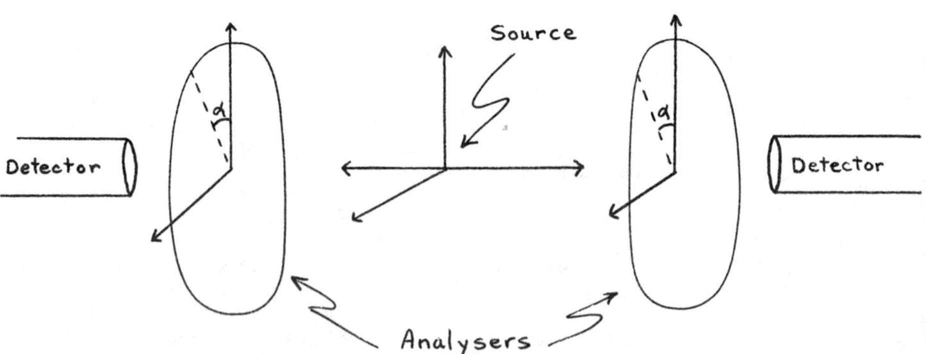

There is an important condition which is part of an EPR situation which I have so far not mentioned. The measurements on the two separated systems are done at the same time and with such a spatial separation that, given the correctness of special relativity, there is no way one measurement could causally influence the other. This reasonable assumption, known as locality, is crucial to the whole affair.

The EPR argument runs: There is a significant correlation between the measured properties of the separated systems. The correlation cannot be brought about by the measuring process itself, because that would violate the locality assumption. Hence, the properties must have already been there to be measured all along. The ψ-function does *not* contain all of the information that it should contain. A complete description of nature would include extra factors (i.e., "hidden variables") which are causally responsible for the correlation found as the results of measurement. In short, we must posit a common cause.

The reasoning is entirely plausible, and, known as the EPR Paradox, has been a thorn in the side of conventional interpretations of quantum mechanics for quite some time. But things have become vastly more complicated due to an amazing argument by J.S. Bell. What Bell did was to prove a rather straight forward result now known as Bell's inequality. Consider the EPR situation once again, only this time with the analysers rotated to different angles. What are being measured this time are the various components of spin. When the two analysers are at the same angle we get a perfect correlation in the results. But when they are measuring different components of spin the correlation is different. Spin has three spatial components and for each of these the values can be up or down. I will use L and R for the particles which go to the left and right respectively, and ↑ and ↓ to indicate that the spin for any given coordinate is up or down. The symbol L-↑- means that the spin of the particle that went to the left is up for the second coordinate (i.e., the y coordinate of the triple x,y,z).

It is impossible to simultaneously ascertain all of the components of the spin of a single particle; however, by rotating the analysers, it is possible to measure different components of spin for the LR pair. The expression pr(L↑--,R--↓) is the probability that the left particle has spin up for the x axis and R has spin down for the z axis; and so on. Starting from premises which could be called "realism", i.e., that there is a common cause, together with locality, Bell derived the following result which is now called Bell's Inequality: pr(L↑--,R-↑-) ≤ pr(L↑--,R--↑) + pr(L-↑-,R--↑).

Quantum mechanics, without the assumption of a common cause or hidden variables, makes quite a different prediction. It says that sometimes the inequality is violated. Consequently, we have the makings of a crucial test (at least as crucial as any test can be). The Bell inequality has made it possible to test matters empirically and the result seems to be that the inequality is indeed sometimes violated. The upshot, consequently, is this: *There can be no common cause of the phenomenon in question even though there is a correlation.* Let me encapsulate the argument:

1. There is a significant correlation.
2. Locality (No action at a distance).
3. Assume there is a common cause (Hidden variables).
4. ∴ Bell's Inequality.
5. The inequality is violated.
6. ∴ Assumption 3 is false.
7. ∴ There is a correlation but no common cause.

If this is a miracle, concludes van Fraassen, then so be it. But we cannot posit theoretical entities to explain this, he notes, without flying in the face of the best of contemporary physics. The miracle argument and in particular the principle of the common cause lead us to this unacceptable conclusion, and so they must be rejected. And with them, of course, goes the best argument for scientific realism. Van Fraassen writes: "In any case, weakening the principle in various ways (and certainly it will have to be weakened if it is going to be acceptable in any sense) will remove the force of the realist arguments. For any weakening is an agreement to leave some sorts of 'cosmic coincidence' unexplained. But that is to admit the tenability of the nominalist/empiricist point of view, for the demand for explanation ceases then to be a scientific 'categorical imperative'." (van Fraassen 1980, p. 30f) I shall return to this, but let me first consider an apparently different topic.

I now come to the speculative (i.e., fuzzy) part of the paper. (However, since I am dealing with an extraordinary problem, namely the Bell results, I should be allowed an extraordinary solution.) The first thing I wish to assume is the existence of synthetic *a priori* knowledge. To claim the existence of the synthetic *a priori* is to say that there are propositions which are known to be true independently of experience but not *merely* in virtue of the meanings of their terms. Knowledge of colour exclusions is the most popular candidate for the status of synthetic *a priori*. "Anything red all over cannot be blue" is a prime example. The argument for it goes something like this: We know that bachelors are unmarried because the concept of unmarried is "contained" in the concept of a bachelor. This sort of containment (metaphorical though it is) does not obtain in the case of colours. Colour concepts are all primitive; no colour concept is contained in, nor defined in terms of, any other colour. In other words, red is no more conceptually connected to blue than it is to round. Since a red thing can be either round or not round it should be possible for it to be blue as well as not-blue. Obviously, it cannot be; this we know independently of experience. So knowledge of colour exclusions is synthetic *a priori* knowledge. (See Sumner and Woods 1969 for some pros and cons of this sort of argument.)

Consider, now, some odd sorts of correlations. Bachelors are correlated with unmarried males; red things are (negatively) correlated with blue things. In neither of these cases would we be tempted in the least to posit a common cause for the correlation. The principle seems totally inappropriate in these circumstances (though there is nothing in the principle's expression which prohibits its application in such cases). Why is the principle inapplicable here? In the case of the bachelor/unmarried correlation we could say that the principle of the common cause *is* being applied. Any cause given for John being a *bachelor* is trivially the *same cause* (i.e., the common cause) for John being *unmarried*. We never invoke the principle of the common cause in cases of correlations induced by analytic truths simply because we never need to. The principle is always trivially satisfied.

This seems an adequate account of examples such as the bachelor/ unmarried one, but it will not work for synthetic examples. We cannot say that the principle is trivially satisfied in a case where something which is red also happens to be not-blue, for the two terms, red and not-blue, are not interdefinable. The principle of the common cause is *not* satisfied in the case of synthetic *a priori* correlations. We may look for the cause of a thing's being red, but we do not and should not look for a common cause of a thing's being *both* red and not-blue. Yet, given the synthetic nature of colour exclusions, this is a correlation of two distinct facts, and the principle as it stands calls for positing a common cause.

We can only speculate as to why the principle of the common cause is inappropriate to cases of synthetic *a priori* correlations. The best bet, I think, is that the *a priori* nature of the knowledge has much to do with it. (It cannot be the whole story, however, as the analytic case shows.) The thing about synthetic *a priori* knowledge, presumably, is that its origin is not of the usual *physicalistic* causal kind. To know that the ball is red we must be in the right sort of causal connection with it. To know that it is round we must be in a different sort of causal connection with it. But once we know that it is red, then we can infer that the ball is also not-blue. We can pull off the inference without the further experience, that is, without the further physicalistic causal interaction between knower and known.

In consequence of all of this, the naive version of the principle of the common cause will require modification. Let us try this reformulation: *Every a posteriori correlation must be explained by positing a common cause.* By an *a posteriori* correlation I mean one which can *only* be determined to obtain *a posteriori*. The essential thing about an *a priori* correlation is that it is one which is *possible* to determine *a priori*. And for it *no common cause is required.*

The Bell result undermines the naive version of the principle of common cause. But, playing on causality as it does, it can be employed to undermine several cherished modern theories. Causal theories, which are presently in vogue, will all find Bell's result a great difficulty. It is the causal theory of knowledge which I am particularly interested in here. This is the theory which says that to know requires the right sort of physicalistic causal interaction between knower and thing known. The Bell result constitutes a problem for the causal theory of knowledge in the following way: Imagine an observer at one side of the

EPR apparatus. There is no way of predicting whether the incoming particle will have spin up or down. This is because the particle does not
have spin components except when measured. When it is detected an inference can be made to the remote one. However, on pain of contradiction, it cannot be maintained that there is any causal connection between the knower and the thing known, the remote particle. The observer
acquires knowledge of the spin components of the *near* particle by means
of the usual kind of *causal* connection; the causal theory of knowledge
obtains in this case. But the knowledge of the *other* particle, however,
is totally non-causal. There is no physicalistic causal connection between the knower and the spin components of the remote particle. The
causal theory of knowledge is violated in this case.

So what kind of knowledge does one have of the remote particle? I
suggest we interpret things as follows: Counter-intuitive as it may
be, it seems only appropriate to call it *a priori* knowledge. It is
known independently of any (ordinary) experience of the remote particle
(assuming that to experience is to causally interact) and so it roughly
satisfies the characterization of the *a priori*.

Though the causal theory of knowledge is inapplicable in the Bell
case, it does have a wide domain of applicability. It is inappropriate
in the Bell case, I think, in the same way that it is inappropriate for
other synthetic *a priori* examples. Our knowledge of spin correlations
is of a piece with our knowledge of colour exclusions. (Though spin
correlations are not immediately "self-evident" like the colour example,
self-evidence is not a *necessary* condition for being *a priori*.)

The consequence of all this, I now conclude, is that the Bell result
does not constitute being a counter-example to the principle of the
common cause, at least not when it is properly formulated.

The principle of the common cause was initially introduced to provide an argument for scientific realism. The argument went: Whenever
there is a correlation we must introduce a common cause; sometimes this
cause can be located in the observable realm, but not always. Thus,
the imperative to explain is often an imperative to introduce theoretical entities. The miracle arguments of Smart, and others are, in
effect, justifications of this methodological rule of science.

The miracle argument is probably the best and perhaps the only argument for scientific realism; and it is a fairly compelling argument at
that. The principle of the common cause is obviously a good rule of
thumb, but if it is robbed of its complete universality then it will,
as van Fraassen says, lack its force as a methodological imperative.
The anti-realist can, with impunity, ignore the dictum to posit theoretical entities, and maintain that there are unexplained regularities
and correlations which we just have to live with. Since the realist
will be forced by the Bell result to live with them at the level of
microphysics anyway, then there can be no quarrel with choosing to live
with them at the observable level.

Properly formulated, however, the principle of the common cause skirts round the problem case. Moreover, the new version of the principle works *just as well* as the naive version in providing an argument for scientific realism. And it is equally well justified by the miracle arguments.

It could well be that no solution to the problem is required for it may turn out that the Bell result is compatible with hidden variables (a common cause) after all. (See Clauser and Shimony 1978) It does seem unwise (and boring) to pin our hopes on it though, and that is why I have tried a different (and, I hope, interesting) way to save the principle of the common cause which in turn will save scientific realism.

Notes

[1] For helpful comments I am indebted to Kathleen Okruhlik.

References

Bell, J.S. (1964). "On the Einstein Podolsky Rosen Paradox." _Physics_ 1: 195-200.

----------. (1966). "On the Problem of Hidden Variables in Quantum Mechanics." _Review of Modern Physics_ 38: 447-452.

Clauser, J. and Horne, M. (1974). "Experimental Consequences of Objective Local Theories." _Physical Review D_ 10: 526-535.

------------ and Shimony, A. (1978). "Bell's Theorem: Experimental Tests and Implications." _Reports on Progress in Physics_ 41: 1881-1927.

d'Espagnat, B. (1976). _Conceptual Foundations of Quantum Mechanics._ 2nd ed. Reading, Mass.: W.A. Benjamin.

Einstein, A., Podolsky, B., and Rosen, N. (1935). "Can Quantum-Mechanical Description of Physical Reality Be Considered Complete?" _Physical Review_ 47: 777-780.

Reichenbach, H. (1956). _The Direction of Time._ Berkeley and Los Angeles: University of California Press.

Salmon, W. (1978). "Why Ask 'Why?'? An Inquiry Concerning Scientific Explanation." _Proceedings and Addresses of the American Philosophical Association_ 51: 683-705.

Smart, J.J.C. (1963). _Philosophy and Scientific Realism._ London: Routledge and Kegan Paul.

Sumner, W. and Woods, J. (eds.). (1969). _Necessary Truth._ New York: Random House.

van Fraassen, B. (1980). _The Scientific Image._ Oxford: Oxford University Press.

------------------. (1981). "The Charybdis of Realism: Epistemological Implications of Bell's Inequality." Unpublished Manuscript.

Wigner, E.P. (1970). "On Hidden Variables and Quantum Mechanical Properties." _American Journal of Physics_ 38: 1005-1009.

Ronald Laymon

The Ohio State University

Philosophers tend to view the relationship between a particular
scientific theory and its data as representable by means of a single
logical structure. The relationship is not ordinarily viewed as requiring a hierarchy of structures of different logical type. Two recent and excellent books illustrate this tendency. Glymour in Theory
and Evidence (1980) continues the positivist tradition of viewing theories as axiomatic systems which connect with their evidence by means
of ordinary quantification and truth functional connectives. Theory
testing can be understood in terms of the syntactic properties of a
single logical system. In The Scientific Image (1980), van Fraassen
adopts a version of what is sometimes called the "semantic view": a
theory is a set of models of some purely formal system. Evidence constitutes (if confirmatory) a sub-model imbedded in a model of the
equivalence class that is the theory. Like the axiomatic view, the
relationship between evidence and theory is analyzable in terms of the
formal properties of a single logical system. Of course, canonical representation of these sorts is usually admitted to be an idealization
or first approximation of actual practice.

No one would deny that in actual practice the road from data to theory is long and arduous. There are complications due to the need for
auxiliary theories of instrumentation, for justifications of particular
measures of goodness of fit, for reasons establishing that the relevant
ceteris paribus clauses have been isolated and satisfied. Sometimes
the theory to be tested must be used as the theory of the testing instruments. The general question to raise here is: does the idealization of assuming a canonical representation of single logical type (for
each theory) seriously distort our understanding of the nature of science? In this paper I shall consider the specific issue of the possible effect a more realistic analysis of data and theory may have on
argumentation for and against scientific realism.[1]

Anti-realism arguments typically begin with some canonical or near-

PSA 1982, Volume 1, pp. 107-121

canonical representation of scientific theory. This representation is then sometimes transformed in a way that avoids non-observable theoretical entities as, for example, by transforming axioms into their Craig or Ramsey equivalents. Another move is simply to deny a semantics for non-observables. It is then argued that we accept the transformed or restricted canonical form on the basis of Ockham's razor: _ceteris paribus_ choose canonical representations committed to as few entities as possible. Advocates of realism respond by trying to show that the _ceteris paribus_ clause is not satisfied when comparing non-transformed or semantically complete representations with their anti-realist instrumentalist brethren. If, though, our initial canonical representation is modified so as to reflect a more complex connection between data and theory, then it is possible that Ockham's razor will cut differently and that the possibilities for _ceteris paribus_ violations will be greatly changed.

The starlight deflection experimental tests of the General Theory of Relativity will be sufficient to illustrate the type of complexity I have in mind. In order to apply the Field Equations, one must specify the mass-energy distribution of the particular system in question. However, even accepting Newtonian measures of the sun and earth, the Field Equations cannot be brought to bear since their solution, given such a realistic description, is unknown and probably computationally impossible. So an _idealized description_ is employed: the Schwarzschild "solution" assumes a perfectly symmetrical non-rotating sun and no other masses. Together the Field Equations and the Schwarzschild idealization yield a solution for the metric. (For more details, see, e.g., Weinberg 1972, pp. 175-194.)

Before proceeding further in the description of this case, let me isolate a philosophically interesting feature. Strictly speaking, the Schwarzschild description of sun and gravitational field as symmetric and static is false. If viewed as an "initial condition", the predictive inference is therefore unsound. Hence, no conclusion about the truth value of premises is deducible from an unsuccessful prediction. The usual response to such an observation is to object that the Schwarzchild description is "approximately true" or an "adequate idealization". This cannot be denied. However, this piece of ordinary scientific wisdom is no philosophical analysis of these concepts, nor does it specify what justifies the description as adequate or as approximately true. A useful first step toward a philosophical theory of adequate idealizations is to view them as intended antecedents for some species of _counterfactual_: if the sun were static and symmetrical then it would deflect starlight by such-and-such amount. An interest relative possible-world interpretation seems natural here where truth is controlled by (among other things) the Field Equations. Adequacy of idealization corresponds then to some concept of sufficient "nearness" of the possible world to our world. Regardless of how the philosophical analysis goes, there must be some justification of the Schwarzschild solution as being appropriate for the deflection experiment.

The next step in the derivation of an observable prediction is to

compute the constraints on orbital shapes imposed by the metric. The
details of this computation need not concern us here. What is of interest is how the orbital equation, expressed in rather abstract theoretical terms, gets attached to our measuring instruments. What we
read in standard textbooks is the following. If we assume that the
metric "at the points of origin and detection" of deflected photons is
Minkowskian, and if we ignore complications due to quantum effects,
then "there is no question about the meaning of" our relativistic computation of orbit: "it is the azimuthal angle in a system of coordinates within which light rays define lines that are essentially
straight." Hence, we can "relate" our computation "to the shift of
stellar images on photographic plates by the ordinary rules of geometric optics." (Weinberg 1972, p. 191). In effect, the relativistic
description is <u>transformed</u> into a simple Euclidean calculation of the
deflection of starlight compared to what it <u>would have been if</u> the sun
were not present. Thinking in terms of our proposed possible worlds
analysis, what we have is some sort of mapping from a set of possible
worlds controlled by the Field Equations to a set of worlds controlled
now by Euclidean geometry.

There is an additional complication. The purpose of the eclipse is
to enable us to determine stellar positions ordinarily invisible to us.
Such starlight is bent by the sun. After eclipse, these stars are once
again invisible and are not observable in the night sky until several
months later. Hence, there are small changes in observational conditions that need to be corrected for. Differences due, for example, to
parallax and the earth's motion are calculated out by means of Euclidean geometry in order to test the intended comparison. (See figure 1.)

From the point of view of the data collector, the procedure determined by the above analysis and reduction is to photograph stellar
fields at the time of eclipse and again several months later when these
fields are visible in the evening sky. The photographic plates are
then to be superimposed and relative displacements noted. These displacements should provide the desired experimental test. And so they
would if it were not for (among other things) <u>scale distortion</u> caused
by small immeasurable changes in the optical system as it waits the required several months. The basic means of overcoming this distortion
problem is to test the prediction using a weighed least squares measure
of goodness of fit. Particular measures used are justified, in part,
by appeal to a Euclidean theory of the measuring instruments. Since
the theoretical prediction is of a hyperbolic dependence between dis-
placement and distance from the sun, what is sought for is the hyper-
bola which fits the data best assuming a normal distribution of errors
and a relative weighing of error type importance. (An interesting feature of this case is that because of the absence of data at distances
close to the sun, straight lines can be made to fit the data as well as
hyperbolas!) (See figure 2.) Actually, we are not done yet, since
there are implicit <u>ceteris paribus</u> clauses in the measures used of goodness of fit. For example, seriously distorted star images are assumed
indicative of some systematic error in violation of the goodness of fit
randomization assumptions.

<u>Figure 1</u>

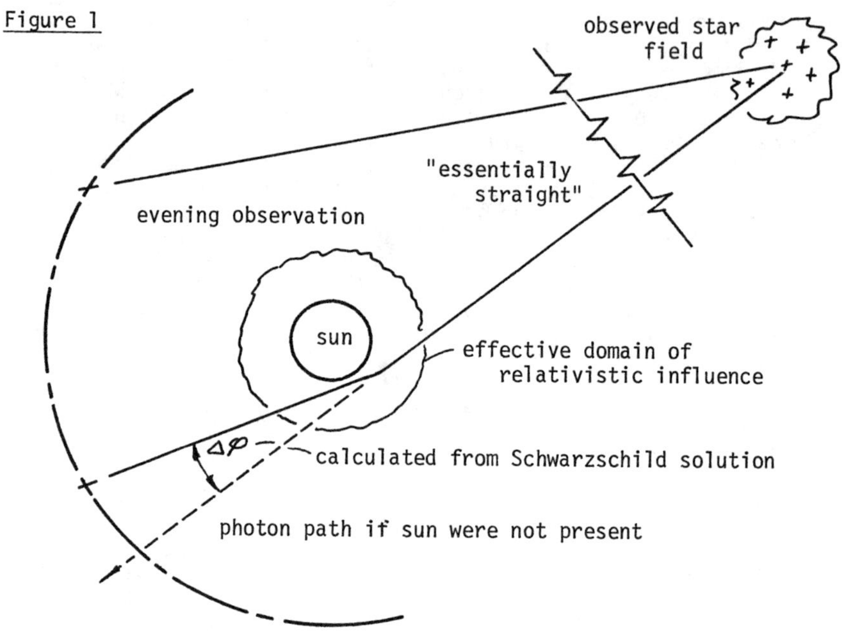

<u>Figure 2</u>

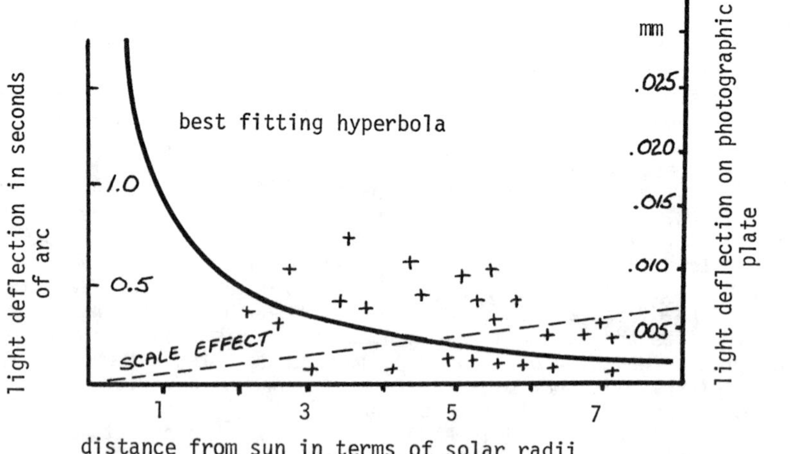

Scale distortion calculated for alteration of 0.1 mm in focal setting for 343 cm. focal length

While there are individual differences, I do not believe that the
general level of complexity illustrated here, on the road from theory
to data, is atypical of the experimental testing of scientific theories.
Let us assume that this is so. Our somewhat impressionistic rendering
yields this hierarchy of logical structures.[2]

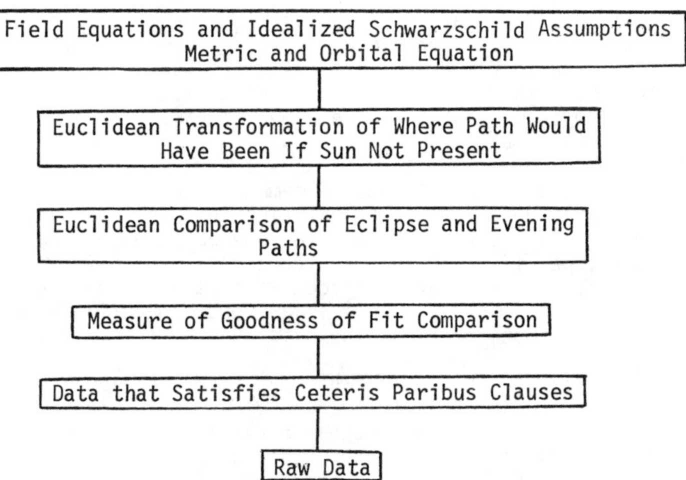

The problem of this paper is to determine what effect consideration
of this hierarchical counterfactual structure leading from data to theory will have on the issue of scientific realism. Prima facie, the
case for instrumentalism seems greatly strengthened. Afterall, everything is treated counterfactually and knowingly so! Therefore, science
does not aim to produce true"theories", but instead opts for calculational convenience and empirical adequacy. This _is_ true. Such an ad-
mission, however, need _not_ stand in the way of realism. Let me explain.

The anti-realist will want to distinguish between two kinds of idealizations. First, there is an idealized treatment of what the anti-realist antecedently accepts as an actual object, usually, something human-
ly observable. Treating the sun or the earth as perfectly spherical
and non-rotating is an example. A theory of _reference_ that ties the
idealized _description_ which will be "true" of some object in some possible world to that "same" object in this world seems a reasonable, and
by now traditional, way to proceed. The anti-realist will also want to
distinguish a second kind of idealization, namely, idealized treatments
of theoretical non-observable objects. But here the anti-realist will
turn instrumentalist and deny reference for the apparent object terms.
Since ordinary ostension does not do duty here, nothing of value seems
lost by giving up the advantages of a referential account.   The problem

112

of explaining the sense in which a theory of unobservables can be said
to be idealized remains. If unobservables do not exist, how can descriptions of them be idealized? Consider Jeans' simple mean free path
version of kinetic theory. While idealized, it is <u>less</u> idealized than
simple collisional theories. For example, Jeans describes the case
this way:

> ...the pressure of a gas was calculated on the assumptions that
> the molecules were infinitesimal in size, and that they exerted
> no forces on one another except when they were actually in col-
> lision. ...neither of these assumptions is true for an actual
> gas in which the molecules are of finite size, and exert forces
> of cohesion on one another even when they are not in contact.
> (Jeans 1940, p. 63).

The surface grammar here seems to demand a semantic account in terms
of more realistic descriptions of existing objects. Before considering the anti-realist options, allow me some additional stage setting.
In particular, we will need some consequences for <u>confirmation</u> that
can be drawn from the hierarchical counterfactual structure connecting
data and theory. Obviously, there is a close connection between our
concepts of confirmation and of the existence of theoretical objects.

Above we asked the question: what justifies the idealizations used
in science? The claim that it is the <u>success</u> of the resultant predictions will not do, since theories never exactly predict the phenomena.
Now it might be claimed that this failure of exact predictive fit can
be accommodated within statistical theories of goodness of fit. While
this is certainly part of the answer, it must be remembered that, typically, idealized theories of the measuring instruments are used to
justify the employed measure of goodness of fit. In the starlight
bending experiments, simple Euclidean analyses were given to justify
data correction and the identification and extent of factors whose random variation would affect experimental outcome. The adequacy of such
analyses gets justified on grounds such as: (1) "at infinity the metric becomes Minkowskian;"; (2) the gravitational field "on the earth's
surface is more than 10^3 times weaker than that of the sun on the sun's
surface." (Weinberg 1972, p. 191). Explicit arguments to demonstrate
the cogency of such grounds are rarely, if ever, given. That such approximations are adequate forms the traditional folklore of a scientific education. My suspicion is that, if pressed, the best that could
be done in the case at hand would be to appeal to the reduction of
General Theory to Newtonian mechanics in special cases, and to point
out analogies of procedure with past Newtonian successes. Synge, in
his book on General Theory, briefly discusses the difficulties of justifying the use of such approximations. In general:

> Approximations based on the neglect of small terms are very
> frequent in mathematical physics, and there is seldom any
> reason to object to them. One feels that if there is any-
> thing wrong, it will show up in some anomaly, and then one
> can revise the theory. (Synge 1960, p. 57).

With respect to the classical tests of General Theory, Synge says:

> The agonist needs no encouragement to work out, as a mathemati-
> cal problem, the geodesics of space-time with the [Schwarzschild]
> metric. The realist, on the other hand, may have some doubts.
> Though convinced of the validity of the geodesic hypothesis for
> very small bodies [e.g., that photons follow geodesics], he may
> wonder just what 'very small' means—are the Earth and Jupiter
> very small? This question cannot be answered until a rational
> theory of the 2-body problem has been developed, and the only
> thing to do is to go ahead with the planetary motion and light
> rays. (Synge 1960, p. 290).

Without delving into the technicalities of the case, the point is
just this: absence of experimental fit within assumed measures of
goodness of fit can be construed as indicative of the inadequacy of
the derivational approximations used. In such cases, the idealizations and approximations can be said to have introduced a bias in the
analysis, that is, a systematic distortion. Given this bias, theoretical prediction had better not satisfy measures of goodness of fit
based on an analysis of random experimental errors. The existence of
derivational bias is the fundamental reason why statistical theories
cannot be a complete account of the concept of adequacy of experimental fit.

A simple example from freshman physics should drive home the point.
If we were to ignore the rotational inertia of the suspension string or
rod of a pendulum, then we would introduce a systematic bias in our prediction. If this inertia is large enough, then a statistical analysis
of random experimental errors cannot help. This was one of the lessons
to be learned from laboratory class: if data were to be fudged, they
had to exceed estimated experimental error and be displaced in a way
that corresponded to the bias introduced by the analysis.

Assuming then the insidiousness of bias due to idealization, our
original problem remains: how are idealizations justified? An obvious
answer is that satisfaction of statistical measures of goodness of fit
justifies the combination of idealized description and fundamental
physical law.[3] Similarly, nonsatisfaction casts doubt on the combination. A Duhemian generalization would be: what is confirmed is the
combination of law and counterfactual idealization; similarly, it is
only the combination that is disconfirmed. This is a cheerful result
for the anti-realist since it restricts confirmation to what is strictly speaking false, i.e., to idealizations and approximations.[4] If,
however, the history of science is to be a normative guide, then this
generalization is to be resisted. This is because in several key episodes in the history of science, fundamental law (or the theoretical
basis of the calculation) was held to be confirmed even though the combination of law and idealization yielded predictions that exceeded estimated experimental errors. Similarly, there was disconfirmation of
law (or theoretical basis) simpliciter when goodness of fit was not
satisfied. My claim is this: sometimes it is possible to argue that

failure of fit is or is not due to the bias introduced by idealization. In the former case, there is confirmation, _even though_ estimated experimental errors are exceeded. In the latter case, there is disconfirmation of law (or theoretical basis).

A particularly clear example is provided by kinetic theory and specific heats. On the basis of spectral lines and other phenomena, it was known that gas molecules have internal structure. Adding structure then to the billiard ball molecules of kinetic theory should have afforded some improvement in the predictions about specific heats. However, it was formally provable that adding degrees of freedom of motion (a consequence of internal structure) would cause a _divergence_ from experimental values. Treating the molecules more _realistically_ would make the predictions about specific heats worse, not better. Hence, there was a disconfirmation of kinetic theory; it was not possible to improve experimental fit; the bias was not eliminable. A similar example is provided by attempts to show that more realistic accounts of the Michelson-Morley experiment would accommodate the null result. Michelson's single ray account is really not an account at all, since it does not entail a finitely-sized shifting fringe pattern. Lorentz was able to give a generalized proof, based on Huygens' principle and the calculus of variations, showing that the null prediction would remain despite improvements in the realism of the description of experimental conditions.

An example of eliminable bias is provided by Newton's response to critics of his _experimentum crucis_. As they noted, color separation and dispersion _is_ observable after the second prism. Newton answered by arguing for this conditional: if the finite size of the apertures is taken into account (and they were not in his published treatment), then an improved prediction will result, one that allows for some color separation and image dispersion. Newton did _not_ actually construct this improved account; he merely argued that it _could_ be done. There are many reasons that block actual constructions of improved accounts: absence of necessary auxiliary theory, absence of required data, absence or impossibility of analytic techniques. Given these reasons, one finds arguments of the sort: _if_ these difficulties were overcome, then improved experimental fit would be possible. Shankland's analysis of Miller's positive Michelson-Morley results provides an example of such argumentation.[5] If adequate theories of expansion and contraction due to thermal disturbances were available, then Special Theory could be made to yield predictions closer to the actual systematic shifts observed.

In compressed slogan form, the view proposed is this: a theory is _confirmed_ if it can be shown that it is possible to show that more accurate but still idealized or approximate descriptions will lead to improved experimental fit; a theory is _disconfirmed_ when it can be shown that such improvement is impossible. Since I have argued for and clarified this account elsewhere, I shall say no more in its support.[6] Assuming its correctness, we can now answer the question of what justifies idealized descriptions. In one sense, I want to deny the meaning-

fulness of the question. Our interest, so far, has been the confirmation or justification of theories. Our problem was that the predictions of theories and idealizations rarely, if ever, are compatible
with estimated experimental error. In order to achieve confirmation,
the blame for this failure of experimental fit must be placed on the
idealized initial conditions. That is, the idealizations are <u>not</u> justified. On the other hand, theory must be logically attachable to descriptions of the world in a way that makes predictions computationally possible <u>in fact</u> and not just in principle. Otherwise, the process of confirmation and disconfirmation cannot even get started.
Idealizations can be said then to be prima facie justified when they
allow for practical computability. <u>If</u> it can be shown that more realistic initial conditions will lead <u>via</u> theory to correspondingly more
accurate predictions, <u>then</u> the original highly idealized initial conditions <u>are</u> justified in the sense that they provided the starting
point for a successful confirmational process. And remember that arguing that improvement is possible need not entail actually constructing
such accounts. (Similarly, although somewhat paradoxically, an idealization is justified if it provides the starting point for a successful
disconfirmation.) This analysis of justified idealization also explains what it means to be <u>approximately true</u>. To say that an idealization is approximately true means that it is justified as above.

I shall now consider the consequences of the above sort of <u>converg-
ing counterfactual</u> theory of confirmation for the issue of scientific
realism. Realist and anti-realist perhaps can agree on this methodological point: proceed <u>as if</u> one were developing ever more accurate
descriptions of an existing reality. Given this agreement, an argument <u>for</u> realism is that cases of successful convergence to better
experimental fit are <u>miraculous coincidences</u> for the anti-realist.
Such an argument obviously will be a close cousin of similar arguments
by Smart (1963), Boyd (1981), and Putnam (1975). There is this difference: whereas they focus on the apparently miraculous nature of scientific reduction and, more generally, similarity across scientific
change, my version stays fixed <u>within</u> the context of a single theoretical <u>program</u>. This restriction makes it somewhat easier to state exactly what it is that is to be explained. At least two aspects of scientific practice within a research tradition require explanation. First,
there is the historical fact that increasingly more accurate and complete descriptions of the apparent objects of reference typically result in more accurate predictions. Second, there is this fact about
the <u>scientific community</u>: <u>convergence</u> to experimental values given
more realistic treatments is seen as <u>confirming</u> the theoretical basis
of these calculations.

Consider Newton's famous moon test. Proposition IV of Book III of
<u>Principia</u> demonstrates this counterfactual: <u>If</u> (a) the earth is sta-
tionary, (b) the earth to moon distance is 60 earth radii, (c) the
moon's period is $27^d7^h43^m$, <u>then</u> assuming inverse square gravitional
attraction, bodies should fall at the surface of the earth at 15 and
1/2 feet per second. Bodies do not fall at this rate. But neither is
the earth stationary, and the earth to moon distance is not 60 earth

radii. These values were known to be <u>outside</u> of the range of reasonable experimental error. Historically, Newton went on to show in various ways that if the counterfactuality of the antecedent were relieved,
then the consequence would approach a value that was, as shown by experiment, <u>correspondingly</u> more accurate.[7] What <u>explains</u> the success of
Newton's modifications is the simple observation that since the earth
and moon are <u>existing objects</u>, then, assuming the correctness of Newtonian mechanics, more accurate input <u>should</u> lead to more accurate output. This explanation would seem agreeable even to the anti-realist.
Such an explanation, however, is rejected in cases where the objects
whose descriptions are apparently improved are non-observables. The
anti-realist response here is that it makes no sense to speak of improving the description of non-existent objects. All that exists is a
heuristic procedure which results in increasingly more accurate predictions. To describe increasing the realism of our treatment of, say,
molecules as just a heuristic, is not an explanation but merely a
transformation of the explanandum. What requires explanation is the
success of a procedure and the very possibility of such success. That
the kinetic theory of Chapman and Enskog has greater empirical adequacy
than Jeans' simple mean free path approach is uncontroversial fact.
The question is why. By denying the notion of increasingly more accurate descriptions of non-observables, the anti-realist shuts himself
off from the most reasonable explanation.

The anti-realist has a powerful objection to this sort of proposed
objection. There were, roughly speaking, increasingly more realistic
theories of phlogiston. (And even if this is not historically as accurate as we would like, we can easily imagine it to have been so.) But
there is no phlogiston; hence, there can be no account of this historical improvement in terms of it. There are several realist responses
to be made here. First, <u>before</u> theory replacement, the existence of
phlogiston is, or so the realist would claim, the best explanation of
improvement. Note that it is <u>not</u> a requirement of realism that our
current theories be true, only that we should make this our aim. And
believing a theory is to believe it to be true. <u>After</u> theory replacement, one explains the success of phlogiston theory in terms of oxygen
theory. Obviously, this cannot be done simply by mapping 'oxygen' into
'phlogiston'. These sorts of explanations, before and after theory replacement, are denied anti-realists.

I shall now turn to the second aspect of progress within a tradition
that requires explanation: the <u>confirmational cogency</u> of improvability
of experimental fit by means of more realistic idealizations. Consider
again Newton's moon test. He was able to show that <u>if</u> more accurate
data were used in the descriptions of earth and moon, then more accurate predictions would result. This is seen as confirmation of Newton's laws. The question is: what <u>justifies</u> such an appraisal? For
both realist and anti-realist the answer is plain. To be a physical
object <u>is</u> to be susceptible to increasingly more accurate and correspondingly more adequate analyses on the basis of correct theories.
Since the earth and moon are physical objects, and since they have been
shown susceptible to <u>improvable</u> analysis on the basis of Newton's laws,

it follows that Newton's laws are confirmed or have received some confirmation. But the anti-realist will deny this explanation of confirmatory value when the objects in question are non-observable. The anti-realist therefore must either bite the bullet and deny that explanation is required in such cases, or provide an anti-realist account of the confirmational cogency of such cases.

Say the anti-realist were to adopt the line of simply <u>defining</u> confirmation to be the following: A theory is confirmed (i.e., is to be accepted as empirically adequate or whatever) when it can be shown that increasing the realism of the initial conditions (where this is judged according to scientific standards) leads to increasingly more accurate predictions. The definition is <u>normative</u>: this is how we should view theory acceptance. And the argument for its acceptance is that it reflects scientific practice (i.e., assume that my examples are typical of good scientific practice). We will let surface grammar be our guide for confirmation, but not for truth. There is an interesting analogy here with causal and necessary-and-sufficient condition theories of natural kinds. The advantage of a causal theory is that it automatically remains consistent with current scientific theory: x is gold if it is <u>relevantly like</u> this sample. If there were a final science, then this advantage would cease and the two linguistic theories would collapse into one. Here's the analogy. The above anti-realist definition of confirmation will always be one step behind current scientific practice.

Consider the following situation. Because of observational and analytic shortcomings, we cannot make our idealized inputs to some theory more realistic and retain computability. However, we can <u>vary</u> them and retain practical computability, but not in ways where the overall realism is comparable or converging. If, in such a case, the predictions exhibit a random-like scatter around experimental values, then we ought to accept the theory as confirmed. This is because a correct theory of observable and unobservable entities will yield, on the basis of descriptions of initial conditions "randomly" distributed about the true, predictions randomly distributed around experimental values. The predictions in such cases are said to be <u>robust</u> with respect to input. An actual instance of this confirmational methodology is given by Wimsatt (1981): Wade's experimental test using laboratory populations of the floor beetle <u>Tribolium</u> of the relative efficacy of individual and group selection.

Now how is the anti-realist to respond to this case? He will have to modify his definition of confirmation in an <u>ad hoc</u> way. Similarly, those who held acids to be proton donors have to modify their definition in an <u>ad hoc</u> way to accommodate scientific change. Stated another way: an anti-realist <u>substitutional semantics</u> for scientific practice may be possible; however, it will be <u>ad hoc</u> and lack the <u>rationale</u> of a realistic referential semantics.

Obviously none of what I have said, by itself, decides the issue of scientific realism. My purpose was the more modest one of showing how argumentation on this issue must change in order to accommodate increasingly more realistic descriptions of the connection between data and theory. This connection, I claim, is based on layers of counterfactual reasoning, where what controls the truth values of the counterfactuals changes from layer to layer.[8] I also tried to show that confirmation of scientific law was possible even though statistical measures of goodness of fit were violated. Such confirmation occurs when it can be shown _possible_ to improve predictions on the basis of more accurate descriptions of the objects of analysis. And to show such possibility need not entail, as the history of science shows, actually constructing better analyses with better predictions. Assuming the correctness of these basic features, it appears that explanation is required of the historical instances where improved predictions resulted from improved descriptions of the objects of analysis. Explanation is also apparently required of why such instances should be seen as confirmatory of the scientific laws used in the analyses. Finally, anti-realists are denied the obvious explanatory benefits of postulation of the existence of non-observable objects. If my proposed analysis is circular, then I claim it to be of larger and more realistic radius than traditional discussions of realism and anti-realism.

Notes

[1]Van Fraassen gives this analysis of scientific realism: "Science aims to give us, in its theories, a literally true story of what the world is like; and acceptance of scientific theory involves the belief that it is true." (1980, p. 8). Since much of what this can mean depends on a theory of canonical form, I shall rest content with this analysis. By realism I will not mean what Laudan (1981) has called _convergent realism_: roughly that mature theories are converging on the truth because they all are referring to the same things.

[2]The notion of a hierarchy of structures connecting data and theory comes from Suppes (1962). For an anti-realist interpretation of the starlight bending experiment, see the dissertation of my student Humphrey (1981).

[3]Because there are so many theoretical layers in the path from theory to data, there is the possibility that mistakes or errors at one level may be fortuitously cancelled by errors at other levels. I ignore this complication in what follows.

[4]This is a cheerful result if one assumes a Russellian theory of reference. Since the idealizations are counterfactual, nothing per se satisfies them. Therefore, there is no question of the existence of the entities apparently referred to by scientists. Science does not aim for truth and its theories are not to be believed as true literal descriptions.

[5]For more details on the cases discussed, and for other related ex-

amples, see my (1977), (1978a), (1978b), (1980), and (1983).

[6]In my (1980). I also discuss in this paper similarities and differences between my views and those of Lakatos and Kuhn. The idea of giving a counterfactual interpretation to the sort of activity discussed was suggested to me by Ron Giere. For work similar to mine, see Koertge (1973) and Wimsatt (1981).

[7]For a more accurate statement of Newton's procedure and more details on his demonstration of universal gravitation see my (1983).

[8]This paper has concentrated on the counterfactual nature of idealizations. I have not dealt with the effect the hierarchical structure of counterfactuals has on confirmation. Obviously, it serves to complicate matters greatly. In the case of the starlight bending experiment, for example, the difficulties at the lower levels of analysis are sufficiently great so that there is no pressure exerted upwards to develop more realistic relativistic analyses. This explains, in part, why the experiment is not seen currently as providing much in the way of confirmational value. See von Klüber (1960) and Earman and Glymour (1980).

References

Boyd, R. (1981). "Scientific Realism and Naturalistic Epistemology. In _PSA 1980,_ Volume 2. Edited by P.D. Asquith and R.N. Giere. East Lansing, Michigan: Philosophy of Science Association. Pages 613-662.

Earman, J. and Glymour, C. (1980). "Relativity and Eclipses: The British Eclipse Expeditions of 1919 and Their Predecessors." _Historical Studies in the Physical Sciences_ 11: 49-85.

Glymour, C. (1980). _Theory and Evidence._ Princeton: Princeton University Press.

Humphrey, S.F. (1981). _An Anti-Realist Conception of Theories in Mathematical Physics._ Unpublished Ph.D. Dissertation, Ohio State University.

Jeans, Sir J. (1940). _An Introduction to the Kinetic Theory of Gases._ Cambridge: Cambridge University Press.

Koertge, N. (1973). "Theory Change in Science." In _Conceptual Change._ Edited by G. Pearce and P. Maynard. Dordrecht: Reidel. Pages 167-198.

Laudan, L. (1981). "A Confutation of Convergent Realism." _Philosophy of Science_ 48: 19-49.

Laymon, R. (1977). "Newton's Advertised Precision and His Refutation of the Received Laws of Refraction." In _Studies in Perception: Interrelations in History and Philosophy of Science._ Edited by P.K. Machamer and R.G. Turnbull. Columbus: The Ohio State University Press. Pages 231-258.

----------. (1978a). "Feyerabend, Brownian Motion, and the Hiddenness of Refuting Facts." _Philosophy of Science_ 44: 225-247.

----------. (1978b). "Newton's _Experimentum Crucis_ and the Logic of Idealization and Theory Refutation." _Studies in History and Philosophy of Science_ 9: 51-77.

----------. (1980). "Idealization, Explanation, and Confirmation." In _PSA 1980,_ Volume 1. Edited by P.D. Asquith and R.N. Giere. East Lansing, Michigan: Philosophy of Science Association. Pages 336-352.

----------. (1983). "Newton's Demonstration of Universal Gravitation and Philosophical Theories of Confirmation." In _Minnesota Studies in the Philosophy of Science,_ Volume 11. Edited by John Earman. Minneapolis: University of Minnesota Press. Forthcoming.

Newton, I. (1687). _Philosophiae Naturalis Principia Mathematica._ London: Royal Society. (As reprinted as _Sir Isaac Newton's Mathematical Principles of Natural Philosophy and his System of the World._ 2 vols. (trans.) A. Motte, revised by F. Cajori. Berkeley: University of California Press, 1973.)

Putnam, H. (1975). "What is Mathematical Truth?" In _Mathematics, Matter and Method._ Cambridge: Cambridge University Press. Pages 60-78.

Smart, J.J.C. (1963). _Philosophy and Scientific Realism._ London: Routledge and Kegan Paul.

Suppes, P. (1962). "Models of Data." In _Logic, Methodology and Philosophy of Science: Proceedings of the 1960 International Congress._ Edited by E. Nagel, _et al._ Stanford: Stanford University Press. Pages 252-261.

Synge, J.L. (1960). _Relativity: The General Theory._ Amsterdam: North Holland.

van Fraassen, B.C. (1980). _The Scientific Image._ Oxford: Clarendon Press.

von Klüber, H. (1960). "The Determination of Einstein's Light-Deflection in the Gravitational Field of the Sun." In _Vistas in Astronomy._ Volume III. Edited by A. Beer. London: Pergamon Press. Pages 47-77.

Weinberg, S. (1972). _Gravitation and Cosmology._ New York: John Wiley and Sons.

Wimsatt, W.C. (1981). "Robustness, Reliability and Multiple-Determination in Science." In _Scientific Inquiry and the Social Sciences: A Volume in Honor of Donald T. Campbell._ Edited by M. Brewer and B. Collins. San Francisco: Jossey-Bass. Pages 124-163.

The Explanatory Import of Dispositions:
A Defense of Scientific Realism[1]

Jon D. Ringen

Indiana University at South Bend

1. Introduction

It is widely assumed that disposition predicates do not designate
events, processes, or states of affairs which could be causal factors
in the production of natural phenomena, yet the fact that an object has
a given dispositional property is frequently taken to help explain the
behavior exhibited by the object to which the disposition can be
ascribed. Considerable philosophical effort has been devoted to the
task of explaining how this could be so.[2] The results run the gamut
from Hume's (1739, p. 224) view that faculties and occult qualities
have absolutely no explanatory import to Armstrong's (1968, p. 88)
recent conclusion that dispositions are causes. Most proposals, however, lie between these extremes. Most contemporary philosophers take
it for granted that disposition ascriptions have some explanatory
import even though disposition predicates do not designate causes.

The instrumentalist, realist, and rationalist analyses of disposition predicates constitute three quite distinct views of this moderate
sort. It will be the burden of this paper to show that only the
realist analysis is adequate to the tasks of empirical science. The
instrumentalist fails to do justice to intuitions concerning the
explanatory import of disposition ascriptions. The rationalist tries
unsuccessfully to locate necessary connections in nature. The realist
provides an account which is intuitively satisfying without introducing
otiose entities into the ontology of empirical science.

Central to each of these analyses of disposition predicates is the
idea that an object has a dispositional property only if some lawlike
conditional sentences are true of the object. In the typical cases,
the antecedent and consequent of the conditional both designate observable events, processes, or states of affairs. Normally, the antecedent
designates some stimulus condition which the object might be in and the
consequent designates some behavior or change of state which the object

---

PSA 1982, Volume 1, pp. 122-133

might exhibit. A crucial feature of the lawlike conditionals is that they must be subjunctive and perhaps counterfactual in character. Typically, their truth or falsity is not a matter of logical necessity (i.e., their truth or falsity is logically contingent). Let us say that an object possesses a subjunctive property (or set thereof) just in case some such subjunctive conditionals are true of it. Derivatively, predicates used to designate subjunctive properties (e.g., magnetic, mass of 20 grams, half-life of 24.1 days, etc.) can be called subjunctive predicates. Using this terminology it is possible to efficiently describe the instrumentalist, realist, and rationalist accounts of dispositions.

2. Three Analyses of Disposition Predicates

On the instrumentalist account, a dispositional predicate can be completely characterized by a (set of) lawlike subjunctive conditional statement(s).[3] An object has a dispositional property if and only if it possesses some subjunctive property. Hence, every subjunctive predicate is a dispositional predicate. Typically, instrumentalists restrict the class of dispositional predicates to those that can be characterized by a set of stimulus-response conditionals.

Instrumentalists rely on the covering-law model of explanation in their account of how disposition ascriptions could have explanatory import (Hempel 1965). So, when unpacked, dispositional explanations exhibit characteristics remarkably like those which, according to the covering-law model, all scientific explanations share. The phenomena explained instantiate the lawlike conditional statements which the specific disposition ascriptions in question entail. This is the source of the intuition that disposition ascriptions have explanatory import. The intuition that disposition explanations are not fully legitimate is said to derive from the fact that the lawlike statements involved are not as general as those which state the basic set of laws in a well-developed scientific theory. Whether dispositional explanations are causal explanations is thought to be a relatively minor matter, since explanation is a matter of subsumption under laws and not primarily a matter of identifying causes.

According to the realist, disposition predicates serve two functions.[4] First, they designate certain subjunctive properties which are possessed by any object to which the predicate is ascribed. Second, they serve as placeholders for terms designating (as yet unspecified) causal factors for manifestations of the subjunctive property in question. Thus, not every subjunctive predicate is a dispositional predicate. At best, a subjunctive predicate designates a component of a disposition and then only if the object having the subjunctive property has some intrinsic characteristic which is a causal factor in producing the behavior to which the consequents of the relevant subjunctive conditionals refer. A disposition ascription does not provide a causal explanation, rather it "marks the spot" where one is (thought likely) to be found. This constitutes the realist response to our basic problem concerning dispositional predicates. Dispositions

are not causes. Disposition predicates do not designate specific
causal factors. Nevertheless, disposition ascriptions have explanatory import precisely because they entail the hypothesis that some as
yet undiscovered and unspecified intrinsic characteristics of the
entity to which the disposition is ascribed are causal factors for
manifestations of the disposition in question.

The rationalist faults both the instrumentalist and the realist for
failing to give an adequate account of the truth conditions for lawlike
conditionals and hence of the "ontological source" of natural
necessity.[5] The rationalist attempts to remedy this fault with a
"nature analysis" of dispositions. On the rationalist analysis, a
disposition predicate designates three things: a subjunctive property,
an unspecified intrinsic characteristic which is a causal factor in
producing manifestations of the subjunctive property in question, and
a nature, the presence of which is the "ontological source" or truth
condition for the subjunctive modality of any lawlike statement true of
the object whose nature it is. Thus, the rationalist account of the
explanatory import of disposition ascriptions is identical with the
realist account except the rationalist claims that any adequate account
of the role of subjunctive properties in scientific explanation must
acknowledge the existence of necessary connections in nature. It must
acknowledge that natural necessity has a source in nature independent
of concepts, propositions, or human beliefs and expectations. In the
rationalist analysis of dispositions, natures are intended to be such
a source.

3. Instrumentalism and Explanation

On the instrumentalist view, the explanatory import of a disposition
ascription derives solely from the subjunctive properties which
constitute the specific disposition in question. For example, the
predicate "... is brittle" is dispositional and hence some lawlike
conditional statements are true of any object to which the predicate is
correctly ascribed. The instrumentalist grants that a piece of glass
which shatters when struck by a stone at a given time may correctly be
said to have shattered <u>because</u> it is brittle. However, the instrumentalist account of the explanatory force of such statements leaves something to be desired. It amounts to saying that the shattering of the
glass on this occasion is simply an instance of the type covered by the
lawlike statements that ascriptions of brittleness entail. Thus, either
(a) this piece of glass broke because at that time it would break if it
were struck or (b) this glass broke when struck because any piece of
glass would break if appropriately struck. This account is unsatisfactory for basically the same reason that any instrumentalist account
of explanation is. It suggests that the causal factors that produce
changes in the world must all be observable phenomena which act as
stimuli for the behavior or changes that observable objects exhibit.
This leaves us with an enormous puzzle, namely, why should different
objects exhibit such a variety of responses to the same stimuli? Why
do bells ring when struck and not shatter?

disposition ascription they warrant will ultimately prove to be correct. Further instance statistics could reveal extrinsic circumstances which account for the differences in behavior. But even if they do not, the differences may not constitute differences in dispositions. Methodological principles to which empiricists routinely appeal could warrant the conclusion that the differences in behavior were not explicable in terms of differences in intrinsic characteristics causally relevant to producing the behavior in question.

Put succinctly, considerations of relative simplicity and of comprehensiveness may favor a theory in which the differences in subjunctive properties which the instance statistics support are taken to be fundamental matters of brute fact. These considerations would constitute empirical scientific reasons for concluding that there are no differences in intrinsic characteristics which could contribute to a causal account of the differences in lawlike behavior in question. The realist can rest assured that there can be empirical grounds for concluding that certain differences in lawlike behavior which cannot be traced to differences in extrinsic circumstances are causally inexplicable (as are differences in inertial mass in Newtonian physics and differences in the electric charge on electrons, neutrons, and protons in classical sub-atomic physics) rather than a result of further intrinsic characteristics which are causal factors in producing the behavior (as are the differences in solubility of salt and sugar sand). Thus, there may be empirical grounds for concluding that certain differences in subjunctive properties do and certain other differences in subjunctive properties do not mark a difference in dispositions. Instance statistics combined with certain general principles of scientific methodology provide warrant for conclusions about which alternative is empirically acceptable. The realist account of the explanatory import of disposition ascriptions is consistent with empiricist epistemology. If Molière's jest has anything like the realist account of dispositions as a target, it quite badly misses the mark. The instrumentalist analysis cannot be preferred to the realist analysis on the grounds that realism is inconsistent with empiricism.

4. Realism, Empiricism, Rationalism, and Natural Necessity

Some philosophers argue that the realist account of the explanatory import of disposition ascriptions is inadequate for the realists' own purposes.[7] These philosophers note that any account of scientific explanation must provide an account of the truth conditions for attributing a subjunctive modality to the conditional statements which such explanations frequently involve. They assert that a properly realistic account must locate these truth conditions in nature and not in concepts, propositions, or human beliefs and expectations. They maintain that these considerations entail that a scientific realist must adopt a "nature analysis" of dispositions. The basic argument is that the causally relevant intrinsic characteristics which any disposition ascription postulates cannot provide the truth conditions required. The supposition that they do starts a vicious infinite regress.

Suppose that the truth condition for the subjunctive conditional "If this lump of sugar were placed in water, then it would dissolve" is that there is some intrinsic characteristic of the sugar lump (e.g., its chemical structure) which is a causal factor in producing dissolution in water. Then since the intrinsic characteristic is a causal factor for dissolution, there must be a lawlike (subjunctive) conditional which is true of the intrinsic characteristic as well (e.g., if an object were to have that characteristic chemical structure, then, everything else being equal, it would dissolve when placed in water). But if causally relevant intrinsic characteristics constitute the truth conditions for lawlike conditionals, there must be an intrinsic characteristic of the chemical structure of sugar by virtue of which the lawlike conditional statement is true of it. This begins a regress which is fatal for any attempt to locate the truth conditions for the modality of lawlike conditionals in the causally relevant intrinsic characteristics of the entity of which the conditional is true. The intrinsic characteristic can serve as the required truth condition only if there is an appropriate lawlike conditional true of it. But this requires another causally relevant intrinsic characteristic be present and so on to infinity. This argument has the clear merit of demonstrating that causally relevant intrinsic characteristics cannot serve the purpose they have been claimed to serve by some realists[8] who seek a mind- and concept-independent source of the necessity involved in laws of nature. The question is whether anything can serve that purpose. Milton Fisk (1970, 1973b) thinks "natures" can. Here, Fisk makes a radical break with both scientific realism and classical empiricism.

Natures are precisely those constituents of the world which provide truth conditions for the necessities which lawlike regularities in nature involve. They are mind- and concept-independent constituents of the world that constitute the truth conditions for the fact that in certain circumstances specific phenomena must occur (with a given probability). The lawful relations among the entities of the world are necessitated by the natures they each (of necessity) have. Fisk's arguments show that in order to serve as the requisite truth condition, an entity's nature cannot be identified with any intrinsic characteristic that is a causal factor in the production of specific lawlike regularities. On Fisk's view, the nature of an object is an entity in its own right which is not just constituted by the capacities and lawlike relations into which the object enters; rather, that the object has these capacities and enters into these lawlike relations is necessitated by the nature of the object in question. In this way, natures are said to provide an account of or ground for natural necessity (Fisk 1973b, p. 207).

Fisk's account is peculiar in a number of ways. First, it is not a causal account of anything nor is it an account in terms of regularities, motives, or intentions, so it is not explanatory in any scientifically familiar way. Indeed, Fisk argues that the possibility of the familiar types of scientific explanations depends on an account of natural necessity such as his. His account is logically prior to the

scientific account. It is in fact intended to be a metaphysical
account of the "ontological basis" of natural necessity. In addition,
it is a metaphysical account which requires entities in the natural
world over and above those which are identified as causal factors or
explanatory variables in the best scientific theories we could ever
have.

Fisk's attempt to demonstrate the existence of such entities repays
some scrutiny. He begins from the premise that lawlike statements of
science exhibit a subjunctive modality. He takes scientific realism to
involve the assumption that the lawlike sentences of science are statements about the world, and hence to entail that there must be something
in the world which makes it true that these lawlike statements have the
requisite subjunctive modality. Since Hume's (1739, 1748) celebrated
critique of Locke (1690) and the ideas of power and necessary con-
nexion, empiricists have held that it is something about the status
conferred on the sentences in question (e.g., by virtue of being part of
or a consequence of some well-confirmed and appropriately systematized
theory) that makes it true that they have the requisite subjunctive
modality.[9] Fisk argues that the empiricists' "intentional notion of
explanation ... is characteristic of the instrumentalist rather than
the scientific realist." (1973b, pp. 202-3). Fisk goes on to show by
impeccable reasoning that if one accepts this assessment of the empiricists' intentional notion of explanation, it follows that there must be
entities in the world over and above those which our best theories
identify as causal factors in the processes that make up our world.
Some self-proclaimed realists (such as Armstrong and Harré) have clearly been confused about that and it is a merit of Fisk's work to have
unmasked that confusion. Nevertheless, Fisk fails to address one
crucial question: why should a realist be inclined to reject the
empiricists' intentional notion of explanation? Until he does, he has
failed to show that "natures" are required to provide truth conditions
for the lawlike statements of science.

Independently of this gap in Fisk's argument, there appear to be at
least two reasons for a scientific realist to be sceptical about Fisk's
account of natural necessity. The first is that it violates the basic
ontological principle of scientific realism, namely that the constituents of the world are exactly those that play a causal role in the production of observable events.[10] As we have seen, Fisk's "natures" are
not causal factors and hence cannot play a causal role in the production
of anything, yet they are alleged to be constituents of the world.
Hence, a scientific realist cannot consent to the existence of "natures"
without admitting entities which violate his own basic ontological
principles. In addition, as the ontological principle suggests, the
explanations which are of concern to the scientific realist are the
causal explanations of observable phenomena. Explanatory entities whose
presence or absence make no difference in explaining what is observed
are otiose. They have no place in a scientific realist's ontology and
they are unnecessary in his account of scientific explanation. The
burden of proof is on Fisk and other rationalists to show why these conclusions about natures should be rejected. It is difficult to see how

this burden could be successfully carried.[11]

5. Summary

The realist analysis of disposition predicates has considerable advantages over the instrumentalist and rationalist analyses. It provides an intuitively satisfying account of the explanatory import of disposition ascriptions. This account accords well with scientific practice, it makes disposition ascriptions empirically testable, and it does not entail either that dispositions are causes or that any otiose entities need be included in a scientific ontology.

The claim that a piece of salt is disposed to dissolve is empirically testable. The claim entails that some intrinsic characteristic of salt is a causal factor in the dissolution of the salt when placed in water. Chemical investigations have identified the ionic structure of salt as the causal factor involved. But, neither the ionic structure of salt nor any other mind- or concept-independent constitutent of the world need be introduced by the realist to ground the subjunctive implications of ascribing solubility to salt. Scientific realists can join Chauncy Wright[12] in rejecting the search for some kind of "cosmic glue" to hold the (causally effective) constituents of the universe together. The consistent realist can, with good conscience, locate the source of the modality of all lawlike statements in their status as parts or consequences of some more or less explicitly formulated and well-confirmed scientific theory.[13]

Notes

[1]Support from Indiana University Faculty Fellowships is gratefully acknowledged as are helpful comments from my colleagues in the Philosophy Department at Indiana University at South Bend.

[2]See, for example, the excellent collection of essays in Tuomela (1978).

[3]See, for example, the views developed in Carnap (1936) and Hempel (1965).

[4]The realist account I defend is closest to that of Levi and Morgenbesser (1964). Realist analyses are, however, also presented by Armstrong (1968, 1969, 1973), Harré (1970a, b), Mackie (1977), Pap (1958), and Quine (1960, 1974, 1975).

[5]The most explicit defense of a rationalist account of dispositions is presented by Milton Fisk (1970, 1973b). Armstrong (1968, 1969) and especially Harré (1970a, b) exhibit considerable sympathy for ideas like those which motivate Fisk's account.

[6]Except perhaps Armstrong (1968, 1969).

[7]The arguments I sketch and examine here are based on Fisk (1973b, pp. 204-207).

[8]Such as Armstrong (1968, 1973) and Harré (1970a, b).

[9]See, for example, Pap (1958, pp. 48-50) and Tuomela (1977).

[10]See, for example, Tuomela (1977, p. 2), Popper and Eccles (1977, p. 8), and Alston (1971, p. 374).

[11]Fisk does argue elsewhere that "one cannot consistently engage in inductive thinking while disbelieving in necessary connections in nature." (1973a, p. 385). I do not find Fisk's arguments convincing.

[12]Cited in James (1875, p. 194).

[13]See, for example, the suggestions made in Pap (1958) and the account sketched in Tuomela (1977).

References

Alston, W. (1971). "Dispositions and Occurrences." <u>Canadian Journal of Philosophy</u> 1: 125-154. (As reprinted in Tuomela (1978). Pages 359-388.)

Armstrong, D. (1968). <u>A Materialist Theory of the Mind.</u> London: Routledge and Kegan Paul.

------------. (1969). "Dispositions are Causes." <u>Analysis</u> 30: 23-26.

------------. (1973). "Beliefs as States." In <u>Belief, Truth, and Knowledge.</u> New York: Cambridge University Press. Pages 7-21. (As reprinted in Tuomela (1978). Pages 411-426.)

Carnap, R. (1936). "Testability and Meaning." <u>Philosophy of Science</u> 3: 419-471, 4: 1-40. (Sections 7-10, pp. 439-453, are reprinted in Tuomela (1978). Pages 3-16.)

Fisk, M. (1970). "Capacities and Natures." In <u>PSA 1970. (Boston Studies in the Philosophy of Science,</u> Volume VIII.) Edited by Roger C. Buck and Robert S. Cohen. Dordrecht: Reidel. Pages 49-62.

--------. (1973a). "Are There Necessary Connections in Nature?" <u>Philosophy of Science</u> 37: 385-404.

--------. (1973b). "Capacities and Natures." In <u>Nature and Necessity.</u> Bloomington: Indiana University Press. Pages 229-256. (As abridged and revised by the author for Tuomela (1978). Pages 189-210.)

Harré, R. (1970a). "Powers." <u>British Journal for the Philosophy of Science</u> 21: 81-101. (As reprinted in Tuomela (1978). Pages 211-233.)

---------. (1970b). <u>The Principles of Scientific Thinking.</u> Chicago: The University of Chicago Press.

Hempel, C. (1965). "Dispositional Explanation." In <u>Aspects of Scientific Explanation.</u> New York: The Free Press. Pages 457-463. (As revised by the author for Tuomela (1978). Pages 137-146.)

Hume, D. (1739). <u>A Treatise of Human Nature.</u> London: John Noone. (As reprinted (ed.) L.A. Selby-Bigge. Oxford: Oxford University Press, 1888.)

--------. (1748). <u>Philosophical Essays Concerning Human Understanding.</u> London: Printed for A. Millar. (As reprinted as <u>An Enquiry Concerning Human Understanding.</u> (eds.) Thomas J. McCormick and Mary Calkins. La Salle: The Open Court Publishing Co., 1907.)

James, W. (1875). "Chauncy Wright." Nation 31: 194.

Levi, I. and Morgenbesser, S. (1964). "Belief and Disposition." American Philosophical Quarterly 1: 221-232. (As reprinted in Tuomela (1978). Pages 389-410.)

Locke, J. (1690). An Essay Concerning Human Understanding. London: Thomas Basett. (As reprinted (ed.) Alexander Campbell Fraser. New York: Dover Publications, Inc., 1959.)

Mackie, J. (1977). "Dispositions, Grounds, and Causes." Synthese 34: 361-369. (As reprinted in Tuomela (1978). Pages 99-108.)

Moliere, J. (1673). Le Malade Imaginaire. (As reprinted in Oeuvres Completes. Volume II. (ed.) Maurice Rat. Paris: Librairie de la Gallimard, 1956. Pages 823-910.)

Pap, A. (1958). "Disposition Concepts and Extensional Logic." In Concepts, Theories, and the Mind-Body Problem. (Minnesota Studies in the Philosophy of Science. Volume II.) Edited by H. Feigl, et al. Minneapolis: The University of Minnesota Press. Pages 196-224. (As reprinted in Tuomela (1978). Pages 27-54.)

Popper, K. and Eccles, J. (1977). The Self and Its Brain. New York: Springer-Verlag.

Quine, W. (1960). "Dispositions and Conditionals." In Word and Object. Cambridge: The MIT Press. Pages 222-225.

————————. (1974). "Dispositions." In Roots of Reference. La Salle: Open Court Publishing Co. Pages 8-15. (As reprinted in Tuomela (1978). Pages 155-162.)

————————. (1975). "Mind and Verbal Dispositions." In Mind and Language. Edited by Samuel Guttenplan. Oxford: Clarendon Press. Pages 83-96.

Tuomela, R. (1977). Human Action and Its Explanation. Dordrecht: Reidel.

———————— (ed.). (1978). Dispositions. Dordrecht: Reidel.

Part IV

Probability and Statistical Inference

The <u>Generalization</u> <u>of</u> <u>de</u> <u>Finetti's</u> <u>Representation</u>
<u>Theorem</u> <u>to</u> <u>Stationary</u> <u>Probabilities</u>

Jan von Plato

University of Helsinki

1. Introduction

According to de Finetti's representation theorem, probabilities of exchangeable sequences of events are unique mixtures of Bernoullian probabilities. The exchangeable probability $P(E)$ of the sequence E equals an integral of the form $_0\int^1 P_x(E)\,dF(x)$, where $P_x(E)$ is for each x $(0 \leqq x \leqq 1)$ a Bernoullian measure with the probability x for 'success' and independence between consecutive events, and F a uniquely determined distribution of the parameter x. $P(E)$ equals the expected value of $P_x(E)$. If F has a derivative f, $P(E) = \int P_x(E)f(x)\,dx$. Probabilities can be calculated under the integral sign as if events were independent and had a constant success probability. Therefore there cannot be anything one can do with Bernoullian probabilities, but could not do with the exchangeable ones. de Finetti thinks that he has, because of the unique correspondence between exchangeable measures and distributions over a parameter each value of which gives a Bernoullian probability, shown the unnecessity of the latter. Interpreting the exchangeable probabilities as subjective, and identifying the Bernoullian ones (for the sake of argument) as objective, he suggests that objective probabilities should be eliminated, except as a mode of speech. It is therefore a well motivated problem to question whether de Finetti's suggested reduction could be generalized to larger classes of (according to de Finetti fictive) objective probabilities. This proves to be the case. The most natural and general case is the representation of stationary probabilities as mixtures of ergodic probabilities. The result is conceptually very satisfactory. It has been known in mathematics for fifty years. There is reason to think that it is the most general case where a non-trivial representation result holds. The concept of ergodic probability is the 'best possible' generalization of independent events from the objectivist point of view, as will be explained below. It is therefore also of some importance for the subjectivist view on the foundations of probability that it stands on a par with the concept of Ber-

———————

PSA 1982, Volume 1, pp. 137-144

noullian probability with respect to reduction. My conclusion as to
this reduction of objective probabilities to the subjective ones will
however be contrary to the subjectivist view. For the physical background of the representation of stationary probabilities as unique mixtures of ergodic probabilities offers a situation where the distribution over ergodic measures, corresponding to the F above, has a physical content. Therefore it cannot always be interpreted subjectively,
and the alleged reduction fails in these cases.

2. The Ergodic Decomposition of Stationary Measures

We will consider for the sake of simplicity a space Ω of infinite
binary sequences $\omega = (\omega_1, \omega_2, \ldots)$. Projections x_n give the nth member
$x_n(\omega)$ of the sequence ω. A transformation T is defined through the
equation $x_n(T\omega) = x_{n+1}(\omega)$, and the n-fold iteration T^n of T gives us
$x_{n+1}(\omega) = x_1(T^n\omega)$. T represents the repetition of trials by giving in
each application of T the 'next' result. T removes the sequence ω by
one step, $T(\omega_1, \omega_2, \ldots) = (\omega_2, \omega_3, \ldots)$. For a set $A \subset \Omega$, $T^{-1}A = \{\omega \mid T\omega \varepsilon A\}$.
A probability measure P over Ω is *stationary* if $P(T^{-1}A) = P(A)$. (If
doubly infinite sequences $(\ldots \omega_{-1}, \omega_0, \omega_1, \ldots)$ are used this may be
written $P(TA) = A$, where $TA = \{T\omega \mid \omega \varepsilon A\}$.) It follows that the probability $P(\omega \mid x_{n+1}(\omega) = i_1, \ldots, x_{n+k}(\omega) = i_k)$ is the same for all n. An intuitive
motivation for stationarity therefore is the following. The (uncondi-
tional) probability laws of finite sequences of events are the same ir-
respective of when the repetition of events or trials is started, or,
in other words, the probabilistic laws remain the same for different
times. Sets for which $T^{-1}A = A$ $(TA=A)$ holds are called invariant. A
stationary measure is *ergodic* if invariant sets have either probability
zero or one.

Stationarity is the most general condition under which relative frequencies converge (almost everywhere). Let us call 'heads' the result
$\omega_i = 0$ and 'tails' $\omega_i = 1$. A stationary probability measure gives
measure one to the set of sequences ω for which the limit of relative
frequency, of for example tails $\lim_{n \to \infty} \Sigma \omega_i/n$, exists. More generally, if f
is an integrable function, the limits of *time averages* $\lim_{n \to \infty} \frac{1}{n} \Sigma f(T^i\omega) =$
$\hat{f}(\omega)$ exist. Note that values $\hat{f}(\omega)$ may vary from one sequence to another.
Therefore the existence of $\hat{f}(\omega)$ is a weaker (more general) statement
than a law of large numbers. If the measure P is ergodic, limits of
time averages are the same except for a set of P-measure zero. Ergodicity therefore guarantees the uniqueness of limits of relative frequencies
and their equality with probabilities as in the law of large numbers.

It can be shown that stationary measures over a space form a convex
set (in fact a simplex) the extreme points of which are exactly the er-

godic measures. Given a stationary measure P over Ω, there is a unique measure μ over the set of ergodic probabilities P_E such that $P(A) = \int P_E(A)d\mu$. A stationary non-ergodic measure is an inner point of the convex set, and the measure μ gives 'weights' to the extreme points P_E. There is only one way of doing this, i.e., μ is unique. Suppose now that P is not ergodic, but that there is a subset $A \subset \Omega$ such that for each $B \subset A$, $T^{-1}B = B$ implies that $P(B) = 0$ or $P(A)$. A is called an *ergodic component*. Under this assumption the conditional measure $P(\cdot\,|A)$ is ergodic. The integral representation of stationary probabilities is known as the ergodic decomposition of stationary measures. It amounts to a decomposition of the space Ω into ergodic parts. Limits of relative frequencies and other asymptotic properties are the same within one component, but vary in non-degenerate cases between them. In the simplest case there are only two components A_1 and A_2 $(= \Omega - A_1)$, and the representation is given by $P(B) = P(A_1)P(B|A_1) + P(A_2)P(B|A_2)$. Since for $a_1 = P(A_1)$ and $a_2 = P(A_2)$ it holds that $a_1 + a_2 = 1$ and $0 \leq a_1, a_2 \leq 1$, $P(B)$ is represented as a unique mixture of the ergodic probabilities $P(\cdot\,|A_1)$ and $P(\cdot\,|A_2)$, with the weights a_1 and a_2. They are different from zero and one in non-degenerate cases. The sequences in A_1 have identical asymptotic properties, and likewise for A_2. Specifically, limits of relative frequencies coincide with the respective probabilities within the components but are not identical between them.

Probabilities cannot be directly identified with limits of relative frequencies, since there will in a measure theoretic formulation always be exceptions of measure zero. But the intuitive idea is captured precisely with the notion of an ergodic probability measure. It is in this sense the 'best possible' generalization of probabilistically independent events. The objective interpretation is motivated for the latter by Bernoulli's theorem. It says that probabilities coincide (up to a set of measure zero) with limits of relative frequencies. But since this coincidence is equivalent to ergodicity, the restriction to independence is unnecessary. Laws of large numbers are conceptually special cases of the ergodic theorem. Therefore the basis of frequentist probability is in ergodicity and not in the special case of independence.

de Finetti's original representation theorem is a special case of the ergodic decomposition theorem. Exchangeable sequences are stationary but have in a non-degenerate case (which for them is that of probabilistic dependence) varying asymptotic properties. Independent events correspond to an ergodic measure. This can be seen from the following. Since the law of large numbers holds for Bernoullian measures and since ergodicity is equivalent to the property of identical limiting behaviour of (almost) all sequences, Bernoullian measures are ergodic. With the ergodic decomposition theorem, de Finetti's representation result can accordingly be seen in the following light. Each value of the parameter

140

p characterizing a Bernoullian probability with two possible outcomes
corresponds to an ergodic component which is a set of sequences having
as one of its properties that the limit of relative frequency of tails
is p. de Finetti claims that the objectivist Bernoullian measure is a
fictive entity. But the physical basis of ergodic theory suggests a
different view.

3. The Physical Significance of the Ergodic Decomposition

Ergodic theory has originated as an abstraction from the theory of
classical dynamical systems. Its concepts are therefore applicable to
the latter. These contain a continuous set Ω of states ω. The motion
of the states in the space of states is determined by the system's
equation of motion. We suppose that the equation is independent of
time. The evolution of a state in one unit of time is then given by a
transformation T. Under the physical condition that the equation of
motion is the same for all times, the system has the property that the
Lebesgue (also called natural) measure of sets of states is preserved
under the transformation T, so that the measure is stationary. It follows, as above, that limits of time averages exist. For an integrable
function of states f, $\lim\limits_{n\to\infty}\frac{1}{n}\Sigma f(T^i\omega) = \hat{f}(\omega)$. A central problem of statistical mechanics is to find out in which cases $\hat{f}(\omega)$ is constant, the
same for (almost) all evolutions of the system. This constancy is equivalent to the ergodicity of the system. It is again, as stationarity, at
the same time a probabilistic and a physical condition. For ergodic systems, the limit of a time average of the values of a function f equals
the expected value of f over the space Ω with respect to the natural
measure. It is thus possible to determine limits of time averages for
ergodic systems by the known natural measure.

A function g of the state space Ω is an invariant of motion if its
value $g(\omega)$ remains unaltered in the mechanical evolution of the system,
i.e., if $g(\omega) = g(T\omega)$ $(= g(T^n\omega)$ for any $n)$. The total energy of an isolated dynamical system remains fixed, and is the same for all states.
It is a constant invariant over the state space. It is required for the
ergodicity of a system to hold that it has no other invariants of motion (which are not functions of the total energy) that would be constants in a set of states of positive measure. Such sets may arise in
the following way. An invariant constant over Ω is a multiple, i.e., a
function, of total energy. Suppose g is an invariant independent of the
total energy. For some parameter a, there are sets A_1 and A_2 such that
for $\omega \varepsilon A_1$, $g(\omega) \leqq a$ and for $\omega \varepsilon A_2$, $g(\omega) > a$. The sets A_i are invariant (i.e.,
$\omega \varepsilon A_i$ is equivalent to $T\omega \varepsilon A_i$) because $g(\omega) = g(T\omega)$ by the invariance of
g. An invariant function thus defines a collection of invariant sets
within each of which it has a constant value. If at least one of these
disjoint sets has positive measure, the system fails to be ergodic,
since by assumption there is a set A such that $P(A) > 0$ and $T^{-1}A = A$. And
since the invariant is independent of total energy, $P(A) < 1$. It follows

from the existence of invariants of motion constants in sets of positive measure that limits of time averages fail to be unique (almost) everywhere in Ω. It is possible that the components arising from the existence of invariants of motion are ergodic. In this case, which is exactly the case of ergodic decomposition, the limits of for example relative frequencies are unique within the components but differ between them. The method of determining them as in the previous section fails in this case. It marks a failure of the frequentist idea of probabilities as unique limiting frequencies as well.

4. de Finetti's Suggested Reduction of Objective Probabilities

In the light of the ergodic decomposition theorem, we may understand de Finetti's representation result as follows. It is unknown what the limit of relative frequency is, but exchangeability guarantees its existence (being a special case of stationarity). This 'ignorance', in subjectivist terms, is taken into account by forming a mixture over the different possible values. If we interpret subjectively the weights of the different hypotheses as to the 'true' value of the objective probability, we have what can be called an *a priori* distribution without any physical significance. It is however a different case if the ergodic decomposition arises from the existence of invariants of motion of a dynamical system. Their existence depends solely on the physical properties of the system's equation of motion, irrespectively of our ability to identify the invariants. The weights of the decomposition have a physical interpretation, so that their distribution is not *a priori* in the sense of being only related to 'ignorance'. A striking case is offered by the ergodic decomposition of (the particle density distribution of) a statistical mechanical system into different phases, when, under suitable values of thermodynamic parameters, the system is divided into for example a liquid and a gaseous part. Molecular motion obeys different statistical laws in different phases, but within a phase the statistical laws are the same. And, as was already noted, the weights of the components are in this case physical probabilities (relative volumes of the phases), rather than subjective *a priori* ones as de Finetti would have it.

5. Bibliographical Note

The best account of de Finetti's theorem and its subjectivist interpretation is de Finetti (1937). Extensions of the theorem can be found in de Finetti (1938), Link (1980), and Diaconis and Freedman (1980). An account of the ergodic decomposition theorem along the lines of Choquet's integral representation theory is given in Jacobs (1963), and a different treatment in Farrell (1962). An ergodic decomposition theorem was established already fifty years ago by von Neumann (1932, p. 617), in his "Zur Operatorenmethode in der klassischen Mechanik". The theorem appears in connection with de Finetti's representation result in Ryll-Nardzewski (1957), and in Freedman (1962).

A stationary non-ergodic system, if left to itself, will remain in one ergodic component. The probabilities of the other components become operative only if an interference is made with the system. This corresponds to *randomization* as in Feller (1971, p. 53 ff. and p. 228).

The concepts and significance of the ergodic theory for the interpretation of probability is discussed in my (1982a, 1982b). The problem of this paper is suggested in my (1981), where more general ideas suggested by de Finetti's reduction also are discussed. A detailed presentation of the results of this contribution will appear in *Synthese* under the title "The significance of the ergodic decomposition of stationary measures for the interpretation of probability".

References

de Finetti, B. (1937). "La prévision: ses lois logiques, ses sources subjectives." _Annales de l'Institut Henri Poincaré_ 7: 1-68. (As reprinted as "Foresight: Its Logical Laws, Its Subjective Sources." (trans.) H.E. Kyburg. In _Studies in Subjective Probability._ Edited by H.E. Kyburg and H.E. Smokler. New York: Wiley, 1964. Pages 95-158.)

——————————. (1938). "Sur la condition d'équivalence partielle." _Actualités Scientifiques et Industrielles_ 737: 5-18. (As reprinted as "On the Condition of Partial Exchangeability." (trans.) P. Benacerraf and R. Jeffrey. In Jeffrey (1980). Pages 193-205.)

Diaconis, P. and Freedman, D. (1980). "De Finetti's Generalizations of Exchangeability." In Jeffrey (1980). Pages 233-249.

Farrell, R.H. (1962). "Representation of Invariant Measures." _Illinois Journal of Mathematics_ 6: 447-467.

Feller, W. (1971). _An Introduction to Probability Theory and Its Applications,_ Volume II. New York: Wiley.

Freedman, D. (1962). "Invariants Under Mixing Which Generalize de Finetti's Theorem." _Annals of Mathematical Statistics_ 33: 916-923.

Jacobs, K. (1963). _Lecture Notes on Ergodic Theory._ 2 vols. Aarhus: Mathematical Institute of Aarhus University.

Jeffrey, R. (ed.). (1980). _Studies in Inductive Logic and Probability,_ Volume II. Berkeley: University of California Press.

Link, G. (1980). "Representation Theorems of the de Finetti Type." In Jeffrey (1980). Pages 207-231.

Ryll-Nardzewski, C. (1957). "On Stationary Sequences of Random Variables and the de Finetti's Equivalence." _Colloquium Mathematicum_ 4: 149-156.

von Neumann, J. (1932). "Zur Operatorenmethode in der klassischen Mechanik." _Annals of Mathematics_ 33: 587-642. (As reprinted in _Collected Works,_ Volume II. (ed.) A.H. Taub. Pages 307-362.)

von Plato, J. (1981). "Reductive Relations in Interpretations of Probability." _Synthese_ 48: 61-75.

——————————. (1982a). "Probability and Determinism." _Philosophy of Science_ 49: 51-66.

--------------. (1982b). "The Method of Arbitrary Functions." _The British Journal for the Philosophy of Science_ (Forthcoming).

On After-Trial Criticisms of Neyman-Pearson Theory of Statistics[1]

Deborah G. Mayo

Virginia Polytechnic Institute
and State University

1. Introduction

Whether it is due to the incompleteness of information, inaccuracies of measurement, or stochastic nature of phenomena, a great deal of scientific inference requires probabilistic considerations. In carrying out such inferences, the statistical methods predominantly used are from the Neyman-Pearson Theory of statistics (NPT). Nevertheless, NPT has been the target of such severe criticisms that nearly all philosophers of induction and statistics have rejected it as inadequate for statistical inference in science. If these criticisms do in fact demonstrate the inadequacy of NPT, then a good portion of statistical inference in science will lack justification. Because of the seriousness of such a conclusion, it is important to carefully consider whether critics of NPT have succeeded in demonstrating its inadequacy.

In this paper I attempt to (1) clarify what I take to be the key issues around which the major criticisms against NPT revolve; and (2) argue that such criticisms fail to provide grounds for rejecting NPT as inadequate for science. I begin by drawing a fundamental distinction between the conception of the aims of statistical inference underlying the NPT, and the conceptions underlying rival views. Corresponding to the distinction between conceptions of statistical inference is a fundamental distinction between the criteria appropriate for judging the adequacy of statistical theories. Clearly, what is adequate for accomplishing the aims of NPT need not be adequate for accomplishing a fundamentally distinct set of aims. I maintain that criticisms of NPT involve judging NPT on the basis of criteria fundamentally distinct from those appropriate for NPT. By showing that NPT fails to satisfy these criteria, I claim, they succeed only in showing that accomplishing the aims underlying these criteria is incompatible with accomplishing the aims underlying NPT criteria. Unless there is reason to think

PSA 1982, Volume 1, pp. 145-158

that this poses a problem for NPT, it does not warrant inferring that
NPT is inadequate--whether the inadequacy is merely being too narrow
in its applicability or, in the extreme case, being totally refuted.

That is, what I call a weak claim (WC): <u>NPT fails to satisfy
criterion C</u>, only warrants inferring a strong claim (SC): <u>NPT is
inadequate</u>, if one accepts the additional premise (P): <u>NPT is adequate
only if it satisfies criterion C</u>, for some specified criterion C.
To justify accepting (P) it must be shown either that (I) NPT claims
to satisfy criterion C, or (II) NPT should satisfy criterion C. I
argue that criticisms against NPT can be taken as providing grounds
for (I) only by misconstruing the aims of NPT, and as grounds for (II),
only by presupposing the correctness of a conception of statistics
fundamentally alien to NPT, which effectively begs the question against
it. Failing to substantiate the needed premise (P), I conclude,
prevents criticisms of NPT from going beyond a weak claim (WC): that
NPT fails to satisfy a given criterion C. I am not thereby claiming
that such criticisms serve no purpose; they serve an important function
in delimiting the aims which one can rightly expect NPT to accomplish.
Nor do I claim to have shown that NPT really is adequate for science;
this requires a positive argument showing the appropriateness of NPT
aims. What I do claim to have shown is that the widespread rejection
of NPT as inadequate by philosophers is premature; the inadequacy of
NPT has not been demonstrated.

2. Error Probabilities vs. E-R Measures

The view considered most plausible by most philosophers of
induction and statistics, is that a theory of statistical inference,
like a theory of deductive inference, should serve to assess the
relationship between evidential claims and conclusions. Since, in
the statistical case, the relationship between evidence (i.e., data) and
a given conclusion may be weaker than deductive validity, a theory of
statistics, on this view, should seek to provide a way of measuring
<u>degrees</u> of the evidential relationship between them. To this end,
measures of <u>evidential-relationship</u> (E-R measures) have been developed.
Theories of <u>statistics based on such E-R measures</u> may be referred to
as <u>E-R theories</u>. Carnap's confirmation measure, (subjective) Bayesian
measures of degrees of belief, Fisher's fiducial probabilities,
Hacking's measure of support, and Kyburg's epistemological probabilities,
are only a few of the many E-R measures upon which E-R theories have
been based. A theory of statistics will be adequate for the aim of
E-R theories to the extent that it provides (absolute or relative) E-R
measures that adequately express the degree of the evidential strength
that <u>specific</u> data affords <u>specific</u> claims of interest.

In contrast, NPT views the aim of a theory of statistics to be that
of providing <u>general</u> procedures or rules for making statistical
inferences; where these general procedures are guaranteed to have
sufficiently low probabilities for leading to various erroneous
inferences.[2] NPT inferences are not assertions about E-R measures,
but about certain properties of a population of interest, called

<u>parameters</u>. Such inferences are based on the result of an
<u>experimental trial</u> consisting of taking a sample of the population
and observing some property of this sample, called a <u>statistic</u>. For
example, one may observe the proportion of 'heads' (a statistic) in
a sample of tosses of a coin to make inferences about the proportion
of 'heads' in the (hypothetical) population of all tosses of the coin
(a parameter). It will facilitate our discussion to outline two
central types of NPT inferences about a parameter θ on the basis of
observing statistic S; <u>hypotheses tests</u>, and <u>confidence interval</u> (CI)
<u>estimates</u>.

<u>General NPT Procedures</u>: A hypotheses testing procedure consists of a
rule which specifies which of the possible values of statistic S are
to result in rejecting a specified hypothesis H (about θ) in favor of
some alternative hypothesis $\overline{H}$. These values form the <u>rejection region</u>,
RR. Hence, a testing rule takes the form: Reject H iff S is in RR. A
general rule for confidence interval (CI) estimation, which I refer to
as a <u>CI estimator</u>, indicates the specific CI estimate that should result
from each possible value of S. A typical CI estimator has this form:
Estimate that the true value of θ is in the interval $[S - c, S + d]$,
i.e., estimate that $(S - c \leq \theta \leq S + d)$.

<u>Specific Inferences</u>: Once the trial is carried out, and S is observed
to have specific value s, these general rules yield specific inferences.
For example, suppose s is in RR. Then the test concludes: Reject H
in favor of $\overline{H}$. The CI estimator concludes: $(s - c \leq \theta \leq s + d)$.

<u>NPT Criteria</u>: The specific inferences that result from a rule will
vary with different values of S; some erroneous, others correct. A
testing rule is adequate only if it is possible to guarantee, <u>before</u>
the trial is made, that (regardless of the true value of θ) the
probability it will lead to erroneously rejecting H is no more than
some appropriately small number, called the <u>size</u> of the test. In
addition, it should be able to guarantee that it will correctly accept
H with suitably high probability, called the <u>power</u> of the test.
Similarly, a CI estimator is adequate only if it has an appropriately
large probability for leading to correct CI estimates, called the
<u>confidence level</u> (CL) of the estimator. Size, power, and CL's are
examples of <u>error probabilities</u>, and NPT inferences are judged adequate
on the basis of the error probabilities of the rules from which they
are generated.

3. Before-Trial Criteria vs. After-Trial Criteria

Unlike E-R measures, error probabilities hold only for general
inference rules <u>before-the-trial</u> is made. Given the frequency view
of probability underlying NPT, a rule's error probability is the relative frequency with which it will lead to correct (or incorrect) inferences in a sequence of (similar or very different) applications of the
rule. For example, if the CI estimator considered above has a CL equal
to .95, we can assert <u>before-the-trial</u> that, for any value of θ,
$P[(S - c \leq \theta \leq S + d)/\theta] = .95$. But <u>after-the-trial</u> is made, it is not

correct to substitute S with the observed value s in this assertion.
This is clearly seen in the case where the result of the substitution
is $P[(1 \leq \theta \leq 3)/\theta] = .95$, and $\theta = 5$. For, the probability that the
resulting CI (i.e., $(1 \leq \theta \leq 3)$) is correct is 0, not .95. NPT is
intended for inferences where the parameters of interest are to be
treated as constants. Hence, a specific NPT inference is either
correct or not; i.e., it is true either 100% of the time or 0% of
the time. So, the only probability that a specific inference may
have is one of the trivial ones; 1 or 0.

It follows that error probabilities do not serve to express the
degrees of probability, support, confirmation or any other E-R
measure, that may be assigned to specific inferences. It is not
surprising, then, that inferences that are adequate according to
NPT criteria about error probabilities may not be adequate according
to criteria about E-R measures. But critics of NPT maintain that
criteria about E-R measures are what matter in analyzing specific
inferences, after-the-trial, i.e., after-trial analysis. As such,
they conclude that NPT is inadequate for after-trial analysis. Such
criticisms may be referred to as after-trial criticisms of NPT. The
most serious criticisms of NPT tend to be cases of after-trial criticisms
of the following form: First, (1) an E-R measure is selected as
appropriately assessing the extent of the evidential strength that
specific data affords specific claims. Second, (2) a criterion is
set out, based on the E-R measure selected, for judging the adequacy
of inferences after-the-trial; call it criterion C. Third, (3) an
example is constructed in which an inference that is more satisfactory,
according to NPT (before-trial) criteria about error probabilities,
is less satisfactory according to (after-trial) criterion C.

From this one may infer what I have termed a weak claim (WC):

(WC): NPT fails to satisfy (after-trial) criterion C.

If, in addition, one holds premise (P):

(P): Satisfying (after-trial) criterion C is necessary for NPT to be
adequate (for after-trial analysis),

then one may infer the strong claim (SC):

(SC): NPT is inadequate (for after-trial analysis).

The problem with after-trial criticisms arises only if they are taken
as grounds for inferring (SC). I argue that no such inference is
warranted since they fail to provide adequate grounds for accepting
(P). It may be suggested that (P) is intuitively obvious, and hence,
requires no justification, as in the following remarks of Cederic
Smith: "Clearly what is wanted is a continuously variable measure of
how probable the various hypotheses are, in the light of the data, and
the NPT fails to provide this. One must conclude that it is not an
appropriate theory of inference." (Smith 1977, p. 74). Here, (P)
amounts to requiring that NPT provide an (after-trial) E-R measure
of probability (i.e., a posterior probability). But NPT is based on
the premise that an adequate theory of inference need not satisfy

such a requirement. And if a criticism of NPT is based on assuming
that a basic NPT premise is false, then it is clearly begging the
question against it. In what follows, I argue that after-trial
criticisms of NPT are based on just this sort of assumption.

4. After-Trial Criticisms of NPT Hypotheses Tests

I base my argument upon those types of after-trial criticisms that
are most serious, as well as most influential. As representatives of
these, I take the criticisms raised by three philosophers: Hacking,
Spielman, and Seidenfeld. Hacking's criticism served as a model for
the other two, both of whom attempted to improve upon it. Hacking
tried to show that NPT tests are "suitable for before-trial betting,
but not for after-trial evaluation." (Hacking 1965, p. 99). After-
trial evaluation, on his view ,involves measuring the extent to which
specific data s supports hypotheses of interest. The E-R measure
chosen for this purpose is the likelihood function (LF). The LF of
hypothesis H given specific data s is the probability (or in the
continuous case, the density) of s given that H is true, i.e., $P(s/H)$.
Tests are judged on the basis of the following (after-trial) criterion
of support. [CS]:

[CS]:   A test should reject hypothesis H on the basis of specific
          data s iff there is a rival hypothesis $\bar{H}$ much better supported
          by s, as measured by the LF, i.e., a test should reject H on the
          basis of s iff there is an $\bar{H}$ such that $P(s/\bar{H})$ is much greater
          than $P(s/H)$.

Hacking provides an example where a test that is better according
to NPT (before-trial) error-probabilities, is much worse according to
his (after-trial) criterion [CS]. More specifically, while the test
has a higher probability of correctly accepting $\bar{H}$ in a sequence of
trials (i.e., has a higher power); a specific instance of this
sequence is seen to result in accepting H although H is false, and
hence has no support. That is, the specific instance gives rise to
an s which leads to accepting H although $P(S/H) = 0$. Similarly,
a test which fails miserably on NPT criteria is seen to satisfy [CS].
From this Hacking concludes the strong claim (SC): NPT tests are
inadequate for after-trial analysis. But his argument only provides
grounds for the weak claim (WC): NPT tests fail to satisfy (after-
trial)criterion [CS]. In one sense even (WC) may be questioned; for [3]
Hacking's examples involve tests which are not "best" on NPT criteria;
and best NPT tests, in his examples, do satisfy [CS]. However, other
examples can be constructed that show the incompatibility between NPT
error criteria and Hacking's criterion of support [CS], thus
establishing (WC). But unless there are grounds for supposing (P)
(NPT tests must satisfy [CS]) there is no warrant for inferring (SC)
(NPT tests are inadequate after- the- trial).

Upon examining the examples showing the incompatibility of NPT
criteria and [CS], I think it is clear that, rather than justifying
the needed premise (P),they provide positive grounds for denying (P),

150

and rejecting [CS] as an inadequate after-trial criterion. For,
[CS] permits one to reject a hypothesis in favor of the most ad hoc
hypothesis, formulated (after-the-trial) to perfectly fit specific
experimental result, s. The reason is that, even if the ad hoc
hypothesis is clearly false, [CS] directs one to accept it simply
because it has the maximum value of the LF; and hence, is best
"supported". For example, observing a coin to land 'heads' gives
maximum support to the hypothesis that both faces of the coin are
heads-even when this is known to be false. In this way [CS] frequently
leads to erroneous inferences. Indeed, it can be shown (see Mayo
1981b) to yield a sequence of inferences, 100% of which are wrong,
even in the simplest one parameter case! Birnbaum (1969) demonstrates
this for the two parameter case. In fact, the reason Neyman and
Pearson (e.g., Neyman 1952) explicitly reject an after-trial analysis
based on LF's alone, is that they saw it would prevent the (before-
trial) guarantees of low probabilities of errors which they sought.

Clearly, then, Hacking's criterion [CS] conflicts radically with
the aims of NPT. So, merely assuming (P) (NPT must satisfy [CS])
is tantamount to assuming that NPT should abandon its own aims. And
just such an assumption is necessary if Hacking's argument is to be
taken as grounds for inferring (SC) (NPT tests are inadequate for
after-trial analysis). Although Hacking does provide positive
arguments in favor of his own theory of statistics based on [CS], this
does not prevent his after-trial criticism of NPT from this question-
begging assumption. For his criticism is based on assuming the
superiority of his own theory--a theory to which NPT is radically
opposed. Moreover, while Hacking has since abandoned his likelihood
theory, his after-trial criticism of NPT has still been seen by many
as providing the groundwork for a non-question begging demonstration
of NPT's after-trial inadequacy.

Spielman (1973) claims that by reconstructing Hacking's criticism,
he can "show that NPT is inadequate on its own terms" (p. 202)
and so provide a genuine "refutation" of NPT tests. However, his
after-trial criticism, I argue, is flawed in much the same way as
Hacking's--despite his attempt to avoid such flaws. Like Hacking,
Spielman wants to show that the error probabilities of NPT tests (i.e.,
size and power) are irrelevant "once an experiment is performed, and
a decision that really counts has to be made" (p. 211), (i.e., for
after-trial analysis). According to Spielman, an after-trial analysis
of a specific inference, based on specific data s, requires a measure
of its reliability; and the E-R measure he selects for this purpose
is the (posterior) probability that the specific inference is correct.
The criterion used in judging tests is the following (after-trial)
criterion of reliability [CR]:

[CR]: A test should lead to a specific testing inference (i.e., accept
       or reject) on the basis of specific data s, only if the
       inference is sufficiently reliable, as measured by the
       probability that it is correct.

Spielman considers an example of a NPT test which has
appropriately high probabilities of yielding correct inferences in
a sequence of applications of the testing rule (and hence satisfies
NPT error probability criteria); but which cannot guarantee that
specific instances of this sequence will yield inferences with
high probabilities of being correct (and hence fails to satisfy
his (after-trial) criterion [CR]). In other words, Spielman shows
that a testing rule,based on observing statistic S,may have a high
probability (before-the-trial) of leading to correct inferences;
while the probability (after-the-trial) of a specific observed value,
s, yielding a correct inference may not be high. This entails the
weak claim (WC): NPT fails to satisfy (after-trial) criterion [CR].
But when Spielman goes on to infer the strong claim (SC) (NPT tests
are inadequate) it becomes clear that his criticism of NPT is not
"on its own terms" - contrary to what he had intended. Again, the
problem involves substantiating premise (P) (NPT must satisfy [CR] in
order to be adequate). For, as we saw in Section 3, NPT never intended
to provide an (after-trial) E-R measure of reliability; and it is
explicitly denied that error probabilities apply once a specific s
is substituted for S. Nevertheless, Spielman maintains (I) that NPT
does intend to satisfy his (after-trial) criterion [CR]; or, (II) if
not,it should.

Spielman offers the following argument in support of (I). He
claims (i) that it is "implicit in the conceptual framework of NPT"
(p. 207) that error probabilities of NPT general procedures are
intended to justify their specific applications. And since according
to Spielman (ii) for specific inferences to be justified they must be
sufficiently reliable (in the sense of having sufficiently high
probabilities of being correct), he concludes (iii) that NPT error
probabilities are intended to guarantee that specific NPT inferences
are sufficiently reliable (in his sense). From this it follows (I)
that NPT intends to satisfy his (after-trial) criterion [CR] (and so,
(P) is true). But it has been shown that (WC): NPT fails to satisfy
[CR]. So, since it fails to satisfy the criterion which it intends to
satisfy, NPT is refuted as promised. However, his proposed refutation
does not succeed; for his argument for (iii) is flawed.

The flaw is in assuming premise (ii). For NPT is based on denying
that (ii) is the case. Rather, it holds that specific inferences get
their justification from having arisen from general procedures with
appropriate error probabilities (in a given long run sequence). Hence,
in assuming (ii) Spielman is assuming the correctness of a criterion
that radically conflicts with the aim of NPT. And in calculating
degrees of probability of specific inferences to be other than 0 or 1,
he makes use of probabilities that are invalid on the frequency view
underlying NPT. Hence,(iii), and so (I) is false; and there is no
longer any basis for regarding error probabilities as "dangerously
misleading."(p. 202). Only by misinterpreting them as (after-trial)
E-R measures of reliability do they appear misleading.

152

Spielman admits that if NPT is only interested in guaranteeing low
error probabilities in long run sequences, then (I) is false and he has
not succeeded in refuting NPT. Nevertheless, he maintains that having
shown that (WC): NPT tests fail to satisfy his (after-trial) criterion
of reliability [CR], "I have shown that [NPT] is too narrow to bother
refuting." (p.214). But, he has not told us why failing to satisfy
[CR] is a deficiency for NPT. That is, he has not shown (II) that NPT
should satisfy [CR]. Yet in claiming that unless it does, then NPT is
"too narrow to bother refuting," Spielman is presupposing (II); and is
thereby begging the question against NPT. Graves (1978) provides a
good discussion of these and other points concerning Spielman's after-
trial criticism.

5. After-Trial Criticisms of NPT of Confidence Intervals (CI's)

Seidenfeld (1979) and (1981), like Spielman, sets out an after-trial
criticism of NPT by reconstructing the argument given by Hacking.
Seidenfeld's criticism shares the same thrust as those of Hacking and
Spielman; namely, that (before-trial) error probabilities of NPT may
fail to indicate the degree of support (Hacking), reliability (Spielman)
or confidence (Seidenfeld) that should be assigned to specific infer-
ences--as measured by an appropriate (after-trial) E-R measure. But,
rather than direct his criticism at error probabilities of NPT tests,
Seidenfeld directs it at error probabilities of NPT of confidence
intervals (CI's); in particular, confidence levels (CL's). He claims
he will show that "it seems reasonable to say, before knowing the data,"
that a CI estimator with a CL equal to p will lead to a correct CI
estimate with probability p; "however, having seen the value x, it may
be unreasonable to maintain the probability statement or use it to
express a degree of confidence in the interval generated by the [CI
estimator]." (Seidenfeld 1979, pp. 56-57).

But, as was noted in Section 3, (and as Neyman and Pearson[4] have
repeatedly warned) applying a CL to a specific interval estimate leads
to absurdities. For, parameter θ is viewed, by NPT, as a constant, and
probabilities are viewed as frequencies in some sequence. Clearly, it
makes no sense to say that the frequency with which θ is contained in
a specific interval [a,b] is, say, .95. A specific CI: $(a \leq \theta \leq b)$ is
either always true or always false; thus, the only probabilities it may
be assigned are 0 and 1. So, to show that CL's fail to provide (after-
trial) measures of probability, is just to show that they fail to
provide a type of (after-trial) evaluation that is illegitimate from
the point of view of NPT. This is precisely what the after-trial
criticisms of NPT tests were seen to amount to.

In order to avoid just this flaw, Seidenfeld develops a clever
strategy: He will mount his after-trial criticism without using any E-R
measure that is illegitimate from the point of view of NPT. The only
probabilities, compatible with the frequency view of probability, that
can be assigned to specific estimates are 1 and 0--according to whether
it is known to be correct or not. Interval estimates that are known to
be correct (i.e., known to contain the true value of θ)are referred

to by Seidenfeld as <u>trivial intervals</u>. Seidenfeld notes that "even
on Neyman's conception of probability there is an acceptable probability
for the trivial intervals. They carry a known probability 1."
(Seidenfeld 1981, p. 283). Hence, by basing his criticism on trivial
intervals, he hopes to carry out his strategy.

   Seidenfeld judges NPT of CI's on the basis of the following (after-
trial) <u>criterion of triviality</u> [CT]:

[CT]:   A CI estimator should not yield specific CI estimates with CL's
        that conflict with their known probabilities. In particular,
        a EI extimator with CL less than 1 should not yield trivial CI
        estimates (i.e., estimates known to have probability 1.)

Seidenfeld considers an example where the CI estimator that is "best"[5],
according to NPT (before-trial) criteria, is inferior to one deemed less
than best by NPT, according to his (after-trial) criterion [CT]. Its
inferiority lies in the fact that it yields trivial intervals more
often, while sharing the same CL of .95; that is, it yields estimates
whose CL (i.e., .95) more often conflicts with the known probability
that the estimate is correct (i.e., 1). From this Seidenfeld infers
the weak claim (WC): NPT of CI's fails to satisfy (after-trial) criterion
[CT].

   Even this inference may be questioned, it seems; for it is arguable
that, in the example Seidenfeld considers, NPT would actually recommend,
not the estimator Seidenfeld criticizes, but an alternative estimator
which does not yield any trivial intervals; and hence, satisfies [CT].
I argue this in detail in Mayo (1981a).[6] Admittedly, the basis for
recommending the alternative interval involves <u>informal</u> criteria about
informativeness. So, Seidenfeld's example can still be taken to show
that satisfying the <u>formal</u> NPT criteria need not result in satisfying
criterion [CT]; and I assume this is all the weak claim (WC) is meant
to assert. The real problem arises when Seidenfeld's argument is taken
as grounds for going beyond (WC), and inferring the strong claim (SC):
NPT of CI's is inadequate for after-trial analysis.

   At some points, Seidenfeld suggests that he only intends to show
(WC); for his only concern is to show "that the N-P theory cannot serve
as an adequate replacement for an inductive logic" (Seidenfeld 1979,
p. 37), where an "inductive logic" is taken to require some (after-
trial) E-R measure. And having shown (WC), that NPT of CI's fails to
satisfy [CT], he has shown that CL's fail to provide valid E-R measures
of probability or confidence. For, (WT) entails that an estimate in
which one has 100% confidence may have a CL less than 100%. Nevertheless,
Seidenfeld's concern with using only an E-R measure that is valid for
NPT clearly suggests that he intends his argument to show some genuine
flaw within NPT itself. He claims that "it is my goal in part, to
strengthen Hacking's evaluation by showing that N-P best tests lead to
clearly inferior confidence intervals, based on [the NPT criteria of
"bestness"] alone." (p. 49). And this clearly implies that his goal

154

is to show (SC): NPT of CI's is inadequate. However, these CI's
are judged "clearly inferior" only in that they fail to satisfy
Seidenfeld's criterion [CT]. To accomplish the goal of showing NPT
of CI's leads to CI's that are "clearly inferior" on the basis of NPT
criteria, he would have to justify premise (P): Failure to satisfy
(after-trial) criterion [CT] renders NPT of CI's inadequate. I will
argue that Seidenfeld's argument entails (P) only if NPT criteria are
either misconstrued or rejected.

To justify (P), it must be shown either that (I): NPT claims to
satisfy [CT]; or, (II) if not, it should. In support of (I),
Seidenfeld appears to reason as follows: Since Neyman suggests that
assigning an (after-trial) probability to a specific estimate is the
"theoretically perfect solution" (Neyman 1937, p. 258), it would seem
that when an (after-trial) probability is known--as in the case of
trivial intervals, Neyman would want the NPT of CI's to provide it.
Hence, Neyman would want to assign probability 1, and not a probability
less than 1, to trivial intervals, i.e., he would want to satisfy
criterion [CT]. It follows that (I).

Firstly, what Neyman personally would want is not the same as what
NPT is logically capable of. NPT is intended to provide an adequate
theory of statistics without (after-trial) E-R measures. While
assigning (after-trial) probability 1 to trivial intervals is compatible
with the frequency theory of probability; such assignments are not
strictly a part of NPT of CI's. Neyman specifically notes that
within the theory of CI's, "we have decided not to consider [them]"
(Neyman 1937, p. 263); for it is simpler to just assert the trivial
interval itself. According to Seidenfeld, the problem with CI
estimators failing to satisfy his criterion [CT], is "the tension
between the confidence level (less than 100%) and a known probability
(of exactly 100%)." (Seidenfeld 1981, p. 282). But, there is no such
"tension" from the point of view of NPT. For, CL's always refer to the
(before-trial) probabilities that CI estimators will lead to errors in
a sequence of applications. And there is nothing contradictory, or
even problematic, about having a specific estimate, which is known to
be true (i.e., a trivial estimate), arise from a general estimating
procedure which is known to lead to correct inferences less than 100%
of the time (i.e., its CL is less than 1). It appears problematic
only by misinterpreting CL's as providing (after-trial) E-R measures
of confidence or probability. Seidenfeld's claim that failure to
satisfy [CT] leads to trivial intervals being asserted "at strictly
less than 100% confidence" (p. 281) involves such a misinterpretation.
For the result of applying a CI estimator is just an assertion that θ
is in the specific interval formed. Nothing is said about the degree
of confidence which is to be attached to this assertion. Admittedly,
the word 'confidence' encourages this sort of misinterpretation--but
this is not a problem for NPT when it is correctly interpreted. Since
CL's are not intended to provide (after-trial) measures of probability,
even if such a probability is known; it follows that (I) is false.

Nor can Seidenfeld's argument be taken as grounds for inferring that
(II): NPT should satisfy criterion [CT]. For, his own example shows
that if the aim is to satisfy his criterion [CT], the result will be
to recommend CI's that fail miserably on NPT criteria. And a CI will
fail to be adequate, for the type of after-trial analysis that NPT is
interested in, unless it satisfies these criteria. Hence, although
the E-R measure Seidenfeld uses is legitimate on the frequency view,
the after-trial criterion [CT] upon which he judges NPT of CI's involves
after-trial considerations that are incompatible with the view of
after-trial analysis underlying NPT. Failing to provide independent
grounds for the correctness of his criterion [CT] prevents Seidenfeld's
argument from legitimately showing (II). To merely assume (II) is
tantamount to assuming that the aims of NPT should be rejected in favor
of the aim underlying E-R theories; namely, to provide an after-trial
measure of the evidential strength that specific data affords specific
conclusions. Without grounds for either (I) or (II), premise (P) is un-
substantiated. Hence, Seidenfeld's argument can not be taken as grounds
for inferring (SC): NPT of CI's is inadequate (for after-trial analysis).
Moreover, Seidenfeld's criticism may be seen to follow the pattern of
argument found in criticisms raised by Fisher (1956), Jaynes (1968), and
Lindley (1971); and each, I claim, involves the same kind of problem.[7]

6. Conclusion

Each of the after-trial criticisms of NPT has been seen to involve
judging statistical inferences on the basis of an (after-trial) criterion
which reflects a very different view of the aim of statistical inference
than the one embodied in NPT. For, underlying each criterion is the view
that a theory of statistics should provide an expression of the extent
of the evidential strength that specific data affords specific conclusions,
after the trial is made. While such criteria are appropriate for
judging E-R theories, they are inappropriate for judging NPT. Moreover,
it has been shown that, unless NPT criteria are either misunderstood or
rejected, the criticisms cannot be taken as grounds for thinking that
it is necessary for NPT to satisfy these (after-trial) E-R criteria.
In effect, the criticisms merely show the difference between NPT criteria
and the (after-trial) criteria based on E-R measures. Therefore, it
can be concluded that the after-trial criticisms of NPT fail to
demonstrate the inadequacy of NPT.

Notes

[1] I would like to thank Ronald Giere and Isaac Levi for very helpful
comments on an earlier draft of this paper. I am also grateful to
Henry Kyburg and Teddy Seidenfeld for sharing their responses to a former paper (Mayo 1981a) with me; they were extremely useful in clarifying
the key points of contention underlying the present paper.

156

[2]The distinction I draw between the NPT conception of statistics and
the conception underlying E-R theories is parallel to the distinction
drawn by Giere (1977) between testing and information models.

[3]A "best" NPT test of a given size is one which also has the
maximum power of any other test with the same size (for the given
hypotheses under test).

[4]See especially Neyman (1937, pp. 261-273; 1952, pp. 210-214; and
1977, pp. 118-119).

[5]A CI estimator, CI*, with a CL of p, is "best" according to the NPT
formal criteria if there is no other estimator (for the given estimation
problem) that also has a CL of p, but which has a smaller probability
of yielding estimates containing incorrect parameter values than CI*.

[6]My argument, briefly, is this: Seidenfeld is able to present a NPT
"best" CI estimator, with a CL of .95, that generates trivial estimates
only by considering a case where parameter θ is known to have a specific
upper bound. This is called the truncated case. Since the formal NPT
criteria are not affected by this truncation, Seidenfeld concludes that
NPT still recommends the same interval in the truncated case. I argue
that by making use of the additional available information in this case,
an alternative interval recommends itself; and this alternative CI does
not give rise to trivial intervals.

My basis for claiming that, in the truncated case, this alternative
CI is superior on the basis of NPT principles is not that it satisfies
criterion [CT]--for, NPT does not seek to do so. Rather, I argue, that
NPT recommends the best CI estimator, relative to the type of information in which one is interested; and the alternative CI yields more
appropriate information in the truncated case.

[7]Their arguments are, roughly, the following. Before the trial, a CI
estimator with CL equal to p has a probability of yielding correct CI
estimates equal to p, in a sequence of applications of the estimator.
However, after the trial yields specific data s, s may be seen as a
member of a subset, T, of the original sequence of applications; and the
probability that the CI estimator will yield a correct estimate in this
subset of applications may be known to be q, where p < q. It is argued
that, in such cases, the correct CL to assign a specific estimate based
on s is not p, but q. (In Seidenfeld's example q equals 1). It is concluded that NPT of CI's are inadequate.

But, unless it is assumed that CL's must provide (after-trial) E-R
measures, there are no grounds for this conclusion. For, it is perfectly valid to assign a CL of p to estimates based on s. It would only be
appropriate to assign it a CL of q if it had been decided before the
trial to limit the sequence of applications to those in T. It is up to
the experimenter to decide, before the trial, which sequence of applications is appropriate for evaluating a given inference; it need not include all possible applications. However, once it is specified, the CL
is fixed; it cannot be altered by the specific experimental result.

References

Barnett, V. (1973). _Comparative Statistical Inference._ New York: Wiley.

Birnbaum, A. (1969). "Concepts of Statistical Evidence." In _Philosophy, Science and Method._ Edited by S. Morgenbesser, _et al._ New York: St. Martins. Pages 112-143.

----------. (1977). "The Neyman-Pearson Theory as Decision Theory, and as Inference Theory; With a Criticism of the Lindley-Savage Argument for Bayesian Theory." _Synthese_ 36: 19-50.

Carnap, R. (1950). _Logical Foundations of Probability._ Chicago: University of Chicago Press.

Fetzer, J.H. (1981). _Scientific Knowledge._ Dordrecht: Reidel.

Fisher, R.A. (1956). _Statistical Methods and Scientific Inference._ New York: Hafner.

Giere, R.N. (1976). "Empirical Probability, Objective Statistical Methods and Scientific Inquiry." In _Foundations of Probability Theory, Statistical Inference and Statistical Theories of Science._ Volume II. _(University of Western Ontario Series in Philosophy of Science._ Volume 6.) Edited by W.L. Harper and C.A. Hooker. Dordrecht: Reidel. Pages 63-101.

----------. (1977). "Testing vs. Information Models of Statistical Inference." In _Logic, Laws and Life._ _(University of Pittsburgh Series in the Philosophy of Science._ Volume 6.) Edited by R.G. Colodny. Pages 19-70.

Godambe, V.P. and Sprott, D.A. (eds.). (1971). _Foundations of Statistical Inference._ Toronto: Holt, Rinehart and Winston of Canada.

Graves, S. (1978). "On the Neyman-Pearson Theory of Testing." _British Journal for the Philosophy of Science_ 29: 1-23.

Hacking, I. (1965). _Logic of Statistical Inference._ Cambridge: Cambridge University Press.

Jaynes, E.T. (1968). "Confidence Intervals vs. Bayesian Intervals." In _Foundations of Probability Theory, Statistical Inference, and Statistical Theories of Science._ Volume II. _(University of Western Ontario Series in Philosophy of Science._ Volume 6.) Edited by W.L. Harper and C.A. Hooker. Dordrecht: Reidel. Pages 175-213.

Lehmann, E.L. (1959). _Testing Statistical Hypotheses._ New York: Wiley.

158

Lindley, D.V. (1971). _Bayesian Statistics, A Review._ Philadelphia: Society for Industrial and Applied Mathematics.

Mayo, D. (1981a). "In Defense of the Neyman-Pearson Theory of Confidence Intervals." _Philosophy of Science_ 48: 269-280.

————. (1981b). "Testing Statistical Testing." In _Philosophy in Economics._ Edited by J.C. Pitt. Dordrecht: Reidel. Pages 175-203.

Neyman, J. (1935). "On the Problem of Confidence Intervals." _Annals of Mathematical Statistics_ 6: 111-116.

————. (1937). "Outline of a Theory of Statistical Estimation Based on the Classical Theory of Probability." _Philosophical Transactions of the Royal Society of London_ Ser. A, 236: 333-380. (As reprinted in _A Selection of Early Statistical Papers of J. Neyman._ Berkeley and Los Angeles: University of California Press, 1967. Pages 250-290.)

————. (1941). "Fiducial Argument and the Theory of Confidence Intervals." _Biometrika_ 32: 128-150.

————. (1952). _Lectures and Conferences on Mathematical Statistics and Probability._ 2nd ed. Washington: U.S. Department of Agriculture.

————. (1977). "Frequentist Probability and Frequentist Statistics." _Synthese_ 36: 97-131.

Pearson, E.S. (1962). "Some Thoughts on Statistical Inference." _Annals of Mathematical Statistics_ 33: 394-403. (As reprinted in _The Selected Papers of E.S. Pearson._ Berkeley and Los Angeles: University of California Press, 1966. Pages 276-283.)

Seidenfeld, T. (1979). _Philosophical Problems of Statistical Inference._ Dordrecht: Reidel.

————. (1981). "On After Trial Properties of Best Neyman-Pearson Confidence Intervals." _Philosophy of Science_ 48: 281-291.

Smith, C. (1977). "The Analogy Between Decision and Inference." _Synthese_ 36: 71-85.

Spielman, S. (1973). "A Refutation of the Neyman-Pearson Theory of Testing." _British Journal for the Philosophy of Science_ 24: 201-222.

A Bayesian Argument in Favor of Randomization

Zeno G. Swijtink

Stanford University

The theory of personal probability must be explored with
circumspection and imagination. For example, applying
the theory naively one quickly comes to the conclusion
that randomization is without value to statistics.
This conclusion does not sound right; and it is not right.

L. J. Savage (1961, p. 585).

1. Introduction

Randomization is a generally accepted principle of sound experimental design and common practice among working scientists. But many
philosophers and some theoretical statisticians have rejected it.
Bayesians have most often stated their opposition to randomization in
terms of a decision theoretic argument. Not all Bayesians, however,
could go along with that argument, as the above quotation by Leonard
Savage testifies.

In this paper I will <u>examine the Bayesian decision theoretic
argument against randomization and show why it fails.</u> By means of an
example I will then <u>give a partial justification of randomization in
Bayesian terms.</u> Randomization can lead to an increase in expected
utility.

Section 2 sketches the Bayesian decision theoretic argument against
randomization. Section 3 traces the argument back to the non-Bayesian
theory of general decision functions of Abraham Wald. Wald's analysis
of randomized experimental designs in terms of his randomized decision
functions is <u>the historical origin of the present-day misunderstanding</u>
of a role of randomization. Section 4 argues for a "de-extensionalized"
analysis of random allocation procedures. I give an example of an
experiment to test the efficacy of a drug in which randomization
("de-extensionalized") over matched pairs leads to an increase in
expected utility. This provides us with a justification of

PSA 1982, Volume 1, pp. 159-168
Copyright © 1982 by the Philosophy of Science Association

160

randomization in mere Bayesian terms.

2. The Bayesian Decision Theoretic Argument Against Randomization

The problem of experimental design is to give criteria for selecting
a set of best designs from all possible designs that are relevant to a
certain question. The Bayesian solution identifies a design problem
with an appropriate decision problem and an optimal design with a
decision that has maximum expected utility.

Here I follow the "honest Bayesian analysis of a decision problem"
as given by D. V. Lindley (1971, p. 19-21). Lindley explicitly
discusses randomization and endorses the Bayesian decision theoretic
argument.

An experimental design $\underline{e}$ is a triplet $(X, \Theta, p(x|\theta))$, where X is a
vector space, Θ is a parameter space of indices of probability measures
P_θ over an appropriate σ-field over X. These probability measures are
supposed to be dominated by a σ-finite measure, μ, so that they can be
described by density functions $p(x|\theta)$ with respect to μ in the sense
that

$$(1) \quad p_\theta (A) = \int_A p(x|\theta) \; d\mu(x)$$

where A is any member of the σ-field.

This notion of experimental design can be illustrated by the following example. A scientist working for a pharmacological firm wants to
do an experiment on the side effects of a drug that has already shown
its efficacy in curing a certain disease. He is interested in the
question whether for a certain group of people -patients having some
persistent allergy problems- the side effects are not worse than the
cure. The scientist decides to compare the effects of the drug with
those of a placebo. An experimental design is then specified by
saying who will get the drug, who the placebo, how much of the drug,
over which period of time, etc. The experimenter must choose between
different experimental designs $\underline{e}$ on a common parameter space Θ. The
different designs involve different sets of observations X that are
governed, however, by some common probabilistic law because the
different experimental designs are all thought to give information that
is relevant to the same question. The purpose of the experiment is to
answer some questions about θ.

How is it determined what experimental designs are optimal? Several
Bayesian analyses have been given. I will discuss one of them. As a
Bayesian analysis it presupposes an assessment of the prior probability
distribution $p(\theta, x|\underline{e})$. In our drug testing case it is plausible that
one will make some decision $d(x) = \underline{d}$ that depends on the outcome of the
experiment; for example, the drug might be forbidden for the allergy
group. This decision carries with it a utility, depending on the true
value of the drug. The general form of the utility function is
$U(d(x), \theta, \underline{e})$, allowing for the possibility that one experimental
design will be more expensive than others. An optimal experimental

design is a design with maximum expected utility with respect to the prior $p(\theta, x|\underline{e})$. Hence an $\underline{e}$ that maximizes

$$(2) \quad \int_X \max_d \int_\Theta U(d(x), \theta, \underline{e}) \; p(\theta|x, \underline{e}) \; p(x|\underline{e}) \; d\theta dx.$$

This analysis and all others that have been given, seem to lead to the same verdict for randomization. Lindley writes:

> ... one important consequence is immediately apparent ... . Let
> E_0 ... be the set of experimental designs satisfying ... (2). If
> it contains a single member, then this is the best experiment
> to perform. If it contains more than one member then all e in
> E_0 are equivalent from a Bayesian viewpoint and any may be
> selected. Consequently it is never necessary to randomize in
> experimental design, though randomization over E_0 would not do
> any harm (nor any good). This goes counter to a popular sampling-
> theory canon. On reflection the Bayesian conclusion seems correct
> to me. Certainly I find it hard to see how the fact that a result
> was obtained by randomization rather than by deliberate choice can
> have any effect on the subsequent analysis; in particular, the
> randomization theory of tests seems unconvincing. (Lindley 1971,
> p. 21).

I have quoted this passage in full to point out that the two issues that Lindley brings forward, should be separated. The second one is the rationale of the randomization theory as advocated, for example, by Fisher (1935). One should not expect that these can be justified within the Bayesian framework. That one justification of randomization fails from a Bayesian point of view, does not mean, however, that there is no Bayesian justification for a certain amount of randomization in experimental design. To give such a justification one has to show what Lindley finds hard to see, viz., how the fact that a result obtained by randomization rather than by deliberate choice can have an effect on the analysis.

When I started thinking about randomization some time ago, I hoped that it would give an Archimedean point from which one could put the Bayesian philosophy of statistical inference out of joint. Now I am more impressed with the Bayesian approach and its seemingly unlimited pliability. Donald Rubin (1978) has shown how a Bayesian can make sense of the practice. He shows how a Bayesian analysis is considerably simplified by randomization. It saves the data analyst from having to assess the mechanisms by which unrecorded subject characteristics and treatments can be confounded. Rubin's results show that it might be easier and thus cheaper to randomize but still permits the Bayesian objection that cheap or easy is only one aspect by which one has to judge the quality of an experimental design. Rubin analyzes randomization as it affects inference. For a Bayesian justification of randomization one needs a decision theoretic argument.

162

3. The Origins of the Argument in Abraham Wald

In his book, Savage (1954) acknowledges that some of the basic ideas
of his theory of personal probability are derived from what he calls
"Wald's minimax theory". But in addition to basic ideas such as
'person', 'world', 'state of the world', 'act', 'consequence' and
'decision', Savage adopted the Waldian analysis of randomization.

Wald wanted to unify the Neyman-Pearson theory of hypothesis testing
and estimation and the work of Fisher's school of experimental design.
His decision theory, however, is anti-Fisherian in spirit and
elaboration. It was heavily attacked by Fisher (1955) who claimed that
it had no relevance for science in general and the design of experiments
in particular.

Wald's statistical decision functions encode 'rules for inductive
behavior' in the tradition of Jerzy Neyman. Formally, Wald's theory
is equivalent to the personalistic decision theory of Savage and
Lindley without the prior probability $p(\theta, x \mid e)$. Consequently, decision
rules cannot be compared on the basis of expected utility as in the
Bayesian theory and other criteria have to be formulated.

Wald generalized the Neyman-Pearson theory in several directions.
He allowed for sequential tests in which one can decide on the basis of
past observations whether to accept, to reject, or to make some more
observations. But his main contribution was to show how the von
Neumann-Morgenstern utility theory can bear on problems of inductive
behavior. To this purpose he summarized the different costs and gains
that are connected with a decision by a non-negative loss function.
For our purposes we can look at the loss function as the negative of a
utility function that is bounded from above.

To tailor Wald's theory to our discussion of randomization in
experiments, let us look again at the scientist who wants to test a
drug for its side effects on people that have an allergy problem. Two
decisions are involved: the choice of an experimental design $\underline{e}$ and the
choice of some decision function $\underline{d}$ that maps the sample space into some
set of terminal decisions. I will denote the set of all relevant
experimental designs by $\underline{E}$ and the set of all decision functions by $\underline{D}$.
The loss function will be denoted by $L = L(\underline{d}, \theta, x, \underline{e})$.

Wald's analysis does not presuppose a prior on θ, and other selection
criteria have to be invoked. The one studied by Wald is minimax risk.
The risk function of $(\underline{d}, \underline{e})$ is defined as

$$(3) \quad R(\underline{d}, \underline{e}, \theta) = \int_X L(\underline{d}, \theta, x, \underline{e}) \, p(x \mid \theta) \, dx$$

The pair $(\underline{d}, \underline{e})$ is said to be inadmissible if there is another pair
$(\underline{d}^*, \underline{e}^*)$ with $R(\underline{d}^*, \underline{e}^*, \theta) \leq R(\underline{d}, \underline{e}, \theta)$ for all θ, with strict
inequality for some θ. The minimax risk rule then says: choose $(\underline{d}, \underline{e})$
so that $\max_{\theta} R(\underline{d}, \underline{e}, \theta)$, the risk level of $(\underline{d}, \underline{e})$, is minimal. The
minimax solution is clearly admissible.

The space of available decision functions is enlarged by incorporating so-called randomized decision functions that can provide for lower risk levels. Both the terminal decision rule and the choice of experimental design can be randomized. In this manner, both the mixed tests of Neyman and Pearson and Fisher's randomization designs seem to be covered by the theory of statistical decision functions. Wald showed that the value of the risk function of a randomized decision rule at θ is some weighted average of the values of the risk function of its deterministic components at θ. This implies that the risk level of a randomized decision rule can indeed be less than that of its components.

Wald's calculation of the risk level of a randomized decision function misses, however, the point of randomization. Wald does not take into account that <u>the stochastic processes studied and the statistical law that governs the process may change with randomization.</u> My contention is that this should be no more acceptable to a Bayesian than to Fisher. For Fisher, randomization was performed to justify the modelling assumptions that go into the construction of the class of distribution functions that the statistician considers as possibly governing the stochastic process. To randomize the allocation of treatments is to do something in the world and thus affects the character of the process that is studied. For example, randomization changes the possible causal interactions in an experimental situation. Unrecorded, subliminal cues cannot influence the treatment allocation. One type of experimenter bias becomes impossible.

Wald was not a Bayesian. He did, however, study what he called Bayes solutions for a statistical decision problem. They arise when the statistician assumes a prior distribution $p(\theta|\underline{e}, \underline{d})$. A Bayes solution is then a decision function that minimizes the average risk with respect to the given prior distribution. Wald proved that under some restrictions the class of all Bayes solutions forms a complete class of decision functions. That is, there are no decision functions that are uniformly better than a Bayes solution with respect to some (possibly improper) prior. When Wald put the two ideas together -- randomization and assuming a prior -- he did not realize that the choice of experimental design can contain information about θ, because he thought that he had given a satisfactory analysis of randomization in his general theory. This made it possible to prove that with a prior randomization becomes unnecessary in the sense that if a Bayes solution with respect to a prior distribution exists, there also exists a non-randomized Bayes solution with respect to the same prior. Here we see the birth of the Bayesian decision theoretic argument against randomization. <u>The irrelevance of randomization became a mathematical theorem.</u>

4. Why It Can Pay To Randomize

If one calculates the expected utility of randomly choosing a design and of deterministically or haphazardly choosing a design in the correct manner, it may pay to randomize. The two key observations that lead to

164

an example that shows this are, one, that given the pre-experimental
data only random choice of treatment allocation guarantees that the
allocation pattern contains no information about the parameters of
interest; and two, that human beings are biased selectors. If I
separate haphazardly and without using an objective randomizer, a
hundred carefully matched pairs in a treatment and control group,
getting to know the allocation that resulted might give me evidence
about the underlying true state of illness of the patients. I might,
for example, "believe in the drug" and be eager to prove its efficacy.
A control group that is chosen haphazardly will then likely contain
slightly more ill patients. This is not a case of cheating because
I might very well have reacted to subliminal cues from the patients.
That this scenario is not farfetched has been amply demonstrated by
Robert Rosenthal (1976).

Let us go back to the scientist who wants to determine the side
effects of a drug for a group of allergy patients. I assume for
reasons of simplicity that he has only two patients in his experiment.
They agree over a broad range of relevant pre-experimental data and
the scientist thinks that it is immaterial who gets the drug and who
the placebo; both experiments would give him the same amount of
information. I will call the patients A and B, the experiment where
A gets the drug and B the placebo, e(A) and the other experiment e(B).

At first sight it might appear completely arbitrary which experiment
the scientist chooses to do. But consider the two possibilities (θ_1)
that A is really more ill than B and (θ_2) that A is less ill than
B. The scientist formulates the following decision matrix $(\varepsilon > 0)$

	θ_1	θ_2
e(A)	ε	0
e(B)	0	ε

The utilities reflect the following reasoning. If θ_1 were the case,
the scientist would rather perform experiment e(A) because he wants to
give the drug a hard time before it can be used by the allergy patients.
(With a larger group of matched pairs, considerations like these become
more convincing since for each θ_i one can find experiments that are
clearly better balanced than a lot of others.) Similarly for θ_2, but
now e(B) is the preferable experiment.

In order to calculate the expected utility of both experiments the
scientist has to assess the probability of the two hypotheses θ_1 and θ_2.
Note that it is already implied that the possibility that A is exactly
as ill as B can be neglected. The pre-experimental data, however,
imply that $p(\theta_1) = p(\theta_2) = \frac{1}{2}$.

How should one calculate the expected utility of the two experiments?
The issue of bias becomes relevant. The scientist has to make a
distinction between doing e(A) because the flip of a coin told him to
do so, or doing e(A) because that was "what I ended up doing when I
haphazardly separated the pair because it didn't matter which of the

optimal experiments I performed." Let me call these two acts or decisions er(A) and eh(A). The decision matrix now becomes (I assume that the cost of haphazard and randomized allocations is the same):

	θ_1	θ_2
eh(A)	ε	0
eh(B)	0	ε
er(A)	ε	0
er(B)	0	ε

The relevant probabilities of the states given the acts are ($\eta > 0$):

	θ_1	θ_2
eh(A)	$\frac{1}{2}-\eta$	$\frac{1}{2}+\eta$
eh(B)	$\frac{1}{2}+\eta$	$\frac{1}{2}-\eta$
er(A)	$\frac{1}{2}$	$\frac{1}{2}$
er(B)	$\frac{1}{2}$	$\frac{1}{2}$

The probabilities that I have jotted down reflect the following reasoning. If one obtains an allocation by randomization, seeing the actual assignment pattern obtained gives no information about θ_1 versus θ_2. But if one obtains it haphazardly, the allocation pattern contains information. The scientist may believe in the drug and may have subconsciously reacted to subliminal cues from the patients so that he was able to give the drug to the healthier patient.

If we calculate the expected utilities of the four experiments we get:

$$EU(eh(A)) = EU(eh(B)) = \tfrac{1}{2}\varepsilon - \varepsilon\eta$$

and

$$EU(er(A)) = EU(er(B)) = \tfrac{1}{2}\varepsilon$$

Randomization pays off.

The example as it stands is unsatisfactory on several counts. In the first place, the probability matrix was motivated by saying that the haphazard act eh(A) was a symptom of θ_2, in the sense that θ_2 increases the chances of doing eh(A) as opposed to eh(B). This means that the analysis has the same causal structure as a so-called Newcomb type decision problem where the true state of nature is thought to increase the probability of one of the acts. These cases are still ill understood. See Gibbard and Harper (1978). Fortunately, the analysis can be transformed into a structure where acts and states of nature are independent.

$$\theta_2/1 \longrightarrow eh(A/B)$$

Figure 1

$$\theta_2/1 \searrow e(A/B)$$
$$eh \nearrow$$

Figure 2

166

Figure 1 gives the causal chains of the preliminary analysis of
haphazard allocation. The true state of illness of both patients
causes the scientist to give the drug to the healthier of the two.
"Causes the scientist" is meant in the probabilistic sense of
"increases the chance of".

   In Figure 2, I have identified the acts differently. There are now
two acts open to the decision maker, eh, the generic act of making a
haphazard choice from the set of designs containing e(A) and e(B), and
er, the generic act of making a random choice. The corresponding
matrix shows that this leads to what Savage calls a small world
analysis.

	$\{\theta_1\}$	$\{\theta_2\}$	$\{\theta_1\}$	$\{\theta_2\}$	$\{\theta_1\}$	$\{\theta_2\}$
eh	e(A) / e(B)	e(A) / e(B)	ε / 0	0 / ε	$\frac{1}{2}-\eta$ / $\frac{1}{2}+\eta$	$\frac{1}{2}+\eta$ / $\frac{1}{2}-\eta$
er	e(A) / e(B)	e(A) / e(B)	ε / 0	0 / ε	$\frac{1}{2}$ / $\frac{1}{2}$	$\frac{1}{2}$ / $\frac{1}{2}$
	(outcomes)		(utilities of outcomes)		(probabilities of outcomes)	

Act and state of nature do not lead to a unique outcome, but to a
"gamble". The outcomes of a small world analysis are the acts of the
grand analysis. The states of nature in the small world analysis are
events in the grand analysis, that is, sets of states of nature.

   The analysis of a design problem in terms of generic acts like eh
and er is what I call a "de-extensionalized analysis". It underlines
a point of Campbell and Boruch (1975) that in evaluating randomized
assignments to treatments one has to consider the alternatives. In an
experiment with 100 matched pairs there are 2^{100} ways of allocating the
treatment, but not all of these allocations are individuated for the
decision maker. There are only a few ways in which he can allocate
treatments: by randomization, haphazardly, and in a number of systematic
ways. Some of these procedures, like randomization and haphazard
choice are gambles. They have a chancy outcome pattern, each with a
different causal and probabilistic structure.

   One could try to fill in the factor a under which θ_2 tends to
decrease the probability of eh, but in practice this will seldom lead
to a complete determination of the outcomes.

	θ_1 & a	θ_1 & not-a	θ_2 & not-a	θ_2 & a
eh	e(A) / $(\frac{1}{2}-\eta)\frac{1}{2}$	e(B) / $(\frac{1}{2}+\eta)\frac{1}{2}$	e(A) / $(\frac{1}{2}+\eta)\frac{1}{2}$	e(B) / $(\frac{1}{2}-\eta)\frac{1}{2}$

Still it is instructive to compare this representation with a similar transformation of the preliminary analysis of randomized allocation. It brings out nicely Fisher's point that when we randomize we have to consider a different set of models.

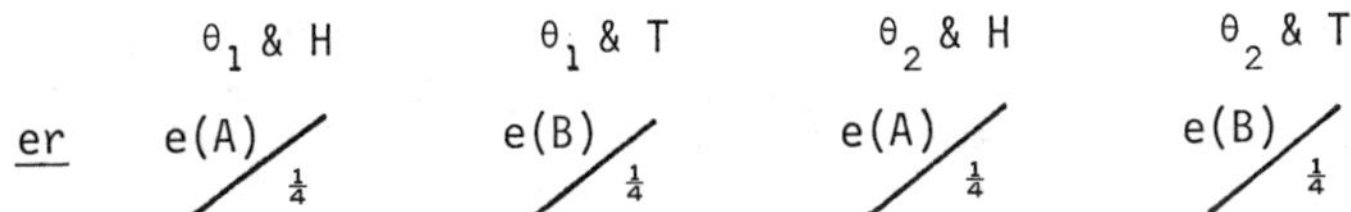

Under randomization, allocations are completely determined by, say, H(eads) and T(ails), with known probabilities. Under haphazard allocation, the relevant event a includes assumptions about the strength of the subliminal cues, the sensitivity of the experimenter to those cues, the particular biases the experimenter is liable to, and much more.

The example as it is, falls short of Lindley's more general format, because no decisions have yet been identified. But it suffices to illustrate my contention that a Bayesian can gain by randomization. And this is as it should be, because experience (e.g., Bennett and Lumsdaine 1975) tells us that randomization combined with stratification or matching gives much better data than haphazard choice combined with those experimental techniques.

References

Bennett, C.A. and Lumsdaine, A. (eds.). (1975). _Evaluation and Experiment: Some Critical Issues in Assessing Social Programs._ New York: Academic Press, Inc.

Campbell, D.T. and Boruch, R.F. (1975). "Making the Case for Randomized Assignment to Treatments by Considering the Alternatives: Six Ways in which Quasi-Experimental Evaluations in Compensatory Education Tend to Underestimate Effects." In Bennett and Lumsdaine (1975). Pages 195-296.

Fisher, R.A. (1935). _The Design of Experiments._ Edinburgh: Oliver and Boyd.

------------. (1955). "Statistical Methods and Scientific Induction." _Journal of the Royal Statistical Society_ Series B, 17: 69-78.

Gibbard, A. and Harper, W.L. (1978). "Counterfactuals and Two Kinds of Expected Utility." In _Foundations and Applications of Decision Theory._ Volume 1. _(University of Western Ontario Series in the Philosophy of Science._ Volume 13.) Edited C.A. Hooker, _et al._ Dordrecht: Reidel. Pages 125-162.

Lindley, D.V. (1971). _Bayesian Statistics, A Review._ Philadelphia: Society for Industrial and Applied Mathematics.

Rosenthal, R. (1976). _Experimenter Effects in Behavioral Research._ 2nd enlarged ed. New York: Irvington Publishers, Inc.

Rubin, D.B. (1978). "Bayesian Inference for Causal Effects: The Role of Randomization." _Annals of Statistics_ 6: 34-58.

Savage, L.J. (1954). _The Foundations of Statistics._ New York: John Wiley & Sons.

------------. (1961). "The Foundations of Statistics Reconsidered." In _Proceedings of the Fourth Berkeley Symposium on Mathematics and Probability._ Volume I. Edited by Jerzy Neyman. Berkeley: University of California Press. Pages 575-586.

------------ _et al._ (eds.). (1962). _The Foundations of Statistical Inference._ New York: John Wiley & Sons, Inc.

Wald, A. (1950). _Statistical Decision Functions._ New York: John Wiley & Sons, Inc.

<u>How to Commit the Gambler's Fallacy and Get Away With It</u>

Davis Baird

University of Arizona

and

Richard E. Otte

University of Arizona and University of Pittsburgh

1. The Gambler's Fallacy

The gambler's fallacy runs like this: We know

1) the bias for heads of a certain coin and flipping device is
    1/2, and

2) flips of the coin with the flipping device are independent.

Given these facts we can easily compute the probability of 4 heads or
fewer turning up in 20 flips, $P(h \leq 4)$, to be

$$(20!(1/2)^{20})/\Sigma_{i=0}^{4}(i!(20-i)!) \cong .0059.$$

Suppose we flip the coin 15 times and get--improbably enough--all
tails. If the coin lands tails up once more in the next 5 flips the
improbable event of 4 or fewer heads in 20 flips, $h \leq 4$, will have
occurred. Since $P(h \leq 4)$ is so small we might be tempted to infer that
all of the next five flips will be heads. If we did not wish to make
the strong claim that <u>all</u> of the next five flips will be heads, we
might at least be tempted to claim that the probability of heads on
each of the next five flips is considerably greater than 1/2. Thus
the gambler betting on red at a roulette table might exclaim after a
string of bad luck--15 black numbers in a row--"If this wheel is fair
the next number is bound to be red!"

Such reasoning is fallacious. To determine $P(h \leq 4)$ we assume that
flips of the coin are independent of each other: the outcome of past
flips does not alter chances on future flips. The probability of

PSA 1982, Volume 1, pp. 169-180

170

heads on the 17th flip is simply 1/2, regardless of whether the first
16 flips were all heads, all tails, or any combination of heads and
tails. Indeed, the probability of getting 5 heads in a row on the
16th, 17th, 18th, 19th and 20th flips is quite small:

$$(1/2)^5 \cong .03.$$

It is not large, as those reasoning a la gambler's fallacy conclude.

This much is uncontroversial, and we shall not dispute it. There
are, however, patterns of reasoning very much like the gambler's
fallacy which we believe are not fallacious. We provide examples of
such patterns of reasoning and argue that they demonstrate that
probabilistic/inductive reasoning is very sensitive to the statistical
model. In particular we exhibit two different statistical models for
coin flipping which are mutually consistent, yet which license contra-
dicting inferences; on one model the gambler's fallacy is a fallacy,
and on the other it may be a legitimate mode of inference.

2. Emergent Probabilities

Our point of departure is a recent paper by Ian Hacking, "Grounding
Probabilities from Below" (Hacking 1980). Hacking begins by asking:

> Is it the case that every stable frequency, correctly represented
> by a mathematical probability, is 'grounded from below' by pro-
> babilities that apply to individuals? That is, does the frequency
> distribution in the population derive from probabilistic facts
> about the individuals that compose it? Or are there some stable
> frequencies that pertain to populations, but do not derive from
> probabilistic facts about members of the population? (Hacking
> 1980, p. 110).

Hacking presents a case for the 'emergentist position': there are
probabilistic laws, manifested by stable frequencies in populations,
where no probabilistic laws cover individuals in the population. Put
another way, Hacking thinks that there may be cases where probabilities
can be correctly assigned to ensembles while no probabilities can be
correctly assigned to individual members of the ensembles in question.
Hacking calls such probabilities, probabilities which are not 'grounded
from below'.

Much of Hacking's argument rests on two examples: suicide
statistics from 19th Century Europe, and fertility statistics from
Germany during the period 1871-1939. The interesting feature of both
of these cases is that stable long run frequencies are found in
populations while no stable frequencies are found in partitions of
these populations. Concerning fertility statistics, Hacking writes:

> Knodel found that the national decline in birth rate is nicely
> mirrored in each administrative district or _Kreis_. The numbers
> are as regular as any which could be hoped for in demography. But,

when we pass to smaller units, such as the village, the uniformity
collapses. Although every <u>Kreis</u> is doing the same thing as every
other, villages within <u>Kreise</u> are all going their own ways, without much in the way of underlying laws. (Hacking 1980, p. 114).

Hacking concludes from this collapse of uniformity that:

> [T]he propensity to limit family size may simply not be a number
> which represents a property of each of the individual couples
> within the <u>Kreis</u>. There is only a very striking property which
> applies to the <u>Kreis</u> as a whole, that the birth rate is declining
> in a systematic and law-like way. (Hacking 1980, p. 115).

When a statistical uniformity appears in a population, there are two
ways to account for it. One way is to ascribe probabilistic properties
to individuals and use results such as the <u>law of large numbers</u> to
explain stable regularities in ensembles of individuals. Another way
is to claim that the uniformity in the population does not arise out
of any probabilistic facts about the individual members of the population, but rather that the probabilities are manifest only at the
level of an ensemble.

We find Hacking's paper suggestive, and shall tentatively grant him
his major point. There are cases where probabilistic properties may
be correctly ascribed to individuals, and, by inferences based on the
probability calculus, to populations; and there are cases where
probabilistic properties may be correctly ascribed only to ensembles
in which no probabilistic properties may be ascribed to individuals
within the ensemble. However, we take issue with Hacking on one of
his closing remarks: "I do not think my distinction makes the
slightest difference to any practical problem of statistical inference."
(Hacking 1980, p. 115). While it is not entirely clear which problems
are practical and which are not, we believe that Hacking's two cases
license different inferences.

3. A Trivial Example

Consider the following fiction. We are gathering census data in
Massachusetts. We happen to be aware of a well established statistical
law that the number of children born per couple--on average--in any
county is 2.44. Let us further suppose that this statistical law is
not grounded from below: no probabilistic properties ascribable to
individual couples or even individual towns grounds the statistical
law. We have completed the census for every town in Middlesex County
except Lexington. The following facts have been revealed:

1) There are 980,000 couples in all of Middlesex County except
    Lexington;

2) There are 2,401,000 children in all of Middlesex County except
    Lexington.

172

We also estimate that:

   3) There are 20,000 couples in Lexington.

From 1) and 2) it is easy to compute the average number of children
per couple in all of Middlesex County except Lexington:

   2,401,000/980,000 = 2.45.

Since it is a statistical law that counties have stable average family
sizes, while towns may have widely varying average family sizes, we
predict the average number of children per couple in Lexington from
that data. Let X be the number of children in Lexington. Then our
statistical law requires that:

   (X + 2,401,000 / 980,000 + 20,000) = 2.44.

Hence,

   X = 2.44(980,000 + 20,000) - 2,401,000 = 39,000.

Accordingly we predict that the average number of children per couple
in Lexington is

   39,000/20,000 = 1.95.

This is different from the county average of 2.44.

   If the statistical law used above was grounded from below then the
reasoning above would commit the gambler's fallacy. Assuming they
exist, let us call the grounds for the statistical law, 'sub-laws';
these sub-laws might stipulate probabilistic facts about the average
number of children per couple in the individual towns. All these sub-
laws together allow us to compute that the average number of children
per couple in Middlesex County should be about 2.44. The important
point is that the County average would be derivative upon individual
town averages. Thus, to draw the analogy with the gambler's fallacy
on the roulette wheel: There is a long run statistical stability to
the effect that only very infrequently will fewer than 5 red numbers
turn up in 20 plays of the wheel. This is analogous to the county law,
which also stipulates a long run statistical stability. The long run
statistical stability of the roulette wheel is derivative upon the
individual probabilities assigned to the individual numbers on the
wheel and the independence of plays; in the same manner, the county
law is derivative upon the 'sub-laws' concerned with the individual
towns. When the gambler commits his fallacy he reasons about the
remaining 5 plays on the basis of the long run statistical stability
in the same way that we have reasoned about Lexington on the basis
of the county law. The gambler commits a fallacy because his reasoning
violates one of the assumptions used to compute the law concerning the
long run statistical stability. Similarly, we violated one of the
sub-laws grounding our county law in making our inference about

Lexington. If our county law is grounded from below, our reasoning commits the gambler's fallacy.

On the other hand, if the county law is not grounded from below by sub-laws about the individual towns, then our reasoning to a conclusion about Lexington does not violate any assumptions used in determining the county law. Consequently, in such cases it is not a fallacy to reason as we did above. Indeed, such reasoning is similar to a very common form of reasoning. We have a law that tells us that planets orbit in ellipses. We gather some data concerning the positions of a particular planet at some particular times, and we fit an ellipse to the observations--as our law bids us. The ellipse then allows us to compute where the planet will be at other times. We claim that when probabilities are not grounded from below, our reasoning concerning Lexington is analogous to the reasoning in the planet-ellipse example. It is not a fallacy at all.

4. A Coin Flipping Model for Emergent Probabilities

Under normal circumstances we assume the following about coin flipping models of probabilistic phenomena:

1) the coin (and perhaps the flipping device) has a property, called bias; the bias, Θ, of a coin [and flipping device] is the probability that the coin will land heads on any flip;

2) distinct flips of the coin are statistically independent of each other.

Let us call a model which has these properties M_b. Alternatively, we can define the following model about coin flipping type phenomena:

1') The coin (and perhaps the flipping device) has a property, called smias; the smias, σ, of a coin [and flipping device] is the probability that the coin will land heads five or more times in 20 flips;

2') distinct 20 flip trials of the coin are statistically independent of one another.

We will call such a model M_s.

If there is a M_b model then there is associated with it an M_s model. If a coin has bias Θ then the coin will have smias

$$\sigma = 1 - 20! \ \Sigma_{i=0}^{4}((\Theta^i(1-\Theta)^{20-i})/(i!(20-i)!)).$$

Since individual flips are independent given a M_b model, 20 flip trials are also independent. Thus, given a M_b model we can find the associated M_s model.

If we admit the possibility of emergent probabilities, the converse need not be true. There may be cases where we can make probabilistic statements about the behavior of a coin on trials of at least 20 flips, while no probabilistic statements can be made about trials of fewer than 20 flips. Perhaps there is some unknown feature of the flipping device which keeps individual flips from being independent, but assures that each 20 flip trial is independent of other 20 flip trials.

Given our preconditioning to think of coins in terms of bias, our smias model appears a bit strange. However, such a model does provide a simple picture of what is going on with Hacking's emergent probabilities. It is perhaps worthwhile to dispel some possible objections by requiring of our smias model some criteria of identity for 20 flip 'individual' trials. By analogy then our individual 20 flip trials operate like Hacking's Kreise. We cannot make probability statements for partitions smaller than the Kreis, or a 20 flip trial, while we can do so for Kreise. Such is a situation where we have a smias model and no bias model to ground it from below.

Hacking says that such emergent probabilities make no difference to statistical inference, but they do. If there is an adequate bias model then reasoning about the remaining 5 flips of a twenty flip sequence on the basis of the previous 15 flips commits the gambler's fallacy. However, if there is no bias model 'underneath' a smias model, then such reasoning is sound. Given 15 tails in the first 15 flips of a 20 flip trial, if we assume smias of .999 and no bias underneath, we can legitimately infer that the next 5 flips will be heads. This is how to commit the gambler's fallacy and get away with it.

5. Bias, Smias and the Theory of Statistical Tests

Our smias model of coin flipping might appear peculiar. On the contrary, however, we believe that smias or something similar is in fact the model tested when statistical tests are run on hypotheses about bias. To clarify this paradoxical remark: we frame a hypothesis in terms of bias; the test of this hypothesis, however, concerns only properties of the smias model associated with the bias model purportedly under test. While the two possible results of a statistical test of some hypothesis about bias is said to be 'reject' or 'do not reject', all we are really justified in inferring is the rejection or not of the associated hypothesis about smias. An example makes these remarks clearer.

Suppose we wish to test the hypothesis H that a coin has bias greater than 1/2,

H: $\theta \geq 1/2$.

According to the standard theories of statistical testing of either Neyman/Pearson or Fisher, to test we adopt a rejection region, R. The rejection region is a set of possible outcomes; if any of these

possible outcomes occurs we reject the hypothesis. Suppose in the
case at hand we can afford to flip the coin only 20 times. Obviously
the rejection region, R, should include those results where a preponderance of tails came up. The rejection region, R= {0,1,2,3 or 4
heads in 20 flips}, provides a fairly stringent test of H. Stringency, or <u>level of significance</u>, is defined as the probability,
assuming the hypothesis under test, of the rejection region; in this
case the probability of R is about .0059.[1] In a certain sense stringency is the probability that such a test would lead to a mistaken
rejection of a true hypothesis. In this case that probability is small,
59/10,000. It is important to note that in order to perform such a
test we need only know whether or not a result occurred in the rejection region. According to accepted testing theory it is the
information of whether or not R is true, and not the actual result,
that is important for inference.

Alternatively, suppose we wish to test a hypothesis J about <u>smias</u>:

J: $\sigma \geq .9941$.

Our financial constraints remain the same; we can afford only one 20
flip trial. Since we can perform only one 20 flip trial, only two
outcomes are possible: 'success', 5 or more heads in 20 flips, or
'failure', 4 or fewer heads in 20 flips. Obviously the sensible test
rejects J if and only if the result is 'failure'. Such a test is
quite reasonable; like our test for bias above, it has stringency of
.0059.

Obviously the test of H is identical with the test of J. We reject
either hypothesis, H or J, just in case 4 or fewer heads turn up in 20
flips. In a certain sense this is just what we want. If there is a
bias model, then

$H = \Theta \geq 1/2$ if and only if $\sigma \geq .9941 = J$.

However, if there is no bias model in the offing, then the equivalence
fails. But we should note that even if there is no bias model we can
still, in a purely formal sense, test hypotheses about a non-referring
parameter, bias. The only features of the bias model which are tested
are features of the smias model associated with it. When we test
hypotheses about bias we are really concerned only with smias
properties.

One might object: The only reason our test of bias is identical to
a test of smias is that we could afford to flip the coin, conveniently
enough, only 20 times. This is true, however the important point is
that no matter how many times we do flip the coin, we can always define
a smias type model for the appropriate number of flips. Our test of
bias will always be equivalent to a test of an appropriately defined
hypothesis in the associated smias type model.

Our smias/bias example plays on a well known problem: why is the

stringency of a test measured by the probability of the entire
rejection region? Should it not be measured by the probability of the
specific result, e.g., 3 heads in 20 flips, instead of simply R?
Recall that in order to test H, $\Theta \geq 1/2$, we had only to know whether
or not R occurred. While several proposed justifications for this
procedure have appeared, none are entirely satisfactory. We believe
considerations about smias, bias, and emergent probabilities shed
some light on this problem.

There is reason to believe test stringency should not be measured
by the probability of the rejection region. An almost universally
adopted maxim of inductive inference is that the premises of an
inductive argument must include <u>all</u> of the <u>relevant</u> known information
concerning the conclusion of the argument. This requirement of total
evidence was perhaps first made clear by Rudolf Carnap in his 1947
paper on inductive logic; a recent discussion with particular attention to problems of statistical inference is in Seidenfeld 1979.
The most widely accepted view is that what are called 'minimally
sufficient statistics' contain all the relevant information for
inference (Seidenfeld 1979, p. 83). Unfortunately, whether or not the
outcome occurred in the rejection region is not sufficient information
for inferences about bias. Hence, according to the sufficiency
reading of the requirement for total evidence, the occurrence or non-
occurrence of the rejection region is not sufficient information for
inferences about bias. It seems then, that measuring stringency by
the probability of the rejection region is not right.

While the occurrence of R is not sufficient for inferences about
bias, the occurrence of 'failure' is sufficient for inferences about
smias. R is identical to 'failure'. Hence, it seems that the information collected in a test of bias is really only sufficient for
inferences about the associated smias hypothesis. On the standard
theory of testing, we test bias by testing the associated smias.

If bias is a property of the coin, then we might justify the rejection of hypotheses about bias, such as H, on the basis of
insufficient information as follows: The information that a result
in the rejection region occurred is sufficient for hypotheses about
smias. In particular, if R occurs we have good reason to reject J,
$\sigma \geq .9941$. Since we assume our smias probabilities are grounded from
below with bias probabilities, we infer that H if and only if J.
Hence, if we reject J, we should also reject H.

6. Metaphysics vs. Epistemology

We believe that the tension between choosing a bias model and a
smias model reflects a tension that exists between metaphysical
intuitions and epistemological intuitions. Russell states the metaphysical intuitions with characteristic flair:

> The theory of probability is in a very unsatisfactory state,
> both logically and mathematically; and I do not believe that there

is any alchemy by which it can produce regularity in large numbers
out of pure caprice in each single case. If the penny really
chose by caprice whether to fall heads or tails, have we any
reason to say that it would choose one about as often as the other?
Might not caprice lead just as well always to the same choice?...
[W]e cannot accept the view that the ultimate regularities in the
world have to do with large numbers of cases, and we shall have to
suppose that the statistical laws of atomic behavior are derivative
from hitherto undiscovered laws of individual behavior. (Russell
1935, p. 168).

Russell explicitly states here that the emergentist position is unsatisfactory. Russell believes, for metaphysical reasons, that we
must explain statistical regularities by an appeal to deterministic
laws about individuals; statistical laws are dependent upon such laws.
For example, consider the statistical laws of classical kinetic theory.
These laws are based upon deterministic laws about individual molecules.
Their statistical character arises only because of the difficulty in
stating the initial conditions of a system and working out the
appropriate calculations. Evidently Russell believes that all
statistical laws should be explained in this manner.

   Russell wants a 'deeper' deterministic explanation for long run
stability. There are two parts to his worry. One is a general uneasy
feeling regarding indeterministic hypotheses. The other is a belief
that laws operate at the level of individuals--not collectives. More
current sentiment regarding determinism is well expressed by Salmon:

   This view involves an a priori commitment to determinism--... .
   This position seems to me untenable. I do not mean to argue that
   present physical theory is complete and correct, but, rather that
   there is no reason to make an a priori decision as to the nature
   of further physical theories. Perhaps, in the future, improved
   theories will provide a deterministic account of events that
   current theory regards as causally undetermined--but perhaps
   they will not. We should be prepared for the possibility that
   the indeterministic character of physical theory is correct and
   that there are events which are intrinsically improbable, not
   merely improbable in relation to our present incomplete knowledge.
   (Salmon 1971, p. 9-10).

We agree with Salmon, and think his attitude towards determinism is
also appropriate for the second part of Russell's worry. It is an
empirical question as to what level natural laws operate at.

   In our opinion the level at which laws operate should be left open
and not assumed a priori to be that of the individual. This is not
generally accepted, as Hacking's reference to the 'much maligned topic
of emergentism' indicates (Hacking 1980, p. 110). While we seem to
have outgrown our deterministic prejudices, as the acceptance of
Quantum Mechanics and the growing popularity of probabilistic
theories of causality shows, we remain clinging to the second half of

Russell's worry: laws operate at the level of individuals. The result is a metaphysical 'push' away from properties like smias towards bias and beyond.

On the other hand, there is an epistemological reason to be unhappy with a law such as $\Theta = 1/2$. Such a law is completely untestable, except in terms of its consequences for many flips. Suppose that we flip a coin once and it comes up heads; does that give us any reason to either claim the coin is fair, or unfair? Of course not. We accept or reject the hypothesis on the basis of many flips. But in dealing with many flips, we are dealing with smias-type properties. It seems like this gives an epistemological reason to work with smias-type models, and not be concerned with properties such as bias. In a sense then, there is an epistemological 'push' towards smias away from bias.

There appear to be three levels of hypotheses that we are dealing with in this discussion: deterministic laws, bias type properties, and smias type properties. We might sketch the situation as follows:

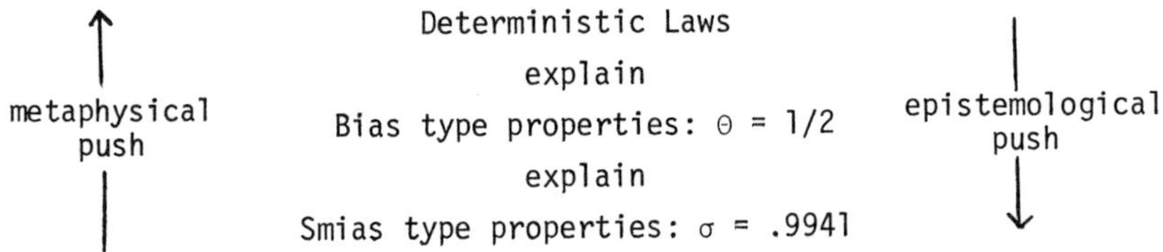

The result is confusion. We have a problematic theory of testing where we do not use all the relevant information when inferring about bias; and we have an inadequate explanation of the regularities observed in ensembles by means of untestable hypotheses about bias.

7. Is the Gambler's Fallacy Really a Fallacy?

We simply are in a position of not knowing for certain whether smias properties are always, ever, or never grounded from below by bias properties. In some cases we may be able to gather some pertinent data. We can test for independence of flips. Such tests are not, however, conclusive. Ultimately our belief in grounding probabilities from below rests upon a metaphysical assumption. This intuition may be very plausible but it is metaphysical, nonetheless. Consequently, we can never be certain whether the gambler's fallacy is really a fallacy, since it also rests upon this metaphysical assumption. Perhaps there is some consolation in this. After all, how could so many gamblers be so wrong?

The later part of our exposition has been given in terms of bias and smias. There are two reasons for this. In the first place, it is helpful to have a concrete model in mind; coin flipping serves nicely in this role. It should be clear though, that smias type probabilities throughout stand for any sort of population probabilities, and bias type probabilities stand for the possible grounds from below that the smias type probabilities might enjoy. Secondly, Hacking discusses binomial coin-like phenomena as cases where Poisson's analysis shows how emergent regularities can be derived from the probabilistic properties of the individuals that compose the population. (Hacking 1980, pp. 112-3). Hacking closes by asking for Poisson's analysis to be extended:

> A useful continuation of Poisson's programme for a law of large numbers would examine the extent to which quantitative frequencies within a collective may derive from purely qualitative and unstructured propensities assigned to individual members of the collective. (Hacking 1980, p. 115).

We second Hacking's proposal for a continuation of Poisson's programme. We would add, however, a request for a better explanation of why we ought to adopt bias-type models even in those cases considered by Poisson.

Notes

[1]This exposition is condensed. The theory of testing is not our principal concern here and we have accordingly not discussed it in detail. Those interested may consult standard texts of Neyman (1952), or Fisher (1956 and 1970), and Cox and Hinkley (1974). Philosophical work on hypothesis tests may be found in Hacking (1965), Seidenfeld (1979), and Baird (1981).

References

Baird, D. (1981). <u>Significance Tests: Their Logic and Early History.</u> Unpublished Ph.D. Dissertation, Stanford University.

Carnap, R. (1947). "On the Application of Inductive Logic." <u>Philosophy and Phenomenological Research</u> 8: 133-148.

Cox, D.R. and Hinkley, D.V. (1974). <u>Theoretical Statistics.</u> London: Chapman and Hall Ltd.

Fisher, R.A. (1956). <u>Statistical Methods and Scientific Inference.</u> New York: Hafner Publishing Company.

------------. (1970). <u>Statistical Methods for Research Workers.</u> 14th ed. New York: Hafner Publishing Company.

Hacking, Ian. (1965). <u>Logic of Statistical Inference.</u> Cambridge: Cambridge University Press.

-------------. (1980). "Grounding Probabilities from Below." In <u>PSA 1980.</u> Volume 1. Edited by P.D. Asquith and R.N. Giere. East Lansing, Michigan: Philosophy of Science Association. Pages 110-116.

Neyman, J. (1952). <u>First Course in Probability and Statistics.</u> New York: H. Holt and Company.

Russell, B. (1935). <u>Religion and Science.</u> London: T. Butterworth.

Salmon, W. (1971). <u>Statistical Explanation and Statistical Relevance.</u> Pittsburgh: University of Pittsburgh Press.

Seidenfeld, T. (1979). <u>Philosophical Problems of Statistical Inference: Learning from R.A. Fisher.</u> Dordrecht: Reidel.

PART V

MEASUREMENT, VERISIMILITUDE AND DECISION

Quantity and Quality: Some Aspects of Measurement[1]

Arnold Koslow

The Graduate School, C.U.N.Y., and Brooklyn College

1. Introduction.

We shall not belabor the importance of having a clear idea of the
difference between classificatory, comparative, and quantitative terms.
Even though some of the goals that were set for theories of measurement
of empirical attributes no longer seem viable, and some do not seem generally applicable[2], there is every reason to believe that suitable general goals for such theories will rely upon a distinction between quantitative and qualitative terms. The distinction seems also to have an
intrinsic interest, even though some of the claims originally made for
it, no longer seem historically accurate, or philosophically defen-
sible.[3] The subject, then, is quantitative terms in theories of the
measurement of empirical attributes like length, mass, charge, and utility. We shall consider two problems. The first concerns a characterization of what counts as a quantitative term of such theories; the
second concerns the description of the relation between the quantitative terms of (say) a theory of length, and the notion of the empirical attribute length which is supposed to be the target or subject
of such a theory.

In what follows, we shall try to characterize afresh, the distinction between quantitative and qualitative terms so that these types of
terms are seen not merely to be different; they are seen to be contrast-
ing types within a family of terms. By relativizing the distinction
to the different ways in which the terms partition their domains, it
can be proven that they fall into mutually exclusive classes. This is,
of course, not a surprising conclusion. Although the literature provides different examples of each type, the provision of different examples doesn't fully support the idea that the types are demonstrably
exclusive. The classic description of a quantitative term (Carnap
1926, 1950; Hempel and Oppenheim 1936; and Hempel 1952) is as a func-
tor (linguistic), or as a numerical-valued function. It is a description, which, despite its obviousness, has its drawbacks. If quanti-

---

PSA 1982, Volume 1, pp. 183-198
Copyright © 1982 by the Philosophy of Science Association

184

tative terms are construed as many-one relations between certain objects and numbers, then there is no obvious contrast between comparative and quantitative terms. Moreover, there are certain debates over
the quantitativeness of a term, where the case for quantitativeness
rests upon non-numerical geometrical or set-theoretical considerations.
Further, there are certain terms like 'x is twice as long as y', which
when expressed in an empirical relational system as xLLy: xL(yoy)&
(yoy)Lx (where 'xLy' holds if and only if x is at least as long as y),
are believed to be quantitative, and the existence of numerical-valued
representing functions for them seems beside the point.[4] There are a
host of other examples which seem to be quantitative, although they
are not explicitly numerical: Lxyz: x is longer than y, by z (xL(yoz)&
(yoz)Lx), the relation of the hypotenuse of a right triangle to its
other sides, and perhaps even the relation of an object x to its equiv-
alence class [x] = {y |yLx & xLy }.[5]

Thus, it would be useful to have a way of characterizing quantitative terms which does not preclude their being numerical, but which
does require them to be numerical.

Our characterization of quantitative terms is somewhat unusual in
that it makes no reference to numbers, concatenation operations, or
ordering relations. There is no possibility of a "proof" that this
characterization is "correct". The most that we can hope for is a wide
agreement over standard examples, and the absence of serious counterexamples. Since there are few structural assumptions in this characterization, it has wide applicability, but is limited in the amount of
detail that it can provide, without supplementary assumptions. Nevertheless, despite the generality, we can show that many questions of
interest, some of which have already received a full and elegant account [Krantz et al] -representation and uniqueness theorems, the notion of an <u>exact</u> quantitative term, and the relation between ordering relations and concatenation operations, can still be raised. Indeed, we shall describe a number of theorems which are provable, even
at this level of generality. Although there is not enough space to
set out their proofs, in most cases, they are fairly straightforward.

The final section of this paper concerns the notion of an empirical
attribute. These, like length, charge, mass, and utility, are the usual targets of empirical relational systems. Indeed, although these
attributes are the foci of such theories, there does not seem to be a
satisfactory description of the role that they play for those theories.
We do not think of these attributes as quantitative terms of an empirical relational system, but as families of predicates satisfying certain conditions which specify on the one hand how these predicates
are related to quantitative terms, and on the other hand, how these predicates are related to the partitions of quantitative terms. We shall
explore what it is for a family of predicates P, to be an <u>underlying</u>
<u>attribute of a quantitative term R</u>. The notion of an underlying at-
tribute has a number of features worth noting: (1) it will make specific how it is that the underlying attribute for a quantitative term
helps to "fix" the partition of that term, (2) it avoids two "extension-

ality" problems which beset the usual proposal that an empirical attribute like length is just a function which maps the domain of the theory to the non-negative real numbers. The first problem is that if length is a function over the domain(or even a class of such functions) then a shift in the domain of the theory automatically induces a shift in the concept of length - if functions are treated extensionally. The second problem concerns the situation, on some domain, when two relations - say 'x is at least as long as y' and 'x is at least as charged as y' are coextensional. In that case, the numerical functions defined over the domain will be the same, and consequently, there will be no distinction between length and charge on that domain. As we shall see, the present construal of such empirical attributes seems to escape these difficulties. (3) There are historical examples, to be found in the writings of Newton, Cotes, and Helmholtz, which seem to lend some support to the present suggestion.[6] (4) The proposal bears a close relation to the determinable-determinate distinction introduced by Johnson (1921, pp. 173-185), and to some remarks suggested by Campbell (1921, p. 118). Indeed, it is not too misleading to say that our proposal treats the empirical attributes as determinables. Lastly, (5), there are a number of theorems which illustrate the various ways in which quantitative terms may be related to each other, if there is a common underlying attribute.

2. Classificatory, Comparative, and Quantitative Terms.

By a partition of a domain D, we shall mean a set of non-empty subsets of D, whose members are mutually exclusive, and whose union is D itself. The proposed characterization of these terms differs from the customary way, in that the distinctions are relativized to a partition. Thus, we shall describe what it is for a relation to be classificatory, comparative, and quantitative, with respect to a partition of its domain.

2.1. Classificatory Terms.

We shall say that an n-place term $'R^n x_1 \ldots x_n'$ is classificatory with respect to a partition M of its domain, if and only if

$$R^n x_1 \ldots x_n \text{ if and only if } x_i \in M_i, \text{ for } i = 1, 2, \ldots, n.$$

Monadic predicates can be accommodated within this scheme. Let M be a two-membered partition, the first member being the set of objects of which 'Px' holds; the second being the set of those for which it fails. We shall say that 'Px' is classificatory with respect to M just in case 'Px & -Py' is classificatory with respect to M. It is easily seen that a term can be classificatory with respect to only one partition.

Theorem 1. If R is classificatory with respect to the partition M, and is also classificatory with respect to the partition N, then M = N.

2.2. Comparative Terms.

Let 'Rxy' be a two-place term.[7] We shall say that 'Rxy' is comparative with respect to a partition $\{M_i\}$ of the domain, if and only if:

(1) If Rxy, and x and x' belong to the same M_i, then Rx'y.;

(2) If Rxy, and y and y' belong to the same M_j, then Rxy'.;

(3) Rxy is weakly connected in D: for any x and y in D, either; Rxy or Ryx.

(4) Rxy is strictly transitive: Rxy is transitive, but it is not anti-transitive(anti-transivity requires that if Rxy and Ryz, then not Rxz).

Conditions (3) and (4) are standard, or nearly so, and (1) and (2)permit the substitution of x' for x (y' for y) when they belong to the same member of the partition. When the partition is discrete - every member is a unit set - then this relativized notion coincides with the usual one. It is easy to verify that if R is comparative with respect to the partition M, then it is also comparative with respect to any weaker partition. Furthermore, the partition I, consisting of the indifference sets $[x] = \{y \mid Rxy \ \& \ Ryx\}$ is the strongest partition with respect to which R is comparative. Thus we have

Theorem 2. Comparative terms, unlike classificatory ones, countenance a certain slack in their partitions: there is a weakest, a strongest, and all partitions in between, with respect to which they are comparative. The following is also easily proven:

Theorem 3. No classificatory term is comparative.[8]

2.3. Quantitative Terms.

We should like to be able to distinguish between the quantitative and the other types of terms in a way that parallels the way in which classificatory terms contrasted with the comparatives. Quantitative terms would then be n-place terms over a domain D, which satisfied certain conditions that could be directly compared with those for classificatory and comparative terms. We suggest the following:

We shall say that the term 'Rxy' is quantitative for a domain D, and a partition $\{M_i\}$ of D, if and only if the following conditions hold for any x, x', y, and y'.

(1) Rxy and Rx'y imply that x and x' both belong to some member M_i of the partition M.

(2) Rxy and Rxy' imply that y and y' both belong to some member M_j of the partition M.

(3) If x and x' both belong to some member of the partition, then Rxy if and only if Rx'y.

(4) If y and y' both belong to some member of the partition, then Rxy if and only if Rxy'.

Intuitively, conditions (1) and (2) state that if x and x' are both re-

lated to the same objects in the same way by some quantitative term R,
then the quantitativeness of R requires that they be grouped together,
or be "alike"(Helmholtz, 1887); they are in the same member of the partition. Conditions (3) and (4) state that objects that are grouped together by the partition, are intersubstitutable - R makes no distinction between them.[9] Several consequences follow immediately:

__Theorem 4__. If 'Rxy' is quantitative over a domain, with respect to a
partition that has more than two members, then it is not comparative[10],
and

__Theorem 5__. If 'Rxy' is quantitative over a domain, with respect to a
partition that has more than two members, and 'Rxy' covers its domain
(for every x, there is a y, such that Rxy, or, for every y there is an
x, such that Rxy)[11], then 'Rxy' is not classificatory.

Thus, except for certain pathological cases, the classificatory,
comparative, and quantitative terms are mutually exclusive types. There
is another result worth noting, which we will have to cite without
proof: quantitative terms determine their partitions uniquely.

__Theorem 6__. If 'Rxy' is quantitative over the domain D with respect to
the partition M, and also quantitative with respect to the partition N,
then M = N (provided that 'Rxy' covers its domain).

Comparative terms are comparative with respect to several partitions; quantitative terms are quantitative with respect to exactly one.
We can describe the members of the partition as __values__ (Suppes (1951)
and Kyburg(1979) call them magnitudes), and say that the value of any
object in the domain is the unique member of the partition to which it
belongs. One difference between comparative and quantitative terms can
be stated this way: as far as comparative terms are concerned, the objects in the domain can have any of several values; quantitative terms,
however, require that each object in the domain has exactly one value.
Thus, without requiring that values be numerical, and without requiring that quantitative terms be explicitly numerical, the difference
between comparative and quantitative terms emerges in a natural manner.

2.4. Examples of Quantitative Terms.

There is an abundance of examples of this kind of term. We shall
list a few, drawn from various sources. (1) If we have an extensive
relational system, using 'xLy' (x is at least as long as y), and an appropriate concatenation operation 'o', then 'xLy & yLx', 'xLLy' (x is
twice as long as y: xL(yoy) & (yoy)Lx), 'xL(m/n)y' (m x's are as long
as n y's) 'Lxyz' (x is longer than y, by z: xL(yoz) & (yoz)Lx) all
count as quantitative. (2) There are also arithmetical examples:
$x + y = 12$ (over N), $(x + y = 20) \& (x - y = 10)$ (over I), $x = y \pmod{n}$
(over a partition that has n values, M_i, where $M_i = \{i, i+n, i+2n, \ldots\}$.
There are (3) geometrical examples which involve congruence and similarity relations, and (4) set-theoretical examples such as RXY: Y is

188

in one-to-one correspondence with the power set of X (each member of
the relevant partition consists of exactly those sets which are in
one-to-one correspondence), RXYZ: $Z = X \cup Y$, where X and Y are disjoint (otherwise the relation is not quantitative), and the partition
is the discrete one.

2.5. An Application: Exact Quantitative Terms.

Using the fact that quantitative terms have uniquely associated partitions, we shall say that one quantitative term, '$Ax_1 \ldots x_n$' is <u>more
definite</u> than another '$Rx_1 \ldots x_n$', if and only if 'A' implies 'R', and
the partition of 'A' is a proper refinement of the partition of 'R'.
We shall say that a quantitative term is <u>exact</u> if and only if there
is no quantitative term which is more definite than it. Thus, if a
term is exact, then any quantitative term which implies it must have
the same partition. There is no way in which a quantitative term that
implies an exact term, can refine that term's partition further.[12]
There is the following result about the set of exact quantitative terms.

<u>Theorem 7</u>. (Decomposition) Either an exact term has a discrete par-
tition or a non-discrete one. The set of all exact terms with non-
discrete partitions can be decomposed into mutually disjoint classes
E_1, E_2, ... such that (1) any two terms in an E_i have the same partition, (2) terms belonging to distinct classes are logically independent, (3) each class E_i is closed under implication, and (4) (closure from above) any exact term which implies a member of E_i, belongs
to E_i.

If we apply this distinction to a standard formulation, L, of an extensive empirical system for length, which uses 'xLy', and a suitable concatenation operation 'o', then we obtian the following result:

<u>Theorem 8</u>. The exact, non-discrete quantitative terms of L are pre-
cisely those quantitative terms of L which partition the domain into
the familiar indifference sets $[x] = \{y \mid xLy \ \& \ yLx\}$.

2.6. Representation and Uniqueness.

Since the notion of a quantitative term, on our account, uses very
few structural conditions, it is not to be expected that very detailed results can be demonstrated. However, several general results
can be proven, which reduce to certain standard results about particular systems, once one uses the particulars of those standard systems. If 'Rxy' is quantitative over a domain, with the partition π,
then let R* be the reduced relation which holds between π_i and π_j of π
exactly when 'Rxy' holds between x and y, for some x in π_i and
some y in π_j. Since 'Rxy' is quantitative, R* is well-defined. Let
us say that and function f which maps the partition to itself is an
<u>R*-endomorphism</u> if and only if it is R*-preserving. That is,

$$R^* \pi_i \pi_j \quad \text{if and only if} \quad R^* f\pi_i f\pi_j.[13]$$

If we assume that R covers its domain, then it follows that there is a unique R*-endomorphism such that all R*-endomorphisms commute with it(or else, such an endomorphism exists for a converse of R*).[14] An example: if the quantitative term is Lxyz, then $L*\pi_i\pi_j\pi_k$ if and only if $\pi_i = \pi_j$ o π_k. Moreover, if f is any L*-endomorphism, then it follows from the quantitativeness of Lxyz, that $f(\pi_i) = f(\pi_j)$ o $f(\pi_k)$.

Let us close this section with a result which might be called a"reshuffling" theorem. It seems to be a generalized version of a uniqueness theorem. Suppose that 'Rxy' [15] is quantitative over the domain D, with the partition π, and that 'Sxy' is quantitative over the domain S, with partition ρ. We shall say that a function ϕ which maps π to ρ is <u>a representing function of R* with respect to S*</u> if and only if

$$R*\pi_i\pi_j \text{ if and only if } S*\phi\pi_i\phi\pi_j .$$

We then have the following theorem:

<u>Theorem 9.</u> (Reshuffling)[16] If ϕ and ψ are two representing functions of R* with respect to S*, then (1) there is a unique S*-endomorphism h, such that $h\phi = \psi$, and (2) there is a unique R*-endomorphism g, such that $\phi g = \psi$.

According to (2), there is a unique way of "reshuffling" the members of ρ, which is S*-preserving, and which transforms ϕ into ψ. According to (1), there is a unique way of "reshuffling" the members of π, which is R*-preserving, and which transforms ϕ into ψ.

There is a good deal more to be said about how the different sets of endomorphisms are related to each other, and what they reduce to in special cases -e.g., when R's partition has the structure of an Archimedean ordered semi-group. Nevertheless, I hope that even with these few remarks, the reader will agree that the interesting and central questions about uniqueness and representation can still be raised, even though the notion of quantitative term which we have described, is not explicitly numerical, and the representing functions are not necessarily numerical-valued.

3. Underlying Attributes of Quantitative Terms.

We want to suggest that empirical attributes such as length, mass, and charge, are not to be identified with any of the quantitative terms in the standard empirical relational systems which study those notions; neither are they to be identified with numerical-valued functions that are defined over the domains of such theories - for the reasons already indicated. Instead, we think of those attributes as related in a special way to quantitative terms - we shall say that they are empirical attributes which underlie quantitative terms. The following account is suggested by some historical examples discussed by Newton, Cotes, and Helmholtz. Here, we can only give the merest sketch of things.

Quantitative terms, as we have seen, have their unique partitions. They therefore serve, among other things, to group members of the do-

main into sets, the members of which are "alike" (Helmholtz 1887). We
cannot identify length with a particular partition -that would lead to
the same difficulties which beset the idea of taking length as a function over the particular domain. However, length has something to do
with the partition of 'xLLy', for example. It provides a way of determining or "fixing" that partition, a way of grouping objects together as "alike", in just the way that 'xLLy' provides, by its partition.

Let $P = \{P_\lambda\}$ be a set of predicates. We shall say that P is an
underlying attribute for the quantitative term 'Rxy' (with partition
π) if and only if:

(1)　Rxy $\Longrightarrow P_\lambda x$ and $P_\mu y$, for some P_λ and some P_μ in P.

(2)　For every P_λ in P, the restriction of P_λ to the domain of R
　　　is some member of R's partition, i.e., $P_\lambda|D = \pi_i$, for some π_i.

(3)　For every π_j in π, there is a P_μ in P, such that $P_\mu|D = \pi_j$.

In the case of 'xLLy', we call families of predicates like P, 'length
given P', and, by 'length' (simpliciter) we mean the weakest of
such families. We shall also say that x has P (length) just in case
there is some predicate in P which holds true of x. We shall say that
x and y have the same P (length) if and only if there is some predicate in P, which holds true of both x and y. If length is construed as
an underlying attribute of the quantitative term 'xLLy', for example,
then it follows that (1) any objects which satisfy 'xLLy', have length,
(2) all objects in the domain have length, (3) if x and x* in the domain have the same length, then xLLy if and only if x*LLy, (4) if x and
x* are twice as long as some y, then x and x* have the same length.
These remarks hold not merely for 'xLLy'; they hold for any quantitative term that has length as an underlying attribute.

The following consequences may perhaps illustrate some of the sys-
tematizing power that is involved in the assumption that an empirical
attribute underlies a quantitative term.

Theorem 10. (Upward Closure) If R and S are quantitative over the
same domain, and R implies S, then every attribute which underlies
S, also underlies R.

Theorem 11. (Sub-Domain) If R is quantitative over a domain D, then
the restriction R* of R to a subdomain D* of D, will also be quantitative. Moreover, every attribute which underlies R will also underly
R*.

Theorem 12. There are quantitative terms over disjoint domains, which
nevertheless have an underlying attribute in common. (Just restrict
R to disjoint subdomains. The restrictions will have all the underlying attributes of the parent term in common.)

There are also a number of consequences which exhibit the connection between underlying attributes of quantitative terms, and the partitions of those terms. We single out one: suppose that R is quantitative over a domain D, with partition π, and R* is quantitative over D*, with partition π*. How are their partitions related? In general, not at all; not even if D is a subdomain of D*. However, if R and R* share an underlying attribute, then the issue is forced.

Theorem 14.(Stability) If R and R* are as above, and D is a subset of D*, and R and R* have an underlying attribute in common, then the partition of R is a refinement of the partition of R*. Thus, objects that are grouped as "alike" by R, will also be grouped as "alike" by R*.

Theorem 15. If R and S are quantitative over a domain D, but partition D differently, then they have no underlying attribute in common.

Theorem 16. If R and S have no underlying attribute in common, then there is no parent quantitative term, T, over a more inclusive domain, which has R and S as its restrictions.

There is much more to the story of underlying attributes. Their unifying power for quantitative terms runs deep. If a quantitative term has an underlying attribute, then there is a canonical form for describing all the quantitative terms on that domain which have its partition.   There is also a type of representation theorem to be had for those quantitative terms that have underlying attributes.

Theorem 17. If R is a quantitative term over the domain D, with partition π, and it has some underlying attribute $S = \{S_\lambda\}$, then every quantitative term which covers its domain can be put into a canonical form $'S^\phi|D'$, where ϕ is a partition preserving function (ppf) of the domain into the set of indexes of S. [17]

Theorem 18.(Representation) If R is a quantitative term, with an underlying attribute $S = \{S_\lambda\}$, then there is a function ϕ mapping the domain onto the set of indexes of S, and a quantitative relation S on the set of indexes, such that ϕ is a representing function of R* with respect to S*. [18]

We have, in brief compass, tried to supplement the standard description of theories of measurable attributes, by providing an account of the quantitative-qualitative distinction for those theories, and, adjoined to that account, a suggestion concerning the sense in which empirical relational systems concern, or have as their focus, an empirical attribute such as length. We have not addressed the question of how our account of quantitative terms is related to concatenation operations, or to ordering relations. There are several systematic things that can be said, which develop ideas due chiefly to Helmholtz(1887) and K.Menger(1959). However, that is another equally long story.

192

Notes

[1]This is part of a monograph, *Quantity and Quality* (Koslow 1982). Some
of these results were prepared for a contribution to the Festschrift
for Richard Bevan Braithwaite (Mellor 1980). Unfortunately, the paper reached the press a little too late for inclusion. This is but a
small token of my gratitude and affection. I would also like to thank
J.Barshay, H.Kyburg Jr.,D.T.Langendoen, D.Rosenthal, and R.Schwartz
for some very helpful comments and discussions on earlier versions of
this study. Although I cannot claim full credit for what is right in
it, I do take full credit for any residual errors and shortcomings
that still dog its pages.

[2]I have in mind two traditions. The first, given its classical formulation by Carnap(1926),Hempel and Oppenheim(1936), and Hempel(1952)
and (1958), regarded the central problem for metrical terms as the
same one which they raised for theoretical terms: how are they introduced? More recently, Hempel(1970) has argued that the question presupposes an untenable position or goal -partly under the influence of
some arguments of Putnam(1962). The second tradition, roughly speaking, takes it as a task to show how to pass from the qualitative to
the quantitative. Adams (1974), **Krantz** *et al*(1971), Scott and Suppes
(1958), and, perhaps most clearly, Suppes (1956, p.88) have advanced
this view. According to it, the vocabulary and assertions of empirical theories of measurement are qualitative. The initial restriction
to minimal qualitative grounds is very plausibly argued by Suppes
(1956,§ I), but the view is less plausible for concepts like length,
mass, and charge. Moreover, if we are correct, empirical relational
systems already have many quantitative terms definable within them,so
it is unclear why the use of empirical relational systems represents
a way of passing from the qualitative to the quantitative.

[3]Carnap (1950, pp. 8-9) has occasionally suggested that the dominant use
of these different types of terms corresponds to certain stages of
scientific research, although that claim is somewhat tempered(p.12),
it is somewhat implausible. He has also, in a rather dark passage
(1966, p.105),written that quantitative concepts, unlike qualitative
ones, are part of our language and not part of nature.

[4]It should not be necessary to use the existence of representing
functions to determine the quantitativeness of a term. It would not be
much of an argument to note that 'xLLy' is quantitative because the
number assigned by a representing function to x, is twice that which is
assigned to y.

[5]Kyburg(1979) has argued, in part on grounds of ontological economy,
that any mapping of objects x into their magnitude [x], is a quantity.
This is demonstrably true, on our account of quantitativeness, in the
following sense: let R be any quantitative term over the domain, with
partition π. Let f be any partition-preserving function (ppf)mapping
the domain to the partition (objects are in the same member of the partition if and only if they are mapped into the same value by f). Then
there exists an associated two-place term $R_f xy: (fx = \pi_i) \& (y \in \pi_i)$,
which is quantitative over D, with partition π.

[6]The exchange between Newton and Cotes over whether the inertia, and the true-volume(the volume of a body when its pores are"discounted")of a body are both measures of the quantity of matter or mass, can be seen as a difference over whether mass in an underlying attribute is shared by the quantitative terms 'same inertia', and 'same true-volume'. If mass were an underlying attribute of both terms, then all bodies of equal volume would have to have the same inertia. This is precisely the point that Cotes challenges. For the exchange, see Newton (1709-1713), and the interesting comment on this issue in Maclaurin (1748, p. 99). For more detail on this debate see Koslow(1965). For some further examples concerning weight, inertia, and thermal equilibrium, see P.Hertz (1977, pp. 103-114).

[7]For the sake of systematic comparison with the other types of terms, we consider an n-place term '$Rx_1 \ldots x_n$' to be comparative if and only if it is comparative in every pair of variables, according to the four conditions on two-place comparatives.

[8]Given our relativization of these notions to partitions, the theorem says that it is impossible for R^n to be classificatory with respect to some partition, and for it also to be comparative with respect to some (possibly different) partition.

[9]The intention is that the standard case is one for which there are distinct objects between which R marks no difference -that is, the partition of R is non-discrete. If the partition were discrete, then, if x and y were in the same member of the partition, they would be identical, and R would be quantitative by default. To remedy this situation, a modification is discussed by Koslow (1982) which requires (1) if the partition is non-discrete, then R is quantitative if and only if conditions (1) - (4) are satisfied, and (2) if the partition is discrete, then R is quantitative if and only if it is the restriction of a term which is quantitative according to (1).

[10]In general, the relevant condition for n-place terms in Theorems 4 and 5, is that '$R^n x_1 \ldots x_n$' has a partition with more than n members.

[11]We have not made the requirement that R cover its domain into a condition for quantitativeness. We prefer instead to take it as an additional assumption when it is needed in certain proofs and examples.The coverage requirement expresses different conditions, depending upon the choice of quantitative term. If $R^* \pi_i \pi_j \pi_k$ is $\pi_i \circ \pi_j = \pi_k$, then coverage expresses closure under concatenation.If the term R^* is $\pi_j \circ \pi_k = \pi_i$, then coverage expresses a solvability condition. Whether coverage of R is to be assumed or not, depends upon the theory in which R is embedded.

[12]Thus any quantitative term with a discrete partition is exact. However, as Theorem 8 shows, there can be exact terms with non-discrete partitions. The idea that exactness precludes a more refined partition, seems to be analogous to von Neumann's notion of a dispersion-free ensemble(1932, 350 ff).

[13]The R^*-endomorphisms are one-to-one mappings on the partition, rather than homorphisms on the domain D. It is different from the usage

in the literature. We hope that the difference will not be confusing.
Usually, endomorphisms are taken to be mappings of the domain to itself, which preserve certain relation(s) R -they are R-endomorphisms.
Since the relations under consideration are quantitative, the corresponding reduced relations R*, defined over the partition of R, are always well-defined. In this study, we are concerned with R*-endomor-
phisms, rather than R-endomorphisms. Thus, the difference in usage.
However, it is easily seen that for quantitative terms, every R-endo-
morphism Θ can be associated with an R*-endomorphism $\Theta*$ (set $\Theta*\pi_i = \pi_j$
if and only if Θx is in π_j for some x in π_i). Conversely,
for every R*-endomorphism $\Theta*$ there are associated R-endomorphisms Θ_G
(Let G be any choice-function on the partition: $G\pi_i$ is in π_i , for every π_i , and let Π be the function which assigns each member of the domain to the unique member of the partition to which it belongs. Finally, set $\Theta_G x = (G\Theta*\Pi)x$.) Endomorphisms of qualitative relational
structures have had their most effective use, so far, in the study
of the meaningfulness of a relation, quantitative or not. Some of the
central results are to be found in the work of Adams, Fagot, and Ro-
binson(1965), Luce(1978), Roberts(1980), Suppes(1959), and Narens(1981).
Unfortunately, Roberts(1979) was not available to me at the time of
writing. It is an open issue whether there are any systematic connections between a relation's being A-endomorphism meaningful, and its
being quantitative.

[14] For n=2, we note that there is a unique function f, that is one-
to-one, such that $R*\pi_i\pi_j$ if and only if $f\pi_i = \pi_j$ if and only if
$ff \pi_i = f\pi_j$ if and only if $R*f\pi_i f\pi_j$. Furthermore, g is an R*-endomor-
phism if and only if (1) $gf = fg$. One direction: for any π_i there
is a π_j such that $R*\pi_i\pi_j$. Now $R*\pi_i\pi_j$ if and only if $R*g\pi_i g\pi_j$ if
and only if $fg\pi_i = g\pi_i$. Therefore $fg\pi_i = gf\pi_i$, for all π_i .
Conversely, if $fg = gf$, then $R*g\pi_i g\pi_j$ if and only if $fg\pi_i = g\pi_j$, if
and only if $gf\pi_i = g\pi_j$ if and only if $f\pi_i = \pi_j$ if and only if
$R*\pi_i\pi_j$. For the case n=3, for example, we prove that there
are a set of functions f_{π_i} mapping the partition to itself (they
however are in general, not endomorphisms of R*) such that $R*\pi_i\pi_j\pi_k$
if and only if $f_{\pi_i}\pi_j = \pi_k$. It can then be shown that h is an
R*-endomorphism if and only if (2) $hf_{\pi_i} = f_{h\pi_i}h$, for all π_i .
It is easy to show that (2) is a generalization of (1). The
functions f_{π_i} are useful in computations. They are in one-to-one correspondence with the members of the partition, and if any two of
then are equal on some member of the partition, then they are equal
everywhere.

[15] We use a two-place quantitative term for convenience of expression.
The remarks which follow, hold for quantitative terms of any finite
place.

[16] This is essentially the content of Luce's Theorem 1(1978), and the
proof can be carried over. Although he considers representations with
respect to numerical structures, it is clear that the proofs of all the
clauses of the theorem go through, whether the representing structures
are numerical or not, provided that the representing functions are one-
to-one, and onto. On our account, the representing functions are as-

sumed to be onto, and the one-oneness follows from the fact that the functions represent R^* with respect to S^*, where R and S are quantitative.

[17] For two-place quantitative terms, Rxy, we let Λ be the set of indexes of S, and let λ^* be the set of all indexes μ, such that S_λ and S_μ yield the same member of the partition, when restricted to D. The λ^* form a partition Λ^* of Λ. Let '$(S^\phi|D)xy$' hold if and only if, for some S_λ and S_μ in S, x is in $S_\lambda|D$, y is in $S_\mu|D$, and $\phi\lambda = \mu^*$, where ϕ is any partition-preserving function mapping D to Λ^* (it assigns the same value to members of the domain D if and only if they belong to the same member of the partition). It is straightforward to check that '$S^\phi|D$' is quantitative over D, with partition π. For any quantitative term Txy over D, with partition π, which covers its domain, there is a unique one-one mapping σ, which maps the partition to itself, such that Txy if and only if $x \in \pi_i, y \in \pi_j$, and $\sigma\pi_i = \pi_j$ .Set ϕ, a mapping of Λ to Λ^* as follows: $\phi\lambda = \mu^*$ if and only if $\pi_i = S_\lambda|D, \pi_j = S_\lambda|D$, and $\sigma\pi_i = \pi_j$. ϕ is a partition-preserving function, and it is easy to verify that Txy if and only if $(S^\phi|D)xy$.

[18] Let '$Rx_1,\ldots x_n$' be a quantitative term over D, with partition π. Using the definition of Λ^* as it was given in footnote 17, we define a relation '$S\lambda_1\lambda_2\ldots \lambda_n$' as holding if and only if for some x_1, $x_2,\ldots x_n$ '$Rx_1\ldots x_n$' holds, and $Rx_1\ldots x_n \implies (S_{\mu_1}x_1)\&(S_{\mu_2}x_2)\&\ldots\&(S_{\mu_n}x_n)$, where each μ_i is in λ_i^*. It is straightforward to verify that S is quantitative over Λ, with partition Λ^*. Lastly, define a mapping ϕ of the partition π onto the set Λ^* as follows: $\phi\pi_i = \lambda_i^*$ if and only if $\pi_i = S_\sigma|D$ and $\sigma \in \lambda_i^*$. The mapping is onto Λ^*, because, for any $\lambda^*, \phi\pi_i = \lambda^*$, where $\pi_i = S_\lambda|D$ .From here it is a straightforward computation to show that $R^*\pi_i\pi_j\pi_k$ if and only if $S^*\phi\pi_i\phi\pi_i\phi\pi_k$.

196

References

Adams, E.W. et al. (1965). "A Theory of Appropriate Statistics." Psychometrika 30: 99-127.

——————. (1974). "Model-theoretic Aspects of Fundamental Measurement Theory." In Proceedings of the Tarski Symposium. (Proceedings of Symposia in Pure Mathematics, Volume XXV.) Edited by L. Henkin, et al. Providence: American Mathematical Society. Pages 437-446.

Campbell, N. (1921). What is Science? London: Methuen.

Carnap, R. (1926). Physikalische Begriffsbildung. Karlsruhe: Braun.

——————. (1950). Logical Foundations of Probability. London: Routledge and Kegan Paul.

——————. (1966). Philosophical Foundations of Physics. (ed.) M. Gardner. New York: Basic Books.

Cohen, R.S. and Elkana, Y. (eds.). (1977). Hermann von Helmholtz: Epistemological Writings. (Boston Studies in the Philosophy of Science, Volume XXXVII.) Dordrecht: D. Reidel.

Helmholtz, H. von. (1887). "Zahlen und Messen Erkenntnistheoretisch Betrachtet." In Philosophische Aufsätze. Eduard Zeller zu seinem fünfzigjährigen Doctor-Jubiläum gewidmet. Leipzig: Fues' Verlag. Pages 17-52. (As translated and reprinted as "Numbers and Measuring from an Epistemological Viewpoint." In Cohen and Elkana (1977). Pages 72-103.)

Hempel, C.G. and Oppenheim, P. (1936). Der Typusbegriff Im Lichte der Neuen Logik. Leiden: A.W. Sijthoff.

——————. (1952). Fundamentals of Concept Formation in Empirical Science. (International Encyclopedia of Unified Science, Volume II, Number 7.) Chicago: University of Chicago Press.

——————. (1958). "The Theoretician's Dilemma." In Concepts, Theories, and the Mind-Body Problem. (Minnesota Studies in the Philosophy of Science, Volume II.) Edited by H. Feigl, et al. Minneapolis: University of Minnesota Press. Pages 37-98. (As reprinted in Aspects of Scientific Explanation. New York: The Free Press, 1965. Pages 173-226.)

——————. (1970). "On the 'Standard Conception' of Scientific Theories." In Analyses of Theories and Methods of Physics and Psychology. (Minnesota Studies in the Philosophy of Science, Volume IV.) Edited by M. Radner and S. Winokur. Minneapolis: University of Minnesota Press. Pages 142-163.

Hertz, P. (1977). "Notes and Comments." In Cohen and Elkana (1977). Pages 103-114.

Johnson, W.E. (1921). _Logic. Demonstrative Inference: Deductive and Inductive._ Part I. Cambridge: Cambridge University Press. (As reprinted New York: Dover, 1964.)

Koslow, A. (1965). _Changes in the Concept of Mass, from Newton to Einstein._ Unpublished Ph.D. Dissertation, Columbia University. Xerox University Microfilms Publication Number 65-9164.

----------. (1982). _Quantity and Quality._ (Unpublished manuscript).

Krantz, D. _et al._ (1971). _Foundations of Measurement,_ Volume I. New York: Academic Press.

Kyburg, H.E., Jr. (1979). "Direct Measurement." _American Philosophical Quarterly_ 16: 259-272.

Luce, R.D. (1978). "Dimensionally Invariant Numerical Laws Correspond to Meaningful Qualitative Relations." _Philosophy of Science_ 45: 1-16.

Maclaurin, C. (1748). _An Account of Sir Isaac Newton's Philosophical Discoveries._ London: Patrick Murdoch.

Mellor, D.H. (ed.). (1980). _Science, Belief, and Behaviour: Festschrift for R.B. Braithwaite._ Cambridge: Cambridge University Press.

Menger, K. (1959). "Mensuration and other Mathematical Connections of Observable Material." In _Measurement: Definitions and Theories._ Edited by C.W. Churchman and P. Ratoosh. New York: Wiley. Pages 97-128.

Narens, L. (1981). "A General Theory of Ratio Scalability with Remarks about the Measurement-Theoretic Concept of Meaningfulness." _Theory and Decision_ 13: 1-70.

Newton, Isaac. (1709-1713). _The Correspondence of Isaac Newton_ Volume V: 1709-1713. (eds.) A.R. Hall and L. Trilling. Cambridge: Cambridge University Press, 1975.

Putnam, H. (1962). "What Theories are not." In _Logic, Methodology, and Philosophy of Science._ Edited by E. Nagel, _et al._ Stanford: Stanford University Press. Pages 240-251. (As reprinted in _Mathematics, Matter, and Method,_ Volume I. New York: Cambridge University Press, 1975. Pages 215-217.)

Roberts, F.S. (1979). _Measurement Theory, with Applications to Decision-Making, Utility, and the Social Sciences._ Reading, Mass.: Addison-Wesley.

------------. (1980). "On Luce's Theory of Meaningfulness." _Philosophy of Science_ 47: 424-433.

Scott, D. and Suppes, P. (1958). "Foundational Aspects of Theories of Measurement." _The Journal of Symbolic Logic_ 23: 113-128. (As reprinted in Suppes (1969). Pages 46-64.)

Suppes, P. (1951). "A Set of Independent Axioms for Extensive Quantities." _Portugaliae Mathematica_ 10: 163-172. (As reprinted in Suppes (1969). Pages 36-45)

----------. (1956). "The Role of Subjective Probability and Utility in Decision-Making." In _Proceedings of the Third Berkeley Symposium on Mathematical Statistics and Probability, 1954-1955._ Pages 61-73. (As reprinted in Suppes (1969). Pages 87-104.)

----------. (1959). "Measurement, Empirical Meaningfulness, and Three-Valued Logic." In _Measurement: Definitions and Theories._ Edited by C.W. Churchmann and P. Ratoosh. New York: Wiley. Pages 129-143. (As reprinted in Suppes (1969). Pages 65-80.)

----------. (1969). _Studies in the Methodology and Foundations of Science. Selected Writings from 1951 to 1969._ Dordrecht: Reidel.

---------- and Zinnes, J.L. (1963). "Basic Measurement Theory." In _Handbook of Mathematical Psychology,_ Volume I. New York: Wiley. Pages 1-76.

von Neumann, J. (1932). _Mathematische Grundlagen der Quantenmechanik._ Berlin: Springer. (As reprinted as _Mathematical Foundations of Quantum Mechanics._ (trans.) R.T. Beyer. Princeton: Princeton University Press, 1955.)

<u>Approximate Generalizations and Their Idealization</u>

Ernest W. Adams

University of California, Berkeley

This paper describes recent work on a theory of <u>approximate</u> <u>generalizations</u> which, following J.S. Mill (1895, Book III Chapter XXIII), are propositions often expressed in the form <u>Most A's are B's</u>, and whose <u>degrees of truth</u> ('probabilities' in Mill's terms) are the <u>proportions</u> of A's that are B's. Earlier work by Ian Carlstrom and myself studied the <u>logic</u> of approximate generalizations — the theory of necessary connections among their degrees of truth — and basic ideas and results of that work will be briefly summarized here, along with a related extension of this study. However our present concern is primarily with two matters having to do with the <u>effects of small</u> <u>changes in the extensions of predicates</u> on degrees of truth. One is the question of which generalizations are <u>continuous</u> in that small changes in the extensions of "empirical" predicates would not entail large changes in the degrees of truth of the generalizations. The methodological importance of continuity resides partly in the fact that only "sufficiently continuous" laws are candidates for express- ing "robust" statistical properties of systems which can be reliably estimated by sampling methods. The other topic has to do with <u>idealizability</u> in a kind of Platonic sense: for which generaliza- tions is it the case that when they are approximately true we can be sure that small changes in the extensions of their predicates would make them exactly true? This question will be discussed with reference to generalizations (axioms and theorems) in <u>theories of</u> <u>fundamental measurement</u>, which typically idealize both by ignoring the fact that the generalizations that they "treat logically" as though they were exactly true are really only approximately true, and, connected to this, by ignoring errors of measurement.

We next summarize informally basic concepts and previous work on the theory of approximate generalizations.

These laws are symbolized as <u>formulas with free variables</u> in first order predicate calculus. To use a non-scientific example whose

PSA 1982, Volume 1, pp. 199-207

predicates are symbolized in the obvious way, the proposition "Most married students live off campus" would be symbolized '(Sx & ($\exists$y)xMy)$\rightarrow$ Ox'. This is to be contrasted with the universal generalization '($\forall$x)((Sx & ($\exists$y)xMy) $\rightarrow$ Ox)'. Our formal language includes the universal generalization — a sentence without free variables — as a limiting case, and its logic is assumed to be the standard bivalent one (hence we are not dealing with a many-valued or "fuzzy" logic of sentences). Though it will not concern us further here, it should be noted that the present symbolization, which is called the absolute symbolization in Adams (1974), is inappropriate in important respects, and a better formulation, which comes closer to Mill's own ideas, is relative. The absolute formulation is much the simpler one, however, and since the issues with which we are concerned arise within it, we will stick to it here.

Degrees of truth of approximate generalizations are defined in measure-structures, which are the basic "semantic structures" of our theory (Carlstrom (1975) seems to be the first to have employed this term, prior to its use by Hoover (1978) and Morgenstern (1979)). These are standard first order models, together with countably additive probability measures defined on σ-algebras of subsets of the models' domains. In our example the measure-structure might consist of a model $\langle D,S,M,O \rangle$, together with a probability measure on a σ-algebra of subsets of D. Measure-structures can be thought of as measured possible worlds, wherein the measure itself provides a means for defining a nearness relation at least among worlds with the same domain and measure on it. This is important in giving precise definitions of continuity and idealization, though we will leave this aside here.

Given an approximate generalization and a measure-structure, the degree of truth of the generalization in the structure is defined as the measure of the set of values of the free variables involved for which the formula is true in the structure (care must be taken to assure that this set of values — the extension of the formula in the model — always has a measure, but that will be left aside). Degree of falsity is 1 minus degree of truth, and the degrees of truth of the limiting case sentences are stipulated to be 1 or 0, according as the sentences are true or false in the model. But note that degree of truth 1 in an approximate generalization is not the same as exact truth, since approximate generalizations can be true almost everywhere in the measure-theoretic sense without being exactly true. This is importantly related to the logic of these propositions.

Generalizations of all standard logical concepts like validity, contradiction, etc., applicable to approximate generalizations can be considered, but previous work by Adams (1974) and Carlstrom (1975) focussed principally on a generalization of entailment, which is plausibly the most natural generalization of that notion that satisfies the formal requirements for being a deduction relation. Roughly, if $\varphi,\psi,\ldots$ are premises symbolized as above, and η is a similar conclusion, we say that $\varphi,\psi,\ldots$ measure-entail η if and only if it is

possible to assure an arbitrarily high degree of truth in η, short of
1, by requiring all premises $\varphi,\psi,\ldots$ to have sufficiently high degrees
of truth, short of 1. Interesting formal questions arise in describing the properties of this relation, which does not reduce straightforwardly to classical entailment. Thus, though it is a necessary
condition for an inference's measure-soundness (for the premises to
measure-entail the conclusion) that the inference that results when
all formulas are universally generalized be classically sound, this is
not generally sufficient. For instance, taking the single premise to
be the quantifier-free formula '(Sx & xMy) $\rightarrow$ Ox' and the conclusion to
be the previously cited formula '(Sx & ($\exists$y)xMy) $\rightarrow$ Ox', we get an
inference which becomes classically sound when both formulas are
universally generalized, but for which the premise has degree of truth
1 while the conclusion has degree of truth 0 in the <u>reals-structure</u>
whose domain is the unit interval [0,1], with the standard interval
measure over that domain, and with the interpretations of the
predicates given by: 'x=x' for 'Sx', 'x=y' for 'xMy', and 'x≠x' for
'Ox' (in this interpretation the premise is equivalent to 'x≠y',
which is "true almost everywhere", while the conclusion is equivalent
to 'x≠x', which is exactly false). This example is suggestive of
general properties of measure-entailment, which are briefly noted
next.

Adams (1974) gives simple syntactic <u>rules of inference</u> for deriving
measure-entailed conclusions, and proves their <u>completeness.</u>
Carlstrom (1975) extends the theory to formal languages including
individual constants and identity (hence, by implication to languages
containing function symbols), and to inferences with infinite sets of
premises, and shows the resulting system to be complete and <u>compact.</u>
This paper also gives a <u>decision-procedure</u> for determining measure-
soundness of <u>quantifier-free inferences,</u> which is important in view of
the significance that continuity considerations give to that class of
inferences. Finally, Carlstrom (1975) shows that measure-entailment
is precisely the relation which preserves "truth almost everywhere" in
the measure-theoretic sense (i.e., truth except on a set of measure
0), which contrasts neatly with classical entailment which preserves
"truth absolutely everywhere". In fact, it is shown that if an
inference is not measure-sound then there exists a <u>reals-structure</u>
<u>counterexample</u> in which all premises have degree of truth 1, while the
conclusion's degree of truth is not necessarily 0, but no greater than

$1-k^{-k}$, where k is the number of free variables in the conclusion.

Even when an inference is measure-sound a question of practical
importance often arises which has been studied recently by Adams
(1981b): given that it can be expected that there are small degrees
of falsity in its premises, we may want to know how great a degree of
falsity is possible in the conclusion under the circumstances. This
leads to the study of <u>transmissible falsity functions</u> of inferences
which give finer-grained characterizations of them than does mere
determination of soundness. The transmissible falsity function,
$\text{Tr}(\alpha,\beta,\ldots)$, of the inference from premises $\varphi,\psi,\ldots$ to the conclusion

202

η is defined for sequences $\alpha, \beta, \ldots$ of non-negative real <u>premise falsity bounds</u> to be the maximum degree of falsity of η in all structures in which the degree of falsity of φ is no greater than α, that of ψ is no greater than β, etc. (if the premises are <u>measure-inconsistent</u> and α, β, etc. are sufficiently small there may be no structures "consistent" with these bounds, in which case $\mathrm{Tr}(\alpha, \beta, \ldots)$ is undefined for those bounds).

In the simplest cases transmissible falsity functions are easy to calculate and agree fairly well with what would be expected <u>a priori</u>. For instance, when all formulas are purely monadic, quantifier-free, and involve just the free variable 'x' (e.g., the <u>transitivity inference</u> from premises 'Ax $\to$ Bx' and 'Bx $\to$ Cx' to the conclusion 'Ax $\to$ Cx'), results of Adams and Levine (1975) apply and entail in particular that if all premises are <u>essential</u> then $\mathrm{Tr}(\alpha, \beta, \ldots) = \alpha + \beta + \ldots$ (unless that sum exceeds 1, in which case $\mathrm{Tr}(\alpha, \beta, \ldots) = 1$ — incidentally, it is the disagreement of this result with considerations of Mill's (1895, p. 422) concerning the transitivity inference that suggests switching to the <u>relative</u> symbolization alluded to earlier). However, even generalizing to monadic and quantifier-free inferences with more than one free variable brings to light a surprising phenomenon.

The "odd" two-premise inference with premises 'Ax $\to$ By' and 'Ax' and conclusion 'By' is measure-sound and has the transmissible falsity function $\mathrm{Tr}(\alpha, \beta) = \alpha/(1-\beta)$ (unless $\alpha + \beta \geq 1$, in which case $\mathrm{Tr}(\alpha, \beta) = 1$), which is obviously not symmetric in α and β. Both premises are essential but the first is much <u>more</u> essential than the second. For instance, if $\alpha = .01$ while $\beta = .5$ (the second premise is "50% false"), the conclusion must still have a degree of falsity no greater than .02. Borrowing a term from Adams (1981a), the second premise can be said to be <u>marginally essential.</u> Adams (1981b, section 5) gives a precise definition of marginal essentialness, and shows that a premise of an inference of this special class is marginally essential if and only if the inference which results when that premise is existentially quantified while the other premises and the conclusion are universally quantified is classically sound. This gives precise expression to one aspect of the parallel discussed by Peterson (1979) between the vague pair of quantifiers 'many' and 'most' and the exact classical pair 'some' and 'all'.

Now we turn to <u>continuity.</u>

Intuitively, an approximate generalization is continuous <u>in</u> one or more of its predicates if small changes in the extensions of those predicates would not entail large changes in the degree of truth of the generalization. We have suggested that continuity of these laws in their "empirical" (or "non-rigid") predicates was a <u>prima facie</u> requirement for their expressing "statistically significant properties of the world" (contrast the statistically significant but not exactly true generalization "Swans are white" with "An even number of swans are white" whose degree of truth would change from 1 to 0 with the change of a single member of the extension of either predicate, and whose truth could not conceivably be "estimated" from sample data on swan coloration). Continuity is also importantly related to idealization because that presupposes that what is exactly true in a nearby "ideal world" must be approximately true in this world, and that is only plausible in the case of generalizations whose degrees of truth change continuously from the ideal world to this world.

Given particular predicates, we are interested in the question of which approximate generalizations are continuous in them. This question is only made precise with the introduction of measures of "nearness" of structures, but following a suggestion of Carlstrom it can be reduced to one concerning a kind of <u>substitutivity of approximate equivalents</u>. The case of our formula '$(Sx \mathbin{\&} (\exists y)xMy) \to Ox$' will give the general idea. This will be continuous, say, in the predicate 'M' if, when 'M' is replaced throughout by a "slightly changed" predicate '<u>M</u>', the <u>approximate biconditional</u>

$$((Sx \mathbin{\&} (\exists y)xMy) \to Ox) \leftrightarrow ((Sx \mathbin{\&} (\exists y)x\underline{M}y) \to Ox)$$

is measure-entailed by '$xMy \leftrightarrow x\underline{M}y$'.

The foregoing allows the theory of measure-entailment to be applied, and Adams (1981b, section 3) uses the <u>completeness theorem</u> of Adams (1974) together with pieced-together Craig interpolations to give intuitively significant necessary and sufficient conditions for continuity. A formula which does not contain a given predicate or predicates within the scope of any quantifier and in which no atomic formula in those predicates involves a repeated variable (e.g., 'xMx' would involve 'x' as a repeated variable) is continuous in those predicates, and furthermore any formula which is continuous in those predicates is measure-equivalent to (it measure-entails and is measure-entailed by) a formula of this <u>canonical continuous form</u>. Our example '$(Sx \mathbin{\&} (\exists y)xMy) \to Ox$' illustrates this. Obviously this is continuous in 'Sx' and 'Ox' but it might be discontinuous in 'xMy' since that occurs within the scope of a quantifier. In fact, the formula <u>is</u> discontinuous in 'xMy' since effecting a "change of zero measure" in the previously cited interpretation of 'xMy', from '$x=y$' to '$x \neq y \mathbin{\&} x=y$', changes the degree of truth of the formula from 0 to 1.

The study of continuity leads in many directions, including to possible formulations of certain ideas of classical metaphysics, but I must confine myself to noting two generalizations of scientific significance. One is a kind of <u>local continuity</u>, in turn related to continuities "subject to constraints", and the other is continuity relative not to changes in the extensions of predicates but to changes in the <u>domains</u> of structures. Necessary and sufficient conditions for all of these have been found, that for continuity under changes in domains being, perhaps not unexpectedly, expressibility in quantifier-free formulas which <u>can</u> have repeated variables in atomic constituents. This relates the present theory to a theory of a more "purely logical" sort of continuity studied by Craig (1965), in which expressibility by quantifier-free formulas also plays a central role.

Turning to <u>idealization</u>, in the theories to be discussed all
predicates are "empirical", and we should want to express generalizations in formulas that are continuous <u>tout court</u>: i.e., they should
be equivalent to <u>canonical</u> formulas which are quantifier-free and
have no repeated variables in atomic constituents.

Let us begin with general observations on <u>idealizability</u>: can it
be assumed that an approximately true law can be made exactly true
by making small changes in the extensions of the predicates involved?
Perhaps surprisingly, not even all <u>canonical</u> laws are consistently
idealizable in this sense, as the case of the "law of approximate
ordering relations" symbolized by the predicate 'xRy', namely
'xRy $\leftrightarrow$ -yRx', makes clear. This can be "as true as you like" but
obviously no change in the extension of 'xRy' can make it exactly
true (this and related examples have led to an as yet undecided
conjecture by Adams and Carlstrom (1979, p. 205), roughly that all
canonical laws which are approximately true can be made exactly true
<u>for all assignments of distinct values to distinct free variables</u>, by
making small changes in the extensions of their predicates, but this
will not concern us further here). Thus, it becomes a significant
question, concerning which we have as yet found no general answer,
whether given systems of approximations are consistently idealizable.
Certain special cases have been determined to be idealizable, which
are worth citing.

One case is related to <u>approximate classification systems</u>, whose
laws are expressible in monadic canonical formulas. The one-variable
case is trivial (for instance, "swans are white" symbolized by
'Sx $\rightarrow$ Wx' could obviously not be highly true unless small changes in
the extensions of 'is a swan' or 'is white' would make it exactly
true), and here one can give simple bounds on the "distance to the
nearest possible world" in which generalizations of this sort would
be exactly true. This finding is generalized, less trivially, to
monadic generalizations with several free variables in Adams (1981b,
section 4).

A much more difficult case, which leads to theories of fundamental
measurement, is the system of <u>laws of ordering</u> expressed by
'(xRy & yRz) $\rightarrow$ xRz' (approximate transitivity) and 'xRy v yRx'
(approximate connectedness). Adams and Carlstrom (1979) prove that
at least all finite structures in which the degrees of falsity of
these laws are at most α are $f(\alpha)$-approximations to structures in

which the laws are exactly true, with $f(\alpha) = 40\sqrt[4]{\alpha}$ (a similar result
is proved for the laws of <u>approximate equivalence relations</u>, which
also relates to approximate classification). Of course this is a
very bad bound, since even if the degrees of falsity of the laws are

no more than 10^{-12}, we can so far only conclude that "the nearest
world in which the laws are exactly true" is a .04-approximation of
the real world (a .04 change in the extension of 'xRy' would make the
laws exactly true), but at least we know we can "bring the ideal

world as close as we want" by making the laws "sufficiently true".
This is in turn connected to two matters of considerable scientific
importance.

The consistent idealizability of the ordering laws entails that
they satisfy an important requirement of systems or "theories" of
approximate generalizations: namely that all theorems (of canonical
form) which are classically entailed by axioms of the theories are
also measure-entailed by them. That is by no means true of all
systems of canonical generalizations (for example, the two "axioms"
'(xRy & yRx & xRz) → zRx' and 'xRy v yRx' classically entail the
canonical "theorem" 'xRy', but they do not measure-entail it), and
if it were not true of the theory of approximate orderings it would
mean that certain classical "theorems" such as second-order
transitivity, '(xRy & yRz & zRw) → xRw', would in fact stand as
independent principals of ordering. Consistent idealizability
implies that that cannot happen (though, trivially, such "non-
canonical theorems" as 'xRx' are not measure-entailed and are
independent principles).

The second matter connected with idealizability has to do with
numerical representability, which is in turn connected to another
kind of idealization in measurement. One main objective of fundamental measurement theories is that of formulating axioms involving
"empirical" predicates, from which it is possible to deduce the
existence of numerical representations of various kinds. In the
ordinal case the aim is to formulate axioms guaranteeing the existence
of an ordinal representation, which is a real-valued function m(x)
satisfying the Ordinal Representation Law 'xRy ↔ m(x) ≤ m(y)'. Exact
satisfaction of the cited axioms of ordering (plus a non-elementary
and non-finite topological assumption) does insure the existence of an
ordinal representation exactly satisfying the Ordinal Representation
Law. Approximate truth of the ordering laws obviously cannot guarantee the existence of an exact representation of the above sort, but
it might guarantee that of an approximate representation, for which
the Ordinal Representation Law would be approximately true. In fact,
that is precisely what consistent idealizability guarantees (at
least in the finite case, in which the topological assumption
mentioned above is vacuous). Here we may say that approximate satisfaction of laws, the exact satisfaction of which insures exact representability, guarantees approximate representability.

Can something analogous be claimed for more complex types of
fundamental measurement such as interval or extensive measurement,
for which the problem of errors of measurement first arises (c.f.,
Adams 1965)? As yet not even the question of consistent idealizability has been settled for typical axiom systems such as those discussed
in Krantz et al. (1971). But even assuming that appropriately chosen
systems of axioms are idealizable, a possibly more serious problem
arises with them, which I will note in closing. This has to do with
the results of Titiev (1969) that the canonical laws of these theories
are not finitely axiomatizable. One may demand that all of an

infinite set of independent laws be approximately true, but unless
all of them are "true almost everywhere" their infinite conjunction
(say in a $L_{\omega_1\omega}$ extension of the theory) may still have degree of

truth 0, and the nearest world in which they are exactly true may
still be "a possible galaxy away". This leads to what I regard as
the central <u>problem</u> of formulating an adequate theory of approximate
fundamental measurement. This is that the presently considered class
of canonical generalizations is of too limited expressive power to
"express the empirical content" of important assumptions which are
implicit in scientific measurement (especially, certain "independence
assumptions" discussed in Adams (1966)), and what is needed is an
enrichment of the formal language, which will make it possible to
deal with them systematically.

But those are all of the exact generalizations, conjectures and
problems of the theory of approximate generalizations for which there
is space here.

References

Adams, E.W. (1965). "Elements of a Theory of Inexact Measurement." _Philosophy of Science_ 32: 205-228.

——————. (1966). "On the Nature and Purpose of Measurement." _Synthese_ 16: 125-169.

——————. (1974). "The Logic of 'Almost All'." _Journal of Philosophical Logic_ 3: 3-17.

—————— and Levine, H.F. (1975). "Uncertainties Transmissible from Premises to Conclusions in Deductive Inferences." _Synthese_ 30: 429-460.

—————— and Carlstrom, I.F. (1979). "Representing Approximate Ordering and Equivalence Relations." _Journal of Mathematical Psychology_ 19: 182-207.

——————. (1981a). "Improbability Transmissibility and Marginal Essentialness of Premises in Inferences Involving Indicative Conditionals." _Journal of Philosophical Logic_ 10: 149-177.

——————. (1981b). "Contributions to a Theory of Laws Admitting Exceptions: the Monadic Case." Project Report, University of California, Berkeley.

Carlstrom, I.F. (1975). "Truth and Entailment for a Vague Quantifier." _Synthese_ 30: 461-495.

Craig, W. (1965). "Boolean Notions Extended to Higher Dimensions." In _The Theory of Models._ Edited by J. Addison, _et al._ Amsterdam: North Holland Publishing Company. Pages 55-69.

Hoover, D.N. (1978). "Probability Logic." _Annals of Mathematical Logic_ 14: 287-313.

Krantz, D. _et al._ (1971). _Foundations of Measurement,_ Volume I. New York: Academic Press.

Mill, J.S. (1895). _A System of Logic._ 8th ed. New York: Harper & Brothers.

Morgenstern, C.F. (1979). "The Measure Quantifier." _Journal of Symbolic Logic_ 44: 103-108.

Peterson, P.H. (1979). "On the Logic of 'few', 'many', and 'most'." _Notre Dame Journal of Formal Logic_ XX: 155-179.

Titiev, R.I. (1969). "Some Modeltheoretic Results in Measurement Theory." _Technical Report No. 46, Psychology Series._ Institute for Mathematical Studies in the Social Sciences, Stanford University.

<u>Truthlikeness for Quantitative Statements</u>

Ilkka Niiniluoto

University of Helsinki

1. Introduction: Galileo vs. Nozzolini

In *A History of the Mathematical Theory of Probability* (1865), I.
Todhunter tells that among the "learned conversations" of the Floren-
tine gentlemen the following question was once proposed: "A horse is
really worth a hundred crowns, one person estimated it at ten crowns
and another at a thousand; which of the two made the more extravagant
estimate?" When consulted on this matter, Galileo pronounced the two
estimates to be equally extravagant, "because the ratio of a thousand
to a hundred is the same as the ratio of a hundred to ten". A priest
named Nozzolini thought that the higher estimate is more mistaken,
since "the excess of a thousand above a hundred is greater than that of
a hundred above ten." (p. 5).

Todhunter's comment on this issue is interesting: it does not appear
to him that "the discussion is of any scientific interest or value". He
suggests that the gentlemen - while renouncing such "frivolities" as
attention to ladies, the stables, or excessive gaming - "might have in-
vestigated questions of greater moment than that which is here brought
under our notice." (p. 6).

Todhunter's negative attitude is explainable by the fact that the
Galileo-Nozzolini controversy does not concern the history of probability theory at all. The former wants to measure errors in terms of pro-
portions, the latter in terms of differences. This issue is, therefore,
a special case of a more general problem concerning comparative judgments of *truthlikeness* (verisimilitude): which of two rival statements
(which both may be false) is closer to the truth? This problem has only
recently received systematic attention from logicians and philosophers[1]
who share the view that a suitable notion of 'truthlikeness' or 'approximate truth' is needed for a realist account of science.

It may seem slightly surprising that the most elaborate modern ac-

———————

PSA 1982, Volume 1, pp. 208-216

counts of truthlikeness apply this notion primarily to first-order generalizations or theories in languages with *qualitative* predicates.
Prima facie it would seem easier to measure the size of the error that
is made by a *quantitative* statement, be it singular (as in the Galileo-
Nozzolini case) or a general law. It has been argued that the quantitative case can in principle be treated just as the qualitative one,
i.e., by using Carnapian attribute spaces (see Carnap, 1971, 1980)
based upon denumerable partitions of the value space of a quantity.[2]
This suggestion - even though it may work well for some special kinds
of quantitative statements - does not seem to be the most natural one
if we are dealing with continuous quantities. Indeed, a more direct approach to the definition of truthlikeness for quantitative laws is out-
lined in this paper.

2. Singular Quantitative Statements

Assume that we are trying to estimate some unknown real number $\theta*$.
Here $\theta*$ may be a physical constant (such as Planck's constant h) or the
value $t(a)$ of some quantity t for a particular object a (where t is
measurable in the interval or the ratio scale).

If θ is an estimate of $\theta*$, then the *distance* of θ from $\theta*$ may be
measured either by

(1) $d_1(\theta,\theta*) = |\theta - \theta*|$

or by

(2) $d_2(\theta,\theta*) = (\theta - \theta*)^2$.

Measures d_1 and d_2 can be employed to give two alternative normalized
definitions for the *degree of truthlikeness* of the singular statement
that the value of $\theta*$ is equal to θ:

(3) $M(\theta,\theta*) = 1/(1+d(\theta,\theta*))$,

where d may be either d_1 or d_2. Thus, $0 \leq M(\theta,\theta*) \leq 1$, and $M(\theta,\theta*) = 1$
if and only if $\theta = \theta*$. Both of these definitions lead to the same comparative notion of truthlikeness[3]:

(4) Estimate θ_1 is closer to the truth than estimate θ_2
    iff $|\theta_1 - \theta*| < |\theta_2 - \theta*|$
    iff $(\theta_1 - \theta*)^2 < (\theta_2 - \theta*)^2$.

If constant $\theta*$ is unknown to us, criterion (4) does not help us in
evaluating the truthlikeness of the estimates. However, as soon as our
beliefs about the location of $\theta*$ can be represented by a probability
distribution or density $g(\theta)$ over the real axis R, then the *expected
distance* of an estimate θ_1 *from the truth* $\theta*$ may be defined by

(5) $\int_{-\infty}^{+\infty} g(\theta)d_1(\theta_1,\theta)d\theta$

or by

210

$$(6) \quad \int_{-\infty}^{+\infty} g(\theta) d_2(\theta_1, \theta) d\theta.$$

It is known that the value of (5) is minimized by choosing θ_1 as the *median* of the distribution $g(\theta)$, and the value of (6) is minimized by the *mean* of $g(\theta)$.[4] By Bayesian standards, these two values are the best point estimates of $\theta*$, when "losses" are measured by (1) or (2). It is also known that, under fairly general conditions, these Bayesian estimates converge with probability one towards the true value $\theta*$, if the probability distribution g is modified by conditionalizing sample information in accordance with Bayes's Theorem.

These considerations may be generalized to n-dimensional quantities by defining the distance of an n-dimensional estimate $<\theta_1,\ldots,\theta_n>$ from the true n-tuple $<\theta_1^*,\ldots,\theta_n^*>$ by the *city block measure*

$$(7) \quad d_1(<\theta_1,\ldots,\theta_n>,<\theta_1^*,\ldots,\theta_n^*>) = \Sigma_{i=1}^{n} |\theta_i - \theta_i^*|$$

or by the *Euclidean measure*

$$(8) \quad d_2(<\theta_1,\ldots,\theta_n>,<\theta_1^*,\ldots,\theta_n^*>) = (\Sigma_{i=1}^{n} (\theta_i - \theta_i^*)^2)^{\frac{1}{2}}.$$

In some case, non-Euclidean distance measures have to be used.[5]

3. Interval Statements

Let $I = [\theta,\theta']$ be an interval of real numbers, and let $\ell(I) = \theta' - \theta$ be its length. The middle point $(\theta - \theta')/2$ of interval I is denoted by $m(I)$. We shall represent simply by I itself the statement that the true value $\theta*$ belongs to I, i.e., that $\theta \leq \theta* \leq \theta'$. When $\theta < \theta'$, statement I is thus an infinite disjunction of singular quantitative statements.

Following the two alternative ways of treating (finite) disjunctions in Niiniluoto (1977), the distance of I from the truth $\theta*$ may be defined either by

$$(9) \quad D_1(I,\theta*) = \gamma \cdot \min_{x \in I} d(x,\theta*) + (1-\gamma) \max_{x \in I} d(x,\theta*) \ ,$$

where $0<\gamma<1$, or by

$$(10) \quad D_2(I,\theta*) = 1/\ell(I) \ \Sigma_{x \in I} \ d(x,\theta*),$$

where d is either d_1 or d_2. Following Niiniluoto (1978), the third possibility would be

$$(11) \quad D_3(I,\theta*) = \gamma \cdot \min_{x \in I} d(x,\theta*) + (1-\gamma)B(I) \max_{x \in I} d(x,\theta*) \ ,$$

where function $B(I)$ is proportional to the length $\ell(I)$ of I. For degenerate intervals $I = \{x\}$, we get $D_1(\{x\},\theta*) = D_2(\{x\},\theta*) = d(x,\theta*)$. Hence, $D_1(I,\theta*) = D_2(I,\theta*) = 0$ if and only if $I = \{\theta*\}$. Moreover, if d_1

is chosen as d in definitions (9) and (10), and if $\gamma=\frac{1}{2}$, the following results are obtained:

(12) If $\theta < \theta' \leq \theta*$ or $\theta* \leq \theta < \theta'$, then
$$D_1(I,\theta*) = D_2(I,\theta*) = |m(I) - \theta*|.$$

(13) If $\theta < \theta* \leq m(I) < \theta'$, then
$$D_1(I,\theta*) = |\theta* - \theta'|/2$$
$$D_2(I,\theta*) = [(\theta* - \theta)^2 + (\theta' - \theta*)^2]/2(\theta' - \theta).$$

(14) If $\theta* = m(I)$, then
$$D_1(I,\theta*) = D_2(I,\theta*) = \ell(I)/4.$$

According to (12), the distance between interval I and a point $\theta*$ outside I is simply the distance between the midpoint $m(I)$ of I and $\theta*$. According to (14), the distance of an interval I from its midpoint $m(I)$ is one fourth of its length $\ell(I)$.

The *degree of truthlikeness* of the interval statement I may now be defined by

(15) $M(I,\theta*) = 1/(1+D(I,\theta*))$.

Hence, $M(I,\theta*) = 1$ if and only if $I = \{\theta*\}$.

Given a probability density $g(\theta)$, the expected distance of interval statement I from the truth is

$$(16) \quad \int_{-\infty}^{+\infty} g(\theta)D(I,\theta)d\theta.$$

A Bayesian theory of interval estimation can be based upon the idea of minimizing the expected loss as defined by (16).

All these ideas can be generalized to the n-dimensional case, where interval I is replaced by a region in the n-dimensional space.

4. Simple Quantitative Laws

Quantitative laws are expressions of functional relations between quantities. By a simple quantitative law we mean a statement of the form

(17) $\forall z[Az \rightarrow u(z) = f(t(z))]$,

which says that, for all objects z of kind A, the value $u(z)$ of quantity u for z is related by function f to the value $t(z)$ of quantity t for z.[6] If the values of quantities t and u are represented in the x-y-coordinate system, then the law (17) says that only those pairs of values are physically possible which lie on the curve

(18) $y = f(x)$.

212

Assume now that the true connection between quantities t and u is expressible by the equation

(19) $y = f^*(x)$.

We may then ask how close to the truth the hypothesis (18) is, i.e., what the degree of truthlikeness of (18) is.

This problem has been discussed in another form by those mathematicians who have studied *metric spaces of functions*.[7] For example, let C be the class of real-valued continuous functions defined on a closed interval $[a,b]$.[8] Then the distance between two elements f and g of C may be defined, e.g., in the following three ways:

(20) $\Delta_1(f,g) = \sup\limits_{a \leq x \leq b} |f(x) - g(x)|$

(21) $\Delta_2(f,g) = \int_a^b |f(x) - g(x)| dx$

(22) $\Delta_3(f,g) = (\int_a^b (f(x) - g(x))^2 dx)^{\frac{1}{2}}$.

Here (21) and (22) can be considered as generalizations of the city block measure (7) and the Euclidean measure (8), respectively. (22) can also be considered as a generalization of the Least Square Difference method in statistics.

We may now propose that the distance between the statements (18) and (19) is measured by the distance between the corresponding functions, i.e., by $\Delta_i(f,f^*)$, where Δ_i ($i = 1,2,3$) is defined by (20)-(22). This gives us three alternative notions of truthlikeness for simple quantitative laws.

(23) $T_i(f,f^*) = 1/(1+\Delta_i(f,f^*))$ $\qquad$ ($i = 1,2,3$).

Hence, $T_i(f,f^*) = 1$ if and only if $f = f^*$, i.e., $f(z) = f^*(z)$ for all $z \in [a,b]$. Comparative judgments of truthlikeness, based upon (23), are definable as follows:

(24) Law $y = f_1(x)$ has more truthlikeness$_i$ than law $y = f_2(x)$
  $\quad$ iff $\Delta_i(f_1,f^*) < \Delta_i(f_2,f^*)$ $\qquad$ ($i = 1,2,3$).

A law may be said to be *approximately true* if and only if its degree of truthlikeness is sufficiently high. Definition (23) gives, therefore, three alternative notions of approximate truth for simple quantitative laws:

(25) Law $y = f(x)$ is approximately$_i$ true within range $[a,b]$ (relative
  $\quad$ to degree δ) iff $\Delta_i(f,f^*) \leq \delta$ $\qquad$ ($i = 1,2,3$).

If $i = 1$, condition (25) guarantees that the statement $y = f(x)$ makes *no* large errors within its range:

(26) Law $y = f(x)$ is approximately$_1$ true in $[a,b]$
  $\quad$ iff $|f(z) - f^*(z)| \leq \delta$ $\qquad\qquad$ for all $z \in [a,b]$.

On the other hand, if i = 2 or i = 3, condition (25) guarantees that the *sum* of the errors that the statement $y = f(x)$ makes in $[a,b]$ is not too large.

These conditions show in what senses it is correct to claim that approximately true laws are "empirically successful".[9] If $y = f(x)$ is approximately$_1$ true in $[a,b]$, then *all* its predictions $f(x_0)$ for various values x_0 within $[a,b]$ are close to the truth. If this law is approximately$_i$ true (i = 2,3) true in $[a,b]$, then its predictions will be *in the average* close to the truth.

5. Generalizations

The approach of Section 4 can be immediately generalized to the case where the value of a quantity u depends upon n other quantities $t_1,\ldots,t_n$. In this situation, the true connection between these quantities can be expressed by the equation

(27) $\quad y = f^*(x_1,\ldots,x_n)$,

which defines a hyperplane in a $(n+1)$-dimensional space. If we restrict our consideration to a finite region in this space, the distance of another equation $y = f(x_1,\ldots,x_n)$ from the true one (27) is definable just as in (20)–(22).

An interesting possibility, familiar from the *method of idealization*[10], is the case where the value of y is claimed to depend only upon some of the arguments $x_1,\ldots,x_n$. We are thus lead to compare equations of the form

(28) $\quad y = g(x_1,\ldots,x_k) \qquad (k < n)$

to the true law (27). As equation (28) defines a hyperplane

(29) $\quad \{\langle x_1,\ldots,x_n,x_{n+1}\rangle \mid x_{n+1} = g(x_1,\ldots,x_k)\}$,

this problem is reduced to measuring the distance of hyperplane (29) from that defined by (27) – which again may be defined in analogy with (20)–(22).

For example, consider three rival theories about the connection between the rest mass (m_0), relative mass (m), and velocity (v) of a physical body:

(30) $\quad m = m_0$

(31) $\quad m = m_0/(1-v^4/c^4)^{\frac{1}{2}}$

(32) $\quad m = m_0/(1-v^2/c^2)^{\frac{1}{2}}$.

Then, assuming that (32) is true, law (31) is closer to the truth than the idealizational law (30).

In this paper, we have not discussed the problem of estimating the degrees of truthlikeness of quantitative laws (cf., Sections 2 and 3). However, in practice this problem is often eliminated in the following manner. It is known that any continuous function can be approximated by a polynomial with an arbitrary high precision.[11] A scientist normally starts with a simple polynomial with unknown coefficients, and then estimates these parameters on the basis of data (cf., Sections 2 and 3). If the law obtained in this way does not survive further tests, the scientist proceeds to consider polynomials of higher degree. Within this procedure, only the estimation of real-valued parameters is needed.

Notes

[1]See Hilpinen (1976), Tichý (1978), Niiniluoto (1977, 1978, 1980, 1982), and Oddie (1978).

[2]See Niiniluoto (1978). The same technique is also applied by Watkins (1978). Miller (1975) discusses truthlikeness in connection with quantities, but his "theories" are simply singular statements about constants or about time-dependent parameters.

[3]This criterion is also independent on the scale in which the quantity in question is measured.

[4]See Ferguson (1967). Cf., Niiniluoto (1982).

[5]Miller's (1975) argument is based upon the fact that all coordinate transformations do not preserve the Euclidean structure of the space. Cf., Niiniluoto (1978) for comments on this issue.

[6]The problems concerning lawlikeness cannot be discussed in this paper.

[7]See Simmons (1963).

[8]On the whole real axis R, the values of (20)-(22) are in many cases infinite. Definition (20) is very close to the notion of approximative truth that Patryas and Krajewski have proposed (cf., Krajewski 1977).

[9]Laudan (1981) criticizes scientific realists for failing to establish this claim.

[10]Cf., Nowak (1980) and Krajewski (1977).

[11]Cf., Simmons (1963).

References

Carnap, R. (1971). "A Basic System of Inductive Logic. Part I." In
Studies in Inductive Logic and Probability, Volume I. Edited by
R. Carnap and R. Jeffrey. Berkeley: University of California
Press. Pages 33-165.

----------. (1980). "A Basic System of Inductive Logic. Part II." In
Studies in Inductive Logic and Probability, Volume II. Edited by
R. Jeffrey. Berkeley: University of California Press. Pages
7-155.

Ferguson, T.S. (1967). Mathematical Statistics: A Decision-Theoretic
Approach. New York: Academic Press.

Hilpinen, R. (1976). "Approximate Truth and Truthlikeness." In Formal
Methods in the Methodology of the Empirical Sciences. Edited by
M. Przelecki, et al. Dordrecht: Reidel. Pages 19-42.

Krajewski, W. (1977). Correspondence Principle and the Growth of
Knowledge. Dordrecht: Reidel.

Laudan, L. (1981). "A Confutation of Convergent Realism." Philosophy
of Science 48: 19-49.

Miller, D. (1975). "The Accuracy of Predictions." Synthese 30:
159-191, 207-219.

Niiniluoto, I. (1977). "On the Truthlikeness of Generalizations." In
Basic Problems in Methodology and Linguistics. (University of
Western Ontario Series in the Philosophy of Science, Volume 11,
Part 3.) Edited by R.E. Butts and J. Hintikka. Dordrecht:
Reidel. Pages 121-147.

--------------. (1978). "Truthlikeness: Comments on Recent
Discussion." Synthese 38: 281-329.

--------------. (1980). "Scientific Progress." Synthese 45: 427-462.

--------------. (1982). "What Shall We Do With Verisimilitude?"
Philosophy of Science 49: 181-197.

Nowak, L. (1980). The Structure of Idealization. Dordrecht: Reidel.

Oddie, G. (1978). "Verisimilitude and Distance in Logical Space." In
The Logic and Epistemology of Scientific Change. (Acta Philo-
sophica Fennica 30.) Edited by I. Niiniluoto and R. Tuomela.
Amsterdam: North Holland. Pages 227-242.

Simmons, G.F. (1963). Introduction to Topology and Modern Analysis.
New York: McGraw Hill.

Tichý, P. (1978). "Verisimilitude Revisited." Synthese 38: 175-196.

Todhunter, I. (1865). <u>A History of the Mathematical Theory of Probability from the Time of Pascal to that of Laplace.</u> (As reprinted New York: Chelsea Publishing Company, 1949.)

Watkins, J. (1978). "Corroboration and the Problem of Content-Comparison." In <u>Progress and Rationality in Science.</u> Edited by G. Radnitzky and G. Andersson. Dordrecht: Reidel. Pages 339-378.

Economics, Risk-Cost-Benefit Analysis, and the Linearity Assumption[1]

K. S. Shrader-Frechette

University of California, Santa Barbara

1. Introduction

Risk assessment, or risk-cost-benefit analysis (RCBA), an offshoot of decision analysis, has been widely touted as a "developing science" (see, for example, Levine 1979, p. 634). Used to help administrators make decisions on how to spend money, this set of procedures is directed at calculating whether the expected benefits from a proposed activity outweigh its expected risks or costs. For example, experts might wish to determine, given the risk of nuclear core melt and the possible resultant release of radioactive materials, whether additional containment (for reactor vessels) would be cost-effective.

In the sixties, RCBA was integrated into the planning and budgeting efforts of many federal agencies in the defense, aerospace, and energy fields. In the seventies, a number of regulatory agencies controlling environmental and occupational hazards likewise began to use the technique. Currently, nearly all US regulatory agencies (with the exception only of the Occupational Safety and Health Administration, OSHA) consistently use RCBA to help determine their policies (Carter 1979, pp. 1324-1325; Starr and Whipple 1980; Gage 1979, p. 13).

Despite the fact that RCBA dominates current US decisionmaking regarding science and technology, several major theoretical problems still beset the technique. First, there is no widely accepted theory of rationality, yet such a theory appears necessary if one is to take RCBA seriously. Moreover, as Arrow (1964) has noted, it would be absurd to attempt to find a theory of social rationality. Secondly, numerous scholars have criticized RCBA on the grounds (1) that politics, not some mathematical-economic theory, determines public policy; (2) that RCBA ignores factors such as the equity of distribution and the incommensurability of various parameters; (3) that its data base is inadequate; (4) that there are methodological disagreements among

PSA 1982, Volume 1, pp. 217-232

218

experts in this area; and (5) that various political, ethical, and
moral attitudes cannot be taken into account in RCBA.

Apart from the ultimate success of various proposals for improving
RCBA, methodological criticism of this technique is important, both
because RCBA is central to policymaking and because, despite its de-
fects, there is no currently accepted alternative to it.[2] This means
that whether RCBA ought, or ought not, to be used in making decisions
regarding science and technology, one is better served by being aware
of its methodological limitations, than by being ignorant of them. One
such methodological limitation is a consequence of an assumption widely
accepted by analysts who follow the method of "observed preferences."[3]

2. The "Fatality Interpretation" of "the Linearity Assumption"

Risk assessors traditionally make the normative assumption, when
predicting and evaluating actual future societal risk, that there is "a
linear relation between risk ["defined as the actual probability per
unit time of a unit cost burden occurring"] and the cost [defined "in
terms of injuries (fatalities or days of disability) or other damage
penalties (expenses incurred) or total social costs (including environmental intangibles)] of that risk" (Starr 1979, p. 23; Starr and Whipple 1980, p. 1116; Starr, Rudman, Whipple 1976, pp. 640-41; see
Philipson 1979, p. 385).[4] I have called this "the linearity assumption". In practice, many (if not most (see Cohen and Lee 1979, p. 707))
risk assessors interpret this assumption to mean that there is a linear
relation between the actual "probability of fatality" and "the value of
risk avoidance" or the "cost of a risk" (Starr and Whipple 1980; pp.
1115-1116; Starr et al. 1976, pp. 636-637).

By adhering to this interpretation of the linearity assumption,[5]
many assessors subscribe to at least three doubtful tenets: (1) that a
unit cost is adequately represented by fatalities: (2) that the value
of risk avoidance is independent of any variable (e.g., benefits),
except probability of fatality; and (3) that there is a linear relation
between societal risk (expressed in terms of probabilities) and the
value of risk avoidance. Since (1) and (2) seem to me to be easily
shown to be implausible, and since (3) has (to my knowledge) not been
adequately assessed in the risk analysis literature, I wish to focus on
this last claim. Specifically, I want to discuss one argument against
tenet (3), viz., that it is highly implausible, since people often have
good reasons for viewing risks of fatality in nonlinear ways. That is,
they frequently have good reasons for placing higher value on risk
avoidance for events having a low probability of fatality than for
those having a higher probability. For example, the public values air-
plane safety more than automobile safety.

Assessors, however, maintain that policy should be made on the basis
of the thesis that risks with a higher probability of fatality are less
acceptable than those with a lower value (Cohen and Lee 1979, p. 720;
Gardenier 1979, pp. 399, 467). Many assessors argue, for example, that
it is "inconsistent for the public both to tolerate 50,000 automobile

deaths per year and yet to be alarmed at generating electricity through nuclear fission (Lave 1979b, p. 541; see Starr and Whipple 1980, p. 1116; Okrent 1979, p. 663; Bazelon 1979, p. 278). Maxey accuses those who view nuclear risks as more dangerous than other, more probable, risks, of having "pathologic fear" and "near clinical paranoia." Numerous authors, including Starr, Whipple, Okrent, Maxey, Cohen, and Lee, maintain that if the public only understood the <u>probabilities</u> (of fatalities) of the risks involved, they would not fear statistically less significant hazards, like LNG (liquefied natural gas) or nuclear accidents, more than some more likely ones (see, for example, Cohen and Lee 1979, p. 707; Häfele 1979, p. 139; Maxey 1979, pp. 410, 417; Starr 1972, pp. 26-27; see also Lave 1979b, p. 484). In other words, since they assume that there ought to be a linear relationship between the actual probability of fatality, and the value of risk avoidance, many assessors do not view the observed societal aversion to certain low-probability risks as evidence against the fatality interpretation of the linearity assumption.[6] Even though a number of good arguments could be brought against this assumption,[7] assessors explain counterexamples as simply a result of the fact that the public does not know the accident probabilities in question.

3. The Argument That Low Probability Events Are Perceived As Likely

To support their claim about the public's alleged ignorance of correct probabilities, assessors often appeal to the notion that, below a certain level (say 1 in 10,000), the public cannot distinguish one very small probability from another. Their argument is that there is an intuitively understandable probability range, within which the public's judgments are consistent with the fatality interpretation of the linearity assumption, and analytic methodology, but that, outside this range (where most of the counterexamples fall), the public "has not learned" to view hazards consistently, by means of this assumption, and instead relies on intuitive, subjective, and incorrect assessments of risks (see, for example, Starr 1972, pp. 26-27; Starr and Whipple 1980, p. 1116; Hollister 1979, p. 57).

According to Starr and Whipple, since low-probability catastrophies lie outside the range of intuitively understandable probabilities, their likelihood is misperceived by the public (1980, p. 1117). To support this thesis about the misperception of low probability events, they note, <u>in general</u>, that extremely unlikely occurrences (winning a lottery, for example) are perceived as quite probable. Starr and Whipple also maintain, <u>in particular</u>, that in the paradigm case of low probability, catastrophic risk, that of nuclear power, extremely unlikely accidents are viewed as quite probable (1980, pp. 1116-1117; Starr 1972, pp. 26-27; Starr 1979, pp. 16-17).

At least two difficulties face this theory about misperception of probabilities, especially of catastrophes. First, not all low-probability events are seen as being highly likely. The fact that the low-probability event of a meteor striking the earth is in fact perceived as a highly unlikely event is precisely why a number of nuclear proponents have used it in an analogy to explain the predicted frequency of a given nuclear accident (see, for example, US NRC 1975, pp. 187-221, esp. Fig. A3.2; see also Häfele 1975, pp. 162; Rowe 1977, p. 320; Yellin 1977, p. 978). Moreover, if some low-probability events are in fact perceived as being extremely unlikely, while others are not, then contrary to Starr's and Whipple's suggestion, the reason for misperceiving a certain low probability appears to be something other than merely whether it falls outside some allegedly understandable probability range.

A second difficulty with this theory about misperception of probabilities is that not all cases of public aversion to low-probability, catastrophic risks can be attributed to the belief that the catastophe has a higher probability of fatality. The <u>reason</u> for society's aversion to certain low-probability events could well be something other than that their likelihood is misperceived, because they fall outside some allegedly understandable probability range. It is quite possible that the actual probabilities of fatalities are <u>accurately</u> perceived, but that other factors, e.g., the low level of benefits arising from the risk, cause societal aversion to low-probability risks. Likewise, it is quite possible that the actual probabilities of fatalities are in fact <u>misperceived</u>, but that the dearth of benefits, and not these misperceived probabilities, play the dominant role in causing societal aversion.

In particular, it is <u>not</u> clear that the public views nuclear fission as having a high probability of causing fatalities, and for this reason places a high value on avoiding nuclear risk. In the same study as that cited by Starr and Whipple, for example, the authors concluded, not that nuclear opponents misperceived risk probabilities, but that they placed a different <u>value</u> on nuclear benefits than did the proponents. Hence this research does not show that misperceived nuclear probabilities cause the risk-cost relationship to be viewed in a nonlinear way.[8]

Although I do not have the time to do so here, it might also be argued that it is epistemologically impossible to provide a clean-cut case for the thesis that catastrophic events, especially those involving relatively new technologies, are <u>erroneously</u> perceived as having a high actual probability of fatalities. This is because a given observed accident frequency is consistent with a fairly wide range of probability values.

4. The Argument That There Is a Distinction Between Perceived and Actual Probabilities

Precisely because they often ignore the <u>range</u> or distribution of acceptable accident probabilities, Starr, Whipple, and others are not warranted in asserting either that certain risks are clearly known to be of a low probability, or that there is always a clear distinction between what they define as <u>objective</u> (or "actual") <u>societal</u> risks and <u>subjective</u> (or merely "perceived") <u>individual</u> risks, as they say there is (1980, pp. 1115-1117; see also Burnham 1979, p. 678). They would have us believe that "misperceived," erroneous probabilities account for the high value placed on safety from certain catastrophic accidents, e.g., at LNG facilities. Using their notion of actual and perceived probabilities to explain apparent anomalies, given this interpretation of "the linearity assumption," however, presupposes that their version of the actual/perceived distinction will hold up in every case. For many reasons, I think it will not.

First, there are numerous problems of actual risk estimation (prior to any alleged evaluation) which simply do not admit of analytical resolution by experts. As a consequence, they can only be handled by means of the intuitive estimates of individuals.[9]

Secondly, it is well-known that individuals discount risks in both the space and time dimension. They are more concerned with events in their own neighborhood than with those in either a distant time or a remote place (see Rowe 1977, p. 130). In fact, if societal (or "actual") risk is taken to be the risk to that refined subset of persons closest to a problem, and if the population is divided into finer and finer risk subsets (e.g., those living within 50 miles of an LNG facility; those bearing higher risks from industrial radiation emissions because of previous medical exposures), then the alleged difference between their definitions of _actual_ societal risk and _perceived_ individual risk disappears.

Thirdly, their version of the actual/perceived risk distinction is dependent upon the assumption that only the _average_ of the aggregated risk probabilities represents the correct, or "actual", probability. Meanwhile, all accurate risk probabilities for particular subsets of persons are assumed to be merely "perceived", nonanalytical, subjective, individual, and potentially incorrect.

Consider the consequences to which the tyranny of such "average" probabilities can lead. Starr, for example, argues repeatedly in his work that "the highest level of acceptable risks which may be regarded as a reference level is determined by the normal US death rate from disease" (1979, p. 14; Starr _et al_. 1976, p. 630). Based on this criterion, he suggests that we ought not necessarily to have been overly concerned about the numerous deaths occasioned by the war in Vietnam. Writing during the peak of the conflict, he coolly affirmed that "the related risk, as seen by society as a whole, is not substantially different from the average nonmilitary risk from disease" (1969, p. 1235). Only in one limited sense is his point correct. Part of what is questionable about it is whether the normal death rate from disease provides a sufficient criterion for judging a risk's acceptability, and whether that criterion ought to be defined merely in terms of the actual probability of fatalities. Also questionable is the interpretation which Starr places on risk acceptability as a direct result of the group whose risk probabilities he chose to aggregate, then average. If he had compared the military risk to US males in the 18 to 24 age group, instead of to males and females of all ages, then obviously he could not have concluded that the Vietnam risk was "not substantially different from the average nonmilitary risk from disease" (1969, p. 1235).   Likewise, because they define "actual" (or correctly perceived) risk, by aggregating and then averaging all risk probabilities, assessors' distinguishing between their senses of actual, and merely perceived, risks is question-begging. Once the relevant subsets of persons are considered, their allegedly perceived risk may be more correct, more "actual", than that defined by the assessors as actual.

The same methodological failing, regarding a risk as negligible merely because it is measured against the average risk to all persons from all activities, occurs throughout much risk assessment literature. Typically, effects of low-level radiation are said to be negligible. They are so small that they are masked by other environmental factors and are perceived as insignificant.   Interestingly, however, the same judgment of insignificance could be made about other causes of death (such as murder by handguns) "which we do regard as of some significance.   If we had no way of distinguishing death by murder from death by natural causes, [then] the death rate from murder could increase manyfold before it became noticeable as an increase in the mortality from all causes." (Lovins, 1977, pp. 921-922).

The point of my argument here is twofold. First, risk probabilities for various subclasses of persons are essential to accurate and equitable risk assessment.   Second, once one begins to consider the actual risks faced by particular classes of individuals, then one has already begun to collapse the assessors' distinction between perceived or individual, and actual or societal, risk probabilities. Although there is often a valid distinction between subjective and objective probabilities, Starr, Whipple, and others have <u>defined</u> "actual" (objective) versus "perceived" (subjective) probabilities in a highly doubtful and question-begging way, and have then relied on this definition to defend their linearity assumption. As a consequence, it seems misleading for Starr, Whipple, and others to attempt to "explain" the anomalies with regard to this interpretation of "the linearity assumption" merely by saying that perceived, not actual, risk probabilities, as they define them, cause the public to view risks in nonlinear and "inconsistent" ways.   Their notion of "<u>actual</u>" probabilities is nothing more than probabilities calculated on the basis of their own theoretical assumptions.

5. The Assumption That Overestimated Probabilities Alone Explain Overvaluation of Certain Risks

However, even if Starr, Whipple, and others were correct in claiming that low-probability events are typically misperceived as likely (section 3), and that there is a clear distinction between actual, as opposed to merely perceived, probabilities (section 4), these facts alone would not prove that observed societal aversion to uncompensated public risks is caused merely by misperceived probabilities of fatalities.   As was suggested earlier, many other factors, such as the distribution of the risk, or its tradeoff with benefits, could equally well explain the anomalies. Moreover, even if there were no counter-examples to the fatality interpretation of "the linearity assumption", and even if the public unanimously judged that the value of risk avoidance was always proportional to the actual probability of fatality, neither this claim, nor any other argument based on consensus, would establish the correctness of the fatality interpretation, although it might establish the political desirability of accepting it.

Apart from the difficulties of arguing for the linearity assumption, on grounds stronger than consensus, it is not clear that Starr, Whipple, Okrent, and other assessors, following the method of observed preferences, are even able to justify their assumption on the grounds of consensus. Their claims about how probability estimates influence societal judgments about acceptable risks are especially vulnerable when one realizes that the nine characteristics hypothesized by various authors to influence judgments of perceived and acceptable risk are highly intercorrelated (Fischhoff et al. 1978, pp. 144, 149). Involuntary hazards, for example, "tend also to be inequitable and catastrophic" (Fischhoff et al. 1980, p. 215). This means that it is especially difficult to determine whether or not society's expressed concern about involuntary risks, for example, is merely an artifact of the excellent correlation between involuntariness and other undesirable risk characteristics. (By using the method of observed, rather than expressed, preferences (see note 10), Starr and Whipple seem to me to be unable to affirm, with assurance, that the characteristic of having a low, misperceived probability (rather than some other characteristic highly correlated with this one) is the major cause for certain risks being viewed as extremely costly.) There are numerous allegedly causal explanations, all consistent with the same "observed" phenomena.

If this is true, then assessors ought to re-examine the claim against them, viz., that their tenet (3) [that there is a linear relation between societal risk (expressed in terms of probabilities) and the value of risk avoidance] is incorrect, especially in the cases in which one is evaluating low-probability catastrophic risks. In any event, their position suggests that a few insights about contemporary RCBA are in order.

6. Some Insights About RCBA

First, because of their reliance on the assumption of a linear relationship between probability of fatality and the value of risk avoidance, and because of their frequent failure to consider parameters other than probability of fatality in estimating risk, many (if not most) assessors underestimate the value component in risk determination. They assume that any preference for a risk, whose probability-of-fatality is statistically higher than that of an alternative, is a result of misperceived probabilities, not a result of a given value system. Secondly, in explaining counterexamples to the assessors' models as those in which probabilities are misperceived, they appear to believe that probabilities can be determined in a wholly objective manner in every situation.[10] Thirdly, to the extent that risk assessors

224

fail to deal with the value-ladenness and subjectivity of their work,
and instead postulate a simple dichotomy between their definitions of
correct vs. incorrect, actual vs. perceived, and expert vs. lay probability estimates, then to that degree are they likely to miss some of
the real sources of controversy over technology and to attempt to define ethical and political issues in purely scientific, logical, or
technical terms.

   Fourthly, many risk theorists appear to be overly concerned with
what they regard as erroneous public decisions. The question is not
how to make the public more rational. It is how to accommodate democratic values within analytic assessment. It is surely not appropriate, when citizens insist on certain things as matters of rights, e.g.,
compensation for technological risks incurred, to miss the ethical
import of such demands, and as a consequence, to dismiss them, a
priori, as irrational exhibitions of a discontinuous preference function (see Burns 1977, p. 1150; Lovins 1977, p. 929). The most we can
ask of a good analytic technique is to clarify our scientific and
ethical options. It is not, and ought not be, a substitute for the
dynamics of the political system.

Notes

[1]An earlier version of this paper was presented on December 1, 1981,
at the Boston Colloquium for the Philosophy of Science. I am grateful
to Professors Joseph Agassi, Robert Cohen, Sheldon Krimsky, and Marx
Wartofsky for comments which enabled me to strengthen this essay.

[2]Many assessors claim that the only alternative to RCBA is to make
decisions intuitively, and that it is virtually impossible for persons
to attempt to make rational decisions in the face of opposing intuitions. If parameters are not somehow reduced to a common denominator,
and then aggregated, Gresham's Law is likely to operate; quantitative
factors will be considered, and nonquantitative parameters will be ignored. This means that not to use RCBA is to run a greater risk than
those resulting from using it. This is because RCBA presupposes some
way of adding commensurable effects, and because this addition is
essential to accurate decision making (Lave 1979a, pp. 181-182). Without it, there is no way to gain "perspective" on the activities in
question, no way to "boil down" vast data into something manageable
(Farmer 1979, pp. 163, 167; Jellinek 1979, p. 71).

[3]The method of revealed or observed preferences consists of experts'
formalizations of past societal policy regarding various risks. Followed by assessors such as Starr and Whipple, this method rests upon
the assumption that past behavior regarding risks, benefits, and their
costs is a valid indicator of present preferences. In other words, the
"best" risk-benefit tradeoffs are defined in terms of what has been
"traditionally acceptable", not in terms of some other (e.g., more
recent) ethical or logical justification. This, of course, involves

the assumption that past behavior is normative, whether it was good or bad, or right or wrong. For this reason, some theorists have argued that the method of "revealed preferences" is too conservative in making consistency with past behavior a sufficient condition for the correctness of current risk policy (see Fischhoff _et al_. 1978, p. 149; Starr 1979, p. 7; Okrent and Whipple 1977, p. 1).

Unlike the method of "revealed preferences", that of "expressed preferences" does not rely on past policy. Developed by assessors such as Fischhoff and Slovic, this approach consists of using questionnaires to measure the public's attitudes toward risks and benefits from various activities. The weakness of this method, of course, is that often what people say about their attitudes toward various risks appears inconsistent with how they behave toward them. Some theorists also view the method as too variable since it takes no account of past societal behavior but only relies on selected responses as to what people say they believe about risks (see Fischhoff _et al_., 1978, p. 149.

[4]Comar (1979, p. 319) and others claim, for example, that to "avoid squandering resources" we must not "reduce small risks while leaving larger ones unattended" (see Starr 1979, p. 12). They say, implicitly, that risks with greater probability ought to be attended to before smaller ones. Likewise Okrent claims that "in view of their statistically smaller contribution to societal risk, major accidents may be receiving proportionately too much emphasis compared to other sources of risk" (1980, p. 372). In making this suggestion, he too appears to propose that something should be viewed as hazardous solely on the basis of its likelihood. Gibson makes this point even more directly when he claims that there is no reason to distinguish single- versus multiple-fatality accidents, so long as the risk is of a certain level (1975, p. 599). In fact, says Rasmussen, viewing risks in this manner is a matter of "consistency in public attitudes" (US NRC 1975, p. 37). This suggests that anyone who does not value risks in such a simple manner may be accused of inconsistency (see Braybrooke and Schotch 1981, pp. 29-30).

[5]What motivates assessors to adopt the fatality interpretation of the linearity assumption and, specifically, to accept tenet (3)? Starr and Whipple, for example, claim that "historically revealed social preferences and costs are sufficiently enduring to permit their use for predictive purposes" (Starr 1969, p. 1232), and that "in such historical situations a socially acceptable and essentially optimum tradeoff of such values has been achieved" (Starr 1969, p. 1233). As Starr himself recognizes, these assumptions may be doubtful, since his method doesn't distinguish what is "best" from what is "traditionally acceptable" (1969, p. 1232). What he fails to recognize, however, is that the method cannot distinguish even what is "traditionally acceptable". Risks are obviously not acceptable just because society has encountered or taken them. Some may have been endured through ignorance, or because of political impotence, or the unavailability of alternatives. Moreover, some risks have only been discovered years after exposure to them because of the difficulties associated with data gathering.

[6]Although society's intuitive evaluations of given risks do not provide sufficient grounds for arguing that these risks _ought_ to be evaluated in a certain way, many risk assessors maintain that _correct_ societal evaluations are consistent with tenet (3) and the linearity assumption. Hence, in response to alleged counterexamples to this assumption, assessors maintain that high public aversion to certain low probability risks does _not_ provide a counterexample to the thesis that "actual" risk probabilities and the value of risk avoidance are linearly related, since the public's aversion in such cases is generated by "perceived" (i.e., incorrect) risk probabilities and not "actual" ones.

[7]For one thing, the restriction of risk to "probability of fatality" is highly questionable, since there are obviously many other "cost burdens", e.g., "decreasing the GNP by a given amount", whose probability also determines the value of avoiding a given risk. Another problem is that the fatality interpretation fails to consider the facts that benefits play a role in risk avoidance, and that the value of avoiding a given risk is rarely simply a function of the probability of fatality, but also a function of the benefits to be gained from taking the risk. A third criticism, the specific one on which I wish to focus, is that the fatality interpretation, and especially tenet (3), run counter to the way that people often evaluate risks, especially low-probability catastrophic risks. As Zeckhauser points out (1975, p. 442), once a new element of risk is announced or imposed, it gives citizens something to think and worry about. But doesn't it seem unlikely that a minor risk, say 1/10 of 1%, would generate _only_ 1/10 of the aversion or anxiety of a 1% risk of the same magnitude? Small probability risks of great possible magnitude, in fact, seem likely to be valued in a way that is not linear. If one admits that the anxiety cost is a substantial portion of the amount that one might pay to avoid a given risk, then the case of anxiety costs may well falsify the thesis of linearity. See Zivi (1979, p. 574), who argues that societal "stress" is "not proportional to the number of lives lost."

[8]The authors concluded that "there were no significant differences" between pro-nuclear and anti-nuclear groups with respect to any beliefs about accident risks (Otway 1977, p. 332). Instead, said Otway, the pro-nuclear group viewed "benefit-related attributes as most important" (p. 331), while the con group assigned highest importance to risk items (p. 331). Moreover, as the research of Fischhoff, _et al._, indicates, nuclear risks were judged by the public, in their psychometric surveys, as having "the _lowest_ fatality estimate [probability times magnitude]" for the 30 activities studied, but the _highest_ perceived risk (1980, p. 192). As a consequence, Fischhoff, _et al._ conclude: "we can reject the idea that laypeople wanted to equate risk with annual fatalities, but were inaccurate in doing so. Apparently, laypeople incorporate other considerations besides annual fatalities into their concept of risk."(Fischhoff, _et al._ 1980, p. 192).

The Fischhoff research also indicates that the nuclear opponents did not perceive the nuclear risk as worth the allegedly low benefits accruing from it (Fischhoff _et al._ 1978, p. 150). If this conclusion is correct, then much controversy might be over the issue of tradeoff, rather than over that of misperceived probabilities. In any case, it provides plausible grounds for rejecting the interpretation, either that the value of risk avoidance is linear with the actual probability of fatality, or that misperceived nuclear probabilities cause this relation to be viewed in a nonlinear way.

[9]Authors of a recent study done at the Stanford Research Institute recently admitted, for example, that analytical techniques couldn't handle probability estimates for terrorist attacks on nuclear installations. They concluded: "we must rely on expert judgment, quantified using subjective probabilities" (cited by Lovins 1977, p. 926). One key reason why probabilities are unalterably subjective is that it is impossible to be sure that a given methodology has accurately accounted for all significant possibilities (Philipson 1979, pp. 415, 419).

[10]Häfele (1975, p. 181), Okrent and Whipple (1977, p. 1-9), and Lowrance (1976, p. 95; see Rowe 1977, pp. 3, 5), for example, all postulate that the probability component, $(p_i (x_i))$ of the utility function, $U = p_i (x_i) \cdot u_i (x_i)$ is "objective," even though the u_i is subjective or evaluational (see Kasper 1980, p. 72). Starr and Whipple likewise assume that there is an objective, wholly accurate probability which is capable of being calculated, before the occurrence of all the events in question. They propose that controversy over technology is largely a matter of conflict between those who know actual risk probabilities and those who know merely perceived risk probabilities (1980, p. 1117). Apart from whether their point about the cause of controversy is true, their theory errs in ignoring the fact that all risk probabilities, even allegedly "actual" ones, involve value judgments, and that "experts' risk assessments are also susceptible to bias, particularly underestimation due to omitting important pathways to disaster." (Fischhoff _et al_. 1980, p. 211).

There are a number of different uncertainties which jeopardize the alleged objectivity of given probability estimates or calculations. Fairley summarizes the errors arising in probability estimates as those of reporting, extrapolation, speculation, definition, and theory (1975, pp. 435-36). Speculative estimates, for example, might err because of "inclusion uncertainty," i.e., uncertainty about relevant consequences or about identification of all recipients of exposure coverage, or because of "specification uncertainty," i.e., uncertainty about whether given probability pathways are mutually exclusive or exhaustive. At a more basic level, there are also parameters (e.g., the probability of sabotage) which are in principle uncertain (see Lovins 1977, pp. 925 ff.; and Kasper 1980, p. 74). In addition, there are numerous modeling errors and assumptions, often resulting from oversimplification of pathways and conseqences, as well as measurement errors and assumptions, often in obtaining exposure magnitudes (Rowe 1977, p. 122).

Perhaps in part as a consequence of these various errors in estimating probabilities, numerous scientists and risk assessors chronically

misjudge sample implications and place too much weight on samples which
are too small (Fischhoff _et_ _al_. 1975, p. 299).   They also   often fall
victim to a number of judgmental biases, such   as   the "availability
bias" (judging a probability on the basis of one's ability to imagine
or remember relevant instances) and the "anchoring bias" (judging a
probability on the basis of how well it confirms one's original judgments (anchor), apart from what new information may reveal) (Fischhoff
_et_ _al_. 1975, pp. 299-303).

References

Arrow, K.J. (1964). _Social Choice and Individual Values._ 2nd ed. New York: Wiley.

Ashley, H. _et al._ (eds.). (1976). _Energy and the Environment: A Risk-Benefit Approach._ New York: Pergamon.

Bauer, R. (ed.). (1966). _Social Indicators._ Cambridge: MIT Press.

Bazelon, D. (1979). "Risk and Responsibility." _Science_ 205: 277-280.

Braybrooke, D. and Schotch, P. (1981). "Cost-Benefit Analysis Under the Constraint of Meeting Needs." Unpublished Manuscript.

Burnham, J. (1979). "Panel: Use of Risk Assessment... ." In Mitre Corporation (1979). Pages 583-689.

Burns, S. (1977). "Congress and the Office of Technology Assessment." _George Washington Law Review_ 45: 1123-1150.

Carter, L. (1979). "Dispute Over Cancer Risk Quantification." _Science_ 203: 1324-1325.

Cohen, B. and Lee, I. (1979). "A Catalog of Risks." _Health Physics_ 36: 707-722.

Comar, C. (1979). "Risk: A Pragmatic De Minimis Approach." _Science_ 203: 319.

Committee on Public Engineering Policy. (1972). _Perspectives on Benefit-Risk Decision Making._ Washington, D.C.: National Academy of Engineering.

Fairley, W. (1975). "Criteria for Evaluating the 'Small' Probability... ." In Okrent (1975). Pages 415-452.

----------. (1979). "Panel: Accident Risk Assessment." In Mitre Corporation (1979). Pages 371-476.

Farmer, F. (1979). "Energy Risks Compared to Other Societal Risks." In Mitre Corporation (1979). Pages 157-172.

Fischhoff, B. _et al._ (1975). "Cognitive Processes and Societal Risk Taking." In Okrent (1975). Pages 291-329.

-------------- _et al._ (1978). "How Safe is Safe Enough?" _Policy Sciences_ 9: 127-152.

-------------- _et al._ (1980). "Facts and Fears." In Schwing and Albers (1980). Pages 181-216.

Gage, S. (1979). "Risk Assessment in Governmental Decision Making."
In Mitre Corporation (1979). Pages 5-14.

Gardenier, J. (1979). "Panel: Accident Risk Assessment. In Mitre
Corporation (1979). Pages 371-476.

Gibson, S. (1975). "The Use of Quantitative Risk Criteria in Hazard
Analysis." In Okrent (1975). Pages 591-608.

Gross, B. (1966). "Preface: A Historical Note on Social Indicators."
In Bauer (1966). Pages ix-xviii.

Häfele, W. (1975). "Benefit-Risk Tradeoffs in Nuclear Power
Generation." In Ashley, et al. (1976). Pages 141-184.

----------. (1979). "Energy." In Starr and Ritterbush (1979). Pages
129-140.

Hollister, H. (1979). "The DOE's Approach to Risk Assessment." In
Mitre Corporation (1979). Pages 35-58.

Jellinek, S. (1979). "Risk Assessment at the EPA." In Mitre
Corporation (1979). Pages 59-71.

Kasper, R. (1977). "Cost-Benefit Analysis in Environmental Decision-
making." George Washington Law Review 45: 1013-1024.

----------. (1979). "Panel: Use of Risk Assessment... ." In Mitre
Corporation (1979). Pages 583-689.

----------. (1980). "Perceptions of Risk and Their Effects on Decision
Making." In Schwing and Albers (1980). Pages 71-84.

Kneese, A. et al. (1979). "A Study of the Ethical Foundations of
Benefit-Cost Analysis Techniques." Unpublished Manuscript.

Lave, L. (1979a). "Discussion." In Mitre Corporation (1979). Pages
186-211.

--------. (1979b). "Panel: Perception of Risk." In Mitre Corporation
(1979). Pages 477-582.

Levine, S. (1979). "Panel: Use of Risk Assessment... ." In Mitre
Corporation (1979). Pages 583-689.

Lovins, A. (1977). "Cost-Risk-Benefit Assessment in Energy Policy."
George Washington Law Review 45: 911-943.

Lowrance, W. (1976). Of Acceptable Risk. Los Altos, CA: William
Kaufmann.

------------. (1979). "Discussion." In Mitre Corporation (1979).
Pages 152-157.

Maxey, M. (1979). Managing Low-Level Radioactive Wastes." In Watson (1979). Pages 400-419.

Mitre Corporation (1979). *Symposium/Workshop...Risk Assessent and Governmental Decision Making.* McLean, Virginia: The Mitre Corporation.

Okrent, D. (ed.). (1975). *Risk-Benefit Methodology and Application.* Los Angeles: UCLA School of Engineering and Applied Science.

---------- and Whipple, C. (1977). *Approach to Societal Risk Acceptance Criteria and Risk Management.* PB-271 264. Washington, D.C.: U.S. Department of Commerce.

----------. (1979). "Panel: Use of Risk Assessment... ." In Mitre Corporation (1979). Pages 583-689.

----------. (1980). "Comment on Societal Risk." *Science* 208: 372-375.

Otway, H. (1977). "Risk Assessment and the Social Response to Nuclear Power." *Journal of the British Nuclear Engineering Society* 16: 327-333.

Philipson, L. (1979). "Panel: Accident Risk Assessment." In Mitre Corporation (1979). Pages 371-476.

Rowe, W. (1977). *An Anatomy of Risk.* New York: Wiley.

Schwing, R. and Albers, W.A. (eds.). (1980). *Societal Risk Assessment.* New York: Plenum.

Starr, C. (1969). "Social Benefits versus Technological Risk." *Science* 165: 1232-1238.

----------. (1972). "Benefit-Cost Studies in Sociotechnical Systems." In Committee (1972). Pages 17-42.

---------- *et al.* (1976). "Philosophical Basis for Risk Analysis." *Annual Review of Energy* 1: 629-662.

----------. (1979). *Current Issues in Energy.* New York: Pergamon.

---------- and Ritterbush, P. (1979). *Science, Technology and the Human Prospect.* New York: Pergamon.

---------- and Whipple, C. (1980). "Risks of Risk Decisions." *Science* 208: 1114-1119.

US Nuclear Regulatory Commission. (1975). *Reactor Safety Study: An Assessment of Accident Risks in US Commercial Nuclear Power Plants.* WASH-1400. Washington: US Government Printing Office.

Watson, J. (ed.). (1979). <u>Low Level Radioactive Waste Management.</u>
    <u>Proceedings. Twelfth Midyear Topical Symposium.</u> Williamsburg,
    Virginia: Health Physics Society.

Whipple, C. (1979). "Panel: Public Perceptions of Risk." In Mitre
    Corporation (1979). Pages 477-582.

Yellin, J. (1977). "Judicial Review and Nuclear Power: Assessing the
    Risks of Environmental Catastrophe." <u>George Washington Law</u>
    <u>Review</u> 45: 969-993.

Zechhauser, R. (1975). "Procedures for Valuing Lives." <u>Public Policy</u>
    239: 419-464.

Zivi, S. (1979). "Panel: Public Perceptions of Risk." In Mitre
    Corporation (1979). Pages 477-582.

Part VI

Philosophy of Physics

Do Virtual Particles Exist?

Robert Weingard

Rutgers University

In a recent paper, Shrader-Frechette (1977) has argued that there
is a serious (and perhaps fatal) unclarity in the notions of particle
and fundamental particle currently used in high energy physics.
Shrader-Frechette focuses on the notion of a virtual particle as one
of the contributors to this unclarity. Among the reasons offered for
this focus are: (1) virtual particles are (in some sense) not observable, and (2) they are introduced merely to balance the "books" so
conservation laws hold.

More recently, Hendrick and Murphy (1981) have argued that Shrader-
Frechette's overall critique of the notions of particle and fundamental particle rests on a number of confusions and in this I am in
complete agreement. Here, I am just interested in their response to
Shrader-Frechette's remarks about virtual particles. In particular,
in response to (2) above they reply that virtual particles are introduced (a) as entities mediating the fundamental forces of nature, and
(b) as entities playing a calculational role in perturbative theoretical methods. With regard to (1), they remark (c) that, "The use of
virtual particles in QED...yields theoretical results in remarkable
agreement with experiment." (1981, p. 459). Here I take them to be
asserting an instance of the general point that if something cannot be
"directly" observed that doesn't mean we cannot have indirect evidence
of its existence.

Now with regard to this general point I think they are, of course,
correct. In raising the question of observability Shrader-Frechette
has simply introduced a red herring. Two simple examples of the use
of indirect evidence in particle physics should refresh us on this
point. Assuming for the sake of the discussion that we are unlikely
to find a quark left over from the big bang sitting on a niobium ball,
quarks cannot be directly observed in the sense of finding a bubble
chamber track or seeing a scintillation counterflash due to the passage of a quark. But even if quarks do not exist as free particles

---

PSA 1982, Volume 1, pp. 235-242

they can be observed indirectly, for example, in the deep inelastic scattering of an electron with a proton. We know that the proton is not a point particle and yet the electron can undergo large angle scattering as if it interacted with a point particle. This and other experiments support the picture of a proton as composed of three quarks, where the above-mentioned scattering can be pictured, to lowest order, as below, where the electron interacts electromagnetically with one of the proton's constituent quarks.

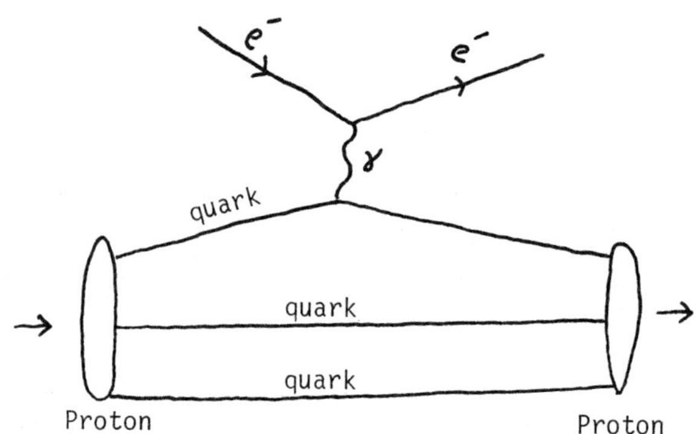

A second example concerns the property of color. Each type of quark comes in 3 colors but quarks can only combine into colorless combinations of 2 or 3 quarks so a colored particle cannot be directly observed. Shrader-Frechette says that color is "in-principle unobservable" and for this reason it is a "very problematic explanation" (1977, p. 432) of the fact that three otherwise identical quarks can form a Δ^{++} because they differ in color. But, just to mention one thing, since the existence of 3 colors triples the number of quarks the phase space available for certain decays is increased, thus implying a larger decay rate than would be expected if quarks did not differ in color.

So using indirect evidence to conclude the existence of (in some sense) not directly observable particles and properties is perfectly alright. But after giving (c) above as their reply to Shrader-Frechette's point about observability, Hendrick and Murphy go on to add in a footnote that, "The question of the ontological status of virtual particles is an important and interesting one, which warrants a more complete account than we can provide in this paper." (1981, p. 459). Does this mean, then, that the remarkable agreement of theoretical results with experiment of the use of virtual particles in Q.E.D. which they mention in (c) does not give (adequate) evidence for the existence of virtual particles (or virtual processes)? And if so, does this call in question their claim (a) above that virtual particles are introduced as entities mediating the fundamental forces of nature (in this case the electromagnetic force)--for if we don't

have (good) evidence for their existence, we don't have (good) reason
for thinking they mediate the fundamental forces.

It is these two questions I wish to comment on in this paper. And
narrowing my scope, I will consider them only with regard to electromagnetic interactions involving electrons and positions.

We will begin by looking at electron-electron scattering. In the
manner of covariant perturbation theory we picture two electrons with
4-momentum p,q in the initial state $|p,q\rangle$ at t= $-\infty$ and we want the
amplitude for a transition to a final state $|p', q'\rangle$ at t= $+\infty$, of two
electrons of 4 momentums p', q' , (suppressing spins). This will be
gotten by applying the S-matrix to the initial state. Expanding S in
powers of the electrons charge e,

$$S = S^0 + S^1 + S^2 + \cdots$$

we get for the amplitude

$$\langle p', q' | S | p, q \rangle = \langle p', q' | \sum_{i=0}^{\infty} S^i | p, q \rangle \tag{1}$$

Each of the $\langle p', q' | S^i | p, q \rangle$ is a sum, the relevant terms of which
correspond to Feynman diagrams. For example, one of the terms in
$\langle p', q' | S^2 | p, q \rangle$ corresponds to the diagram

This diagram is usually described as representing the second order
process of two electrons interacting by exchanging a (virtual) photon
(of 4 momentum k). In this context, our earlier questions reduce to
the question: Is this description literal or metaphorical? That is,
does this diagram represent a process or mechanism whereby two
electrons interact or is it merely an aid to computing the sum (1)?

However, let us first review why the above diagram is usually
described as representing the _process_ of two electrons exchanging a
photon. The reason is that the corresponding term of S^2 contains the
sequence of annihilation and creation operators, $b_{q'}^\dagger\, a_K\, b_q\, b_{p'}^\dagger\, a_K^\dagger\, b_p$.
When applied to the initial state, $|p,q\rangle$, (and ignoring the question
of time orderings) b_q and b_p annihilate it, $b_{q'}^\dagger$ and $b_{p'}^\dagger$ create the
final 2 electron state, while $a_k^\dagger$ creates a one photon state and a_k
destroys it--this being what the virtual (or _internal_) photon line in
the diagram corresponds to. But in a different case, say compton
scattering, the diagram

appears. Here the same operators a and a$^+$ are annihilating and creating initial and final photon states (<u>external</u> lines) while we get a
virtual fermion (internal fermion line). That is, the interpretation
of the annihilation and creation operators as such is independent of
whether they are connected with external or internal lines ("real" or
virtual particle states), depending only on their commutation relations, that the norm of the particle states are non-negative and that
the annihilation operators annihilate the vacuum.

In our case this means that the question of whether or not there
are virtual particles (or more precisely, whether the internal lines
of the Feynman diagrams can be "literally" interpreted) is not a
question about whether there exists a certain kind of particle that
cannot be "directly" observed (unlike the case with gluons). Rather,
it's a question about whether electrons, positrons and photons, which
can be "observed" (as a track in a bubble chamber, as put on a photographic plate) can also occur virtually.

Given what we have just said about annihilation and creation
operators, we can see why it is <u>natural</u> to describe the second order
electron-electron scattering diagram, which corresponds to a term in
the expansion (1), as involving the creation and annihilation of a
virtual photon state. Yet this does not mean we can conclude
straight-away that this is what the diagram represents. Why? The
reason I want to focus on here has to do with the fact that the expansion (1) for the transition amplitude is a superposition. Note
that as far as what is observable in our electron-electron scattering
case, we can picture it as

We have the initial state, the final state and that's all, the circle
representing whatever it is that takes place during the transition.
Now from equation (1), the probability for the transition is given by
(2)

$$\left| \langle p', q' | S | p, q \rangle \right|^2 = \left| \sum_{i=0}^{\infty} \langle p', q' | S^i | p, q \rangle \right|^2$$

(2)

This means that not only do second order diagrams contribute to the transition probability, but higher order diagrams as well, such as, for example (with suitable correction for the infrared divergence)

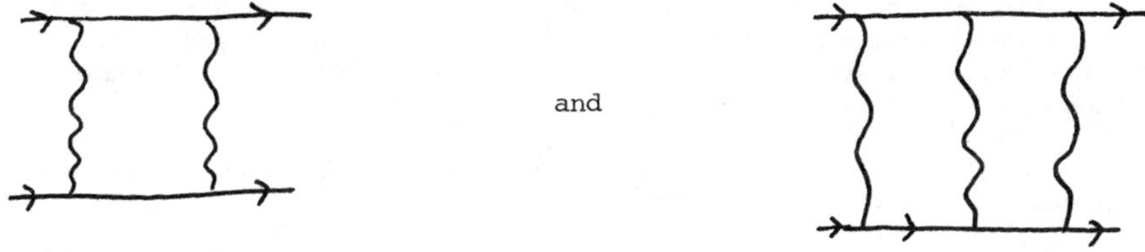

and

4th order   6th order

Even so, if the transition probability were of the form $\sum |\langle p', q'|S'|p,q\rangle|^2$, then the probability of the transition $|p,q\rangle \to |p',q'\rangle$ would be the sum of the probabilities due to the first order diagrams plus the probability due to the second order diagrams, etc. Then if we interpret the diagrams in terms of processes, the transition probability for $|p,q\rangle \to |p',q'\rangle$ would be the sum of the probability due to a first order process plus the probability due to a second order process, etc. But since the transition probability $|\langle p', q'|S|p,q\rangle|^2$ is the square of the sum of the first order, second order, etc., amplitudes, $|\sum \langle p', q'|S'|p,q\rangle|^2$ we cannot make this interpretation.[1] That is, I am here assuming a version of what I take to be the standard view of superpositions,[2] the view which asserts, in elementary quantum mechanics, that if an ensemble is described by a state $\Psi = \sum \phi_i$, where the ϕ_i are the eigenvectors of the property P , then none of the systems in the ensemble have a (sharp) value of the property P . While I think this view is standard, not everyone holds it; for example see Fine (1979) and Cartwright (1976). Here I can only say that I accept it, partly because I see no other way of making sense of non-boltzmanian statistics for identical particles.

But only part of the problem here comes from the standard view of a superposition. To see this let us go back to the second order graphs for electron-electron scattering. Earlier we looked at one graph but in fact two second order graphs contribute to the transition amplitude,

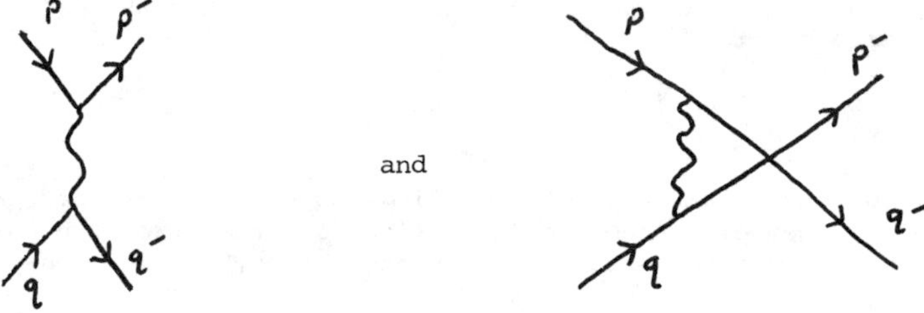

and

So we already have a superposition with the second order term. In
both graphs, however, there is just one internal photon line. Thus,
suppose these two graphs (actually their difference) were the complete
transition amplitude. Then I do not think the fact that it is a
superposition would, by itself, be a bar to interpreting the transition as taking place through the mechanism of the creation and annihilation of a single virtual photon state. This is because there being
one photon would be sharp.

   In the case of electron-position scattering, the two second order
graphs are

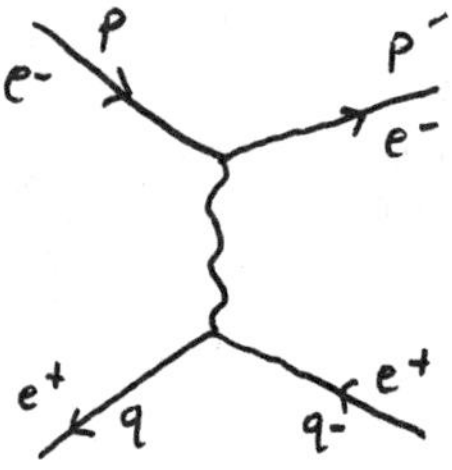

and

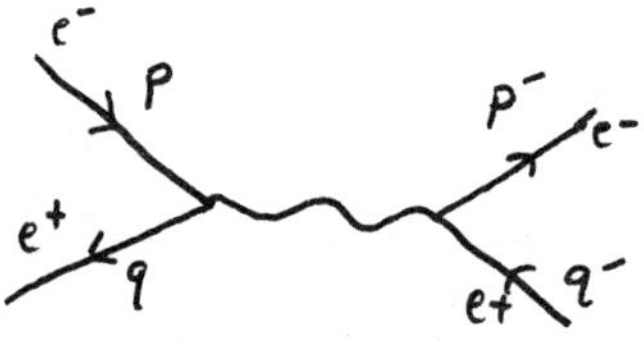

That there is a single photon is still sharp (as far as the second
order term is concerned) although it is perhaps not as easy to think
of the two diagrams (intuitively) as representing a (single) mechanism
for the interaction.[3] In any case, however, when we turn to the
complete transition amplitude for a process, we see that it is a sum
over all relevant diagrams of <u>all</u> orders. According to the standard
view <u>neither</u> <u>the</u> <u>number</u> <u>nor</u> <u>kinds</u> of "virtual" particles will be
sharp. Given this result, it is hard to see how the diagrams in any
literal sense, can be thought of as picturing or representing the
mechanism by means of which transitions such as electron-electron
scattering take place.

<u>Notes</u>

[1]For simplicity, I am ignoring that the transition should actually
be pictured $|P,q\rangle \to S|P,q\rangle$ and then a measurement determines the final
state, since the measurement does not determine that the process took
place via a particular diagram.

[2]The superposition is the sum of those terms of $S|P,q\rangle$ whose inner
product with $|P^-,q^-\rangle$ is nonzero.

[3]Indeed, the single second order diagrams of covariant perturbation
theory are themselves combinations of two diagrams from the point of
view of time ordered perturbation theory. Thus, in electron-positron
annihilation the Feynman diagram

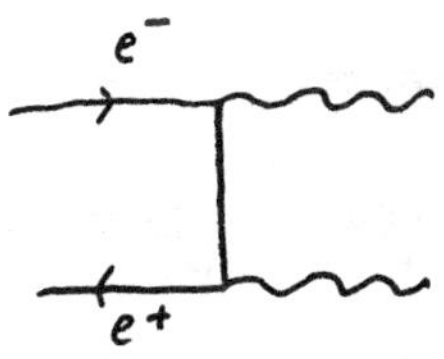

is really a combination of the two time ordered diagrams

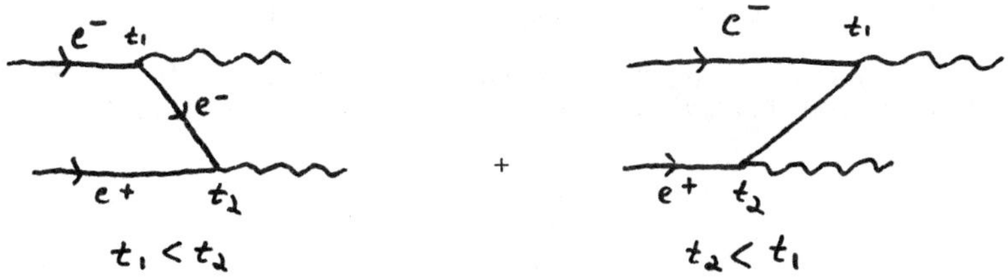

+

In the Feynman diagram, the internal line corresponds to the fermion propagator which combines the electron and positron terms of the time ordered expressions. If we can regard positrons as electrons going back in time then the internal line in the above Feynman diagram would represent a single virtual fermion on a literal interpretation.

242

References

Cartwright, N. (1976). "Superposition and Macroscopic Observation."
In <u>Logic and Probability in Quantum Mechanics.</u> Edited by P.
Suppes. Dordrecht: Reidel. Pages 221-234.

Fine, A. (1979). "How to Count Frequencies: A Primer for Quantum
Realists." <u>Synthese</u> 42: 145-154.

Hendrick, R.E. and Murphy, A. (1981). "Atomism and the Illusion of
Crises: The Danger of Applying Kuhnian Categories to <u>Current</u>
Particle Physics." <u>Philosophy of Science</u> 48: 454-468.

Shrader-Frechette, K. (1977). "Atomism in Crises: An Analyses of the
Current High Energy Paradigm." <u>Philosophy of Science</u> 44: 409-440.

The Logic of Experimental Questions

R. I. G. Hughes

Yale University

1. Introduction

   The experimental procedures of physics assign values to physical
quantities; we measure the temperature of a gas, the luminosity of a
light source, the momentum of a particle. Any pair, $q = (A,\Delta)$,
with A an observable quantity, Δ a Borel subset of the reals, we will
call an experimental question, though some prefer the term proposition.
For example, in lowering my big toe into the bath, I address the experimental question (T,Δ) to the system consisting of my bathwater,
where T is the temperature measured in °F, and Δ is (approximately) the
interval (110,120). Theory tells us what answers we may expect, given
certain initial conditions. A determinist theory like classical mechanics tells us that, under specified conditions, the measurement of a
quantity will yield a particular result with certainty, whereas a statistical theory like quantum mechanics tells us what probability at-
taches to various possible outcomes of a measurement. Thus, given a
full specification of the state of a classical system, every experimental question (A,Δ) will be assigned either 1 or 0; the theory predicts unequivocally whether a measurement of A will or will not yield a
result within Δ; in quantum theory, on the other hand, the state assigns a probability, i.e., a value in the interval [0,1], to each experimental question. In each case we can think of the state of a
system as a function, $w:Q \to [0,1]$, where Q is the set of experimental
questions associated with the system. What state a system is in will
depend on how it has been prepared, that is, on its history.

   Experimental questions are given different mathematical representations in different theories; accordingly, each theory attributes a
specific algebraic structure to the set Q. In the theory of classical
systems, each experimental question corresponds to a subset of a phase
space which is real and finitely dimensional; Q is a field of sets and
has the structure of a Boolean algebra (equivalently, of a complemented
distributive lattice).[1] However, the phase space associated with a

---

PSA 1982, Volume 1, pp. 243-256
Copyright © 1982 by the Philosophy of Science Association

quantum system is a Hilbert space; within it subspaces, rather than
subsets, correspond to experimental questions.[2] The set of subspaces
is ordered by inclusion; given any pair, L, M, of subspaces, there is
a largest subspace included in each of them and a least subspace
which includes them both. Thus in this case too Q forms a lattice.
However, this lattice, as is well known, is not distributive: we may
well have

$$q_1 \lor (q_2 \land q_3) \neq (q_1 \lor q_2) \land (q_1 \lor q_3).$$

Discussions of quantum logic start from an investigation of the algebraic structure of Q within quantum mechanics. For some commentators
this is also where they should end; be that as it may, the account so
far prompts the following question. Are there a priori constraints
which the algebraic structure of the set of experimental questions
dealt with by a physical theory must satisfy?

Notice that there is no reference to any particular physical theory
involved. Of course, a physicist will be led to ask specific questions, i.e., to perform some experiments rather than others, on the
basis of particular theoretical considerations, but, as far as the
present enterprise is concerned, these considerations are irrelevant.
The question is, independently of any theory involved, what can we say
about the set of experimental questions?

A number of writers, among them Finkelstein (1969), Jauch (1968)
and Piron (1972), have suggested that, whatever our physical theory,
the set Q must have the structure of an orthocomplemented lattice. As
they point out, classical mechanics and quantum theory both conform
to this requirement; the lattice for classical mechanics is characterized by the additional assumption of distributivity, and that for
quantum theory by the weaker assumption of weak modularity. However,
on the basis of some straightforward considerations, their accounts
can be seen to be seriously flawed. A critique of their proposals
will lead me to a rejection of the lattice-theoretic approach and to
the endorsement of an alternative.

2. The Ordering of Experimental Questions

As I have said, each experimental question is thought of as an
ordered pair: $q = (A, \Delta)$; its first member A is an observable, a
physical quantity. But at the pre-theoretic stage we can only think
of a physical quantity as that-which-is-measured-by-such-and-such-an-
experiment; each question involves a reference to a specific experimental procedure. I should perhaps emphasize that this doesn't imply
acceptance of Bridgman's operationalist doctrines (see Bridgman
1927). The present analysis does not deny that, within a given
theory, the role of an observable is inadequately characterized by a
description of how it may be measured, but rather looks for constraints
to which, in any theory, the set of all observables must conform.
Prior to a specific theoretic commitment different experimental procedures must be thought of as measuring different physical quantities.
Of course, if two questions turn out to be equivalent, we can think of

them as measuring the same thing, but before we do so we need an
adequate definition of <u>experimental equivalence</u>, and this, it trans-
pires, is not as straightforward as one might think.

I will consider in turn four alternative ways in which a relation
$\leqslant$ might be defined on Q. If this relation can be shown to be transitive and reflexive under one of the definitions, then we may use it
to define an equivalence relation in an obvious way, by writing,

$$q_1 \sim q_2 \quad \text{iff} \quad q_1 \leqslant q_2 \quad \text{and} \quad q_2 \leqslant q_1 .$$

The relation $\leqslant$ will then be a partial ordering on Q (or, more strictly,
on the set $Q/\sim$ of equivalence classes of Q). Defining a partial ordering on Q is a prerequisite for any approach which seeks to show
that Q is a lattice.

Some additional notation will be useful. To any experimental question the answer can be 'Yes' or 'No'. We write '$v(q) = 1$' for 'The
experimental question q receives answer, "Yes"', and '$v(q) = 0$' for
'The experimental question q receives answer, "No"'. Note that a
single experimental result only is involved here, when we assign
values to v. The same experiment can, of course, answer many experimental questions: if I determine that the temperature T of my bathwater lies within Δ, I simultaneously determine that it lies within
any set Γ which includes Δ, and that it does not lie in any set Γ' disjoint from Δ. However, unless I determine the exact value of <u>every</u>
observable for a system at a stroke, v can only be a partial function
from Q to $\{0,1\}$; this point I will return to later.

Here then are the four alternatives.

(1) $q_1 \leqslant q_2$ iff   whenever q_2 is asked of a system
after q_1, if q_1 receives answer
'Yes', then so does q_2.

(2) (Jauch) $q_1 \leqslant q_2$ iff   whenever $v(q_1) = 1$, then
$v(q_2) = 1$.

(3) (Piron) $q_1 \leqslant q_2$ iff   for all states w, $w(q_1) = 1$
implies $w(q_2) = 1$.

(4) (Maczyński) $q_1 \leqslant q_2$ iff   for all states w, $w(q_1) \leqslant w(q_2)$.

3. Alternative (1).

Alternative (1) is not canvassed in the literature. I raise it only
to dismiss it, but as I do so some important points will emerge. The
alternative assumes that every experiment asking a question q is a
filter; that is to say, that systems for which $v(q) = 1$ are separated
out by the experimental procedure so that further experimental tests
may be made on them. To be absolutely precise, it tells us that only

experimental questions asked by filters can be members of the domain
(left field) of $\leqslant$. Note that, we cannot at this stage just demand
that every experiment be _equivalent_ to a filter, since we have no independent way of establishing equivalence. The alternative rules out,
therefore, any experiment in which the system under interrogation is
no longer an independent system after the experiment is performed; it
rules out, for instance, experiments in which electrons or photons
strike a photographic plate or a photomultiplier. Thus it seems to
be unduly restrictive.

Further, it blurs the important distinction between measurement and
preparation: if we try to establish that $q_1 \leqslant q_2$, then the system
of which q_2 is asked has been prepared by the apparatus which asks q_1.
The state of the system after q_1 has been asked will not, in general,
be the same as it was before; there is not even any guarantee that
after the system has passed q_1 once it will do so a second time. Von
Neumann found it necessary to add to quantum mechanics a special postulate, the _projection_ postulate, to the effect that, if an experiment
is performed repeating a measurement which has been made immediately
before it by another, then the second will yield the same result as
the first. By doing so he emphasized that there is no general reason,
independent of particular theories, why this should be the case.[3] In
other words, to adopt alternative (1) would mean that we could not know
a priori that $q \leqslant q$, and hence could not claim that $\leqslant$ was necessarily
a partial ordering on Q.

4. Jauch

(2) $\qquad q_1 \leqslant q_2 \qquad$ iff $\qquad$ whenever $v(q_1) = 1$, then $v(q_2) = 1$.

On this definition $\leqslant$ is transitive and reflexive; thus _prima_ _facie_
it could be used to define an equivalence relation, and hence a partial
ordering of experimental questions. Jauch (1968) sought to show that
Q is a complemented lattice with respect to this ordering.

He defines (p. 76) the _complement_ of any question q as the question $q^{\perp}$, asked with exactly the same apparatus as q, such that
$v(q^{\perp}) = 0$ iff $v(q) = 1$ and $v(q^{\perp}) = 1$ iff $v(q) = 0$. We see that:

(2a) $\qquad (q^{\perp})^{\perp} = q;$

(2b) $\qquad q_1 \leqslant q_2 \qquad$ implies $\qquad q_2^{\perp} \leqslant q_1^{\perp}.$

The existence of the _greatest_ _lower_ _bound_ (infimum) of any set of
experimental questions is asserted by an axiom (p. 74). Writing
'$q_1 \wedge q_2$' for 'inf $\{q_1,q_2\}$', we require that, for any pair of questions
q_1, q_2,

(2c) $\qquad q_1 \wedge q_2 \leqslant q_1 \qquad\qquad\qquad q_1 \wedge q_2 \leqslant q_2$

and, for any q,

(2d) $\qquad q \leqslant q_1$ and $q \leqslant q_2$ together imply $q \leqslant q_1 \wedge q_2$.

Now q_1 and q_2 may be asked with radically different kinds of apparatus; it may, in fact, be impossible to ask them simultaneously. Jauch sees this problem and suggests the following solution. (Note that Jauch's 'propositions' are our 'experimental questions'.)

> Instead of making one pair of measurements, a and b, we construct a filter for the proposition a∧b by using a sequence of alternating pairs of filters for the propositions a and b respectively. ... The proposition a∧b is then true if the system passes this filter, and it is not true otherwise. Although infinite processes are, of course, not possible in actual physical measurements, the construction can be used as a base for an approximate determination of the proposition to any needed degree of accuracy. (p. 75)

De Morgan's Law is then used to define the operation $\vee$ (supremum) on Q:

(2e) $\qquad q_1 \vee q_2 =_{df} (q_1^{\perp} \wedge q_2^{\perp})^{\perp}$.

On Jauch's account, the absurd question F is the question which always receives the answer 'No'. He claims that, for all q,

(2f) $\qquad q \wedge q^{\perp} = F$.

This, with (2a) and (2b) would guarantee that $\perp$ is an orthocomplementation operation on Q.

If the structure $\langle Q, \wedge, \vee, \perp \rangle$ is to be a lattice, Jauch must show us that his construction of the question $q_1 \wedge q_2$ does in fact yield the infimum of $\{q_1, q_2\}$ with respect to $\leqslant$. But it is here where the most obvious problems in Jauch's approach arise.

First of all, it is assumed that to each experimental question there corresponds a filter. We found this objectionable when we considered alternative (1), and the same objections apply here. In particular, Jauch, who elsewhere is at pains to distinguish preparation procedures from measurements, ignores the distinction here. For let us assume that a system passes the filter for $q_1 \wedge q_2$ set up as described. Then _a fortiori_ we know that it has passed the filter for q_1; however, it has passed the filter for q_2 only _after_ emerging from that for q_1. It is only by virtue of quite a strong assumption that we can write

(2c) $\qquad q_1 \wedge q_2 \leqslant q_2$.

There is also an inevitable asymmetry in the construction of $q_1 \wedge q_2$. Although (conceptually at least) we can extend the chain of experiments to infinity, there must be a first member of the chain, either q_1 or q_2. Thus we cannot guarantee that $q_1 \wedge q_2 = q_2 \wedge q_1$, as lattice theory requires.

One could perhaps sidestep these objections by restricting the discussion to theories in which the projection postulate (or some version thereof) holds. But even then the most that could be shown a priori would be that the experiment described was a lower bound for $\{q_1, q_2\}$, i.e., that (2c) held. There is a large assumption made in claiming that (2d) also holds, and that the experiment represents their greatest lower bound with respect to $\leqslant$.

A more general criticism is that Jauch's definition of $\leqslant$ effectively restricts that relation to pairs of questions asked using the same apparatus. Thus it is hard to see what else would figure in an equivalence class of questions containing q apart from q itself. Yet, strictly, it is these equivalence classes that are the elements of Jauch's lattice (p. 74). To see why this restriction exists, consider how we might try to establish whether $q_1 \leqslant q_2$ when the apparatus to ask q_2 differs from that used to ask q_1. If we find that $v(q_1) = 1$, there are two approaches available to us: we can either test then for q_2, or we can wonder what the result of asking q_2 would have been. Neither seems very promising. If we choose the first, we are adopting alternative (1) rather than the alternative Jauch proposes. If we choose the second, we move outside the realm of practical enquiry; we can only guarantee that a system in a certain state would have passed q_2 if all the systems in that state pass q_2. To take the second approach, then, compels us to move to alternative (3). It's worth noting that, in his later work with Piron, this is just what Jauch did.[4]

5. Piron

$$(3) \qquad q_1 \leqslant q_2 \qquad \text{iff} \qquad \text{for all states w,} \quad w(q_1) = 1$$
$$\text{implies} \quad w(q_2) = 1 \, .$$

This is the definition favored by Finkelstein and Piron. (See in particular Piron 1972 .) Clearly the definition makes $\leqslant$ reflexive and transitive. It is also possible to define the infimum of a set of questions with respect to $\leqslant$ (thus defined) which avoids the problems which plagued Jauch's approach. Given two questions, q_1 and q_2, Piron defines $q_1 \cdot q_2$ as the question which consists of selecting one of the two, q_1 or q_2, at random (e.g., by the flip of a coin), and asking whichever is selected. We may agree with him that, in terms of his definition of $\leqslant$,

$$(3c) \qquad q_1 \cdot q_2 \leqslant q_1 \qquad\qquad q_1 \cdot q_2 \leqslant q_2$$

and that, for any q,

$$(3d) \qquad q \leqslant q_1 \text{ and } q \leqslant q_2 \text{ together imply } q \leqslant q_1 \cdot q_2 \, .$$

Thus $\qquad q_1 \cdot q_2 = \inf \{q_1, q_2\}.$

Like Jauch, Piron defines the question $q^{\perp}$ as the question which receives the answer 'Yes' just when q receives the answer 'No', and conversely, and so obtains

(3a) $\qquad\qquad (q^{\perp})^{\perp} = q$

(3b) $\qquad\qquad q_1 \leqslant q_2 \quad$ implies $\quad q_2^{\perp} \leqslant q_1^{\perp}$.

He also introduces a trivial question T, such that, for all states w, $w(T) = 1$. Clearly, for all q,

(3g) $\qquad\qquad q \leqslant T$.

We may also define, though he does not, the absurd question F :

(3h) $\qquad\qquad F =_{df} T^{\perp}$.

But after this, the analysis founders. Piron uses de Morgan's Law to define the connective $\vee$, so that[5]

(3e) $\qquad\qquad q_1 \vee q_2 =_{df} (q_1^{\perp} \cdot q_2^{\perp})^{\perp}$.

Now consider the truth conditions for $q_1 \vee q_2$. We have,

$$w(q_1 \vee q_2) = 1 \quad \text{iff} \quad w(q_1^{\perp} \cdot q_2^{\perp})^{\perp} = 1$$

$$\text{iff} \quad w(q_1^{\perp} \cdot q_2^{\perp}) = 0$$

$$\text{iff} \quad w(q_1^{\perp}) = 0 \text{ and } w(q_2^{\perp}) = 0$$

$$\text{iff} \quad w(q_1) = 1 \text{ and } w(q_2) = 1$$

$$\text{iff} \quad w(q_1 \cdot q_2) = 1 \ .$$

This is an undesirable result. It is forced on us by Piron's analysis, since $w(q_1^{\perp} \cdot q_2^{\perp}) = 0$ just when the experiment $q_1^{\perp} \cdot q_2^{\perp}$, i.e., the experiment which consists in selecting one of $q_1^{\perp}$ and $q_2^{\perp}$ at random, never yields the answer 'Yes'; in other words, just when neither $q_1^{\perp}$ nor $q_2^{\perp}$ ever yields the answer 'Yes'.

Secondly, let us take the apparatus which allows us to ask the question q and perform a series of experiments using it on similarly prepared systems; on each occasion the labels 'Yes' and 'No' are to be assigned to the two possible outcomes of the experiment by some random procedure. Each such experiment is the question $q \cdot q^{\perp}$. Barring cosmic accident or divine intervention, $0 \leqslant w(q \cdot q^{\perp}) \leqslant 1$. But then, since $w(F) = 0$, $q \cdot q^{\perp}$ is not equivalent to the absurd question.

Piron, it is true, explicitly introduces neither the truth value 0 nor the absurd question; none the less he is left with the equally unpalatable result that $T \not\leqslant q \cdot q^{\perp}$.

We may put the matter in terms of truth-functional logic. Effectively, Piron deals with a three-valued logic with diametrical negation; that is, the truth table for $\perp$ is the following (Table A). He also introduces a connective . with the truth value matrix shown in Table B.

Table A:

$\perp$	
1	0
x	x
0	1

Table B:

.	1	x	0
1	1	x	x
x	x	x	x
0	x	x	0

Given these truth functions, it is trivial to show that, for all valuations w, (i) $w(q.q^{\perp}) = x$, and (ii) $w(q_1^{\perp}.q_2^{\perp})^{\perp} = w(q_1.q_2)$.

The proffered structure is thus not the complemented lattice we have been led to expect. Of course, it remains true that $\langle Q, \leqslant , . \rangle$ is a semilattice.

6. Maczyński

(4) $q_1 \leqslant q_2$ iff   for all states w, $w(q_1) \leqslant w(q_2)$.

This relation is a more general version of Piron's; it was suggested by Mackey (1963) and is used by him and by Maczyński (1967) to define an equivalence relation on Q. We can write:

$$q_1 = q_2 \quad \text{iff} \quad \text{for all states w,} \quad w(q_1) = w(q_2) .$$

By using the equality sign "=" to denote equivalence we obviate the need to talk in terms of equivalence classes; in this way $\leqslant$ becomes a partial ordering of Q.

However, there is no obvious way, and probably no way at all, to define a question which is the infimum of a pair of arbitrarily chosen questions with respect to $\leqslant$. Nor does Maczynski try. On his account Q will be an orthomodular poset (which I define below) rather than a lattice.

Following a route parallel to his, we define a complementation operation on Q as Jauch and Piron did.[6] Thus, for all w,

(4j) $w(q^{\perp}) = 1 - w(q)$.

(In this equation and in those that follow, universal quantification over Q is left tacit.)

An orthogonality relation on Q is then defined as follows.

(4k) $q_1 \perp q_2$ iff $q_1 \leqslant q_2^{\perp}$.

Note that (i) $q_1 \perp q_2$ iff $q_2 \perp q_1$,

and that   (ii) $q \perp q$ iff $q = F$, where, as before, F is the absurd question, such that, for all w, $w(F) = 0$.

Also   (iii) $q_1 \perp q_2$ implies that, for all w, $w(q_1) + w(q_2) \leqslant 1$.

Maczyński puts forward the following postulate:

(P1)   If $\{q_i\}$ is a set of pairwise orthogonal questions ($q_i \perp q_j$ if $i \neq j$), then there exists a question r such that

      (i)   for all states w, $w(r) = \Sigma_i w(q_i)$ and

      (ii) $r = \sup \{q_i\}$.

He shows,[7] as a consequence of this postulate, that the ordering is <u>weakly</u> <u>modular</u>, that is:

If $q_1 \leqslant q_2$, then there is a q_3 orthogonal to q_1 such that $q_1 \vee q_3$ exists and $q_1 \vee q_3 = q_2$.

We may also express this modularity condition in terms of compatibility. Two questions, q_1, q_2, are said to be <u>compatible</u> ($q_1 \$ q_2$) if there exist pairwise orthogonal questions, r_0, r_1, r_2 , such that

$$q_1 = r_0 \vee r_1 \qquad\qquad q_2 = r_0 \vee r_2 .$$

The ordering $\leqslant$ is weakly modular provided that $q_1 \leqslant q_2$ implies $q_1 \$ q_2$.

The postulate P1 is thus sufficient to guarantee that $\langle Q, \leqslant, \perp, T, F \rangle$ is an orthomodular poset, defined as follows.

      $\langle Q, \leqslant, \perp, T, F \rangle$ is an <u>orthomodular</u> <u>poset</u> if

      (i)   Q is partially ordered by $\leqslant$;

      (ii)   with respect to $\leqslant$, Q has a maximum element T and a minimum element F ;

      (iii)   a complementation operation, $\perp$, is defined on Q satisfying   (a) $(q^\perp)^\perp = q$

                  (b) $q_1 \leqslant q_2$ implies $q_2^\perp \leqslant q_1^\perp$

                  (c) $q \wedge q^\perp = F$;

      (iv) $\langle Q, \leqslant, \perp, T, F \rangle$ is weakly modular .

I will not here rehearse or comment on Maczyński's arguments for (P1), save to say that they motivate it (Czelakowski's expression) rather than compel assent. His enterprise is slightly different from those we have considered; it is to produce a set of algebraic axioms for the set Q of experimental questions of quantum mechanics, and then to argue for the plausibility of each, rather than to discuss physical theories in general. However, in view of the conclusions I come to in the next section, I would like to point out the striking effect of adding the further postulate (P2), as Czelakowski (1974) has suggested.

(P2)   If $q_1 \$ q_2$, $q_2 \$ q_3$, $q_3 \$ q_1$, then $q_1 \vee q_2 \$ q_3$.

A theorem of Gudder's (1972) tells us that, if Q is an orthomodular poset for which (P2) holds, then Q is isomorphic to a transitive partial Boolean algebra. Of these structures more anon.

7. Partial Structures

I want now to develop a different approach, one that recognizes that a given experiment may yield a number of possible results, in other words, that it may ask a number of experimental questions simultaneously. Let us divide experiments into two kinds. Those of the first kind yield outcomes lying on the continuum of real numbers, or within a small range of the reals. (Typically we write down, "i = 2.21 $\pm$ 0.02 amps " as a measurement of current.) The ranges involved usually overlap. (2.21 $\pm$ 0.02 overlaps with 2.20 $\pm$ 0.02.) Experiments of the second kind yield mutually exclusive outcomes; the yes-no experiments we have been considering are the simplest examples, but the outcomes need not be restricted to two. We could, I suppose, call the two types of experiment analogue and digital respectively. But for any given experiment, of whichever type, we have a set $\{x_i\}$ of possible <u>outcomes</u>. If we close this set under (infinitary versions of) the classical logical operations of conjunction ($\wedge$), disjunction ($\vee$) and complementation ($\perp$), we get a Boolean (σ-) algebra of <u>events</u>. Each event can be thought of as an experimental question. The disjunction of all possible outcomes (the certain event) is the trivial question T, and their conjunction (the impossible event) is the absurd question F .

For any experiment in our manual we can obtain such an algebra, and each state then throws a probability measure onto its elements obeying the usual Kolmogorov axioms. Further, the maxima (the trivial questions) of all these algebras coincide, as do the minima (the absurd questions). The set Q of experimental questions, then, consists of a set of Boolean algebras "pasted together", as it were, at top and bottom, to form what Hardegree and Frazer (1981) call a <u>Boolean atlas</u>. Let us call it $\mathfrak{M}$. On $\mathfrak{M}$, $\vee$ and $\wedge$ are partial operations, defined just for pairs of elements which lie within the same Boolean subalgebra, and each of these Boolean subalgebras contains just the set of questions asked by a particular apparatus. Note also that the ordering on $\mathfrak{M}$ (this is the usual ordering within each subalgebra, defined by: $a \leqslant b$ iff $a = a \wedge b$) is the restriction to pairs of elements of Q lying within a common subalgebra of alternative (4) among those discussed. Now we may extend this ordering by lifting this restriction, so that, as before, for <u>any</u> q_1, q_2 in Q,

(4) $q_1 \leqslant q_2$ iff   for all states w, $w(q_1) \leqslant w(q_2)$.

In terms of this ordering questions occurring in different subalgebras may be equivalent: both T and F are such questions, but for any q_1, q_2 we may say that

$$q_1 \text{ is equivalent to } q_2 \ (q_1 \sim q_2) \text{ iff for all states } w, \ w(q_1) = w(q_2).$$

If q_1 and q_2 are equivalent, it is tempting to regard them as asking the same experimental question. In fact, this has been done in the previous sections without vitiating what was being said. But, as we learn from quantum mechanics, this may be imprecise: q_1 and q_2 may ask questions involving different observables while still being equivalent in the sense described. (This occurs in quantum theory when two distinct Hermitian operators share an eigenvector.)

But consider the case when we have a set $\{q_{a_i}\}$ of questions in the Boolean subalgebra $\mathcal{A}$ of $\mathcal{M}$ and a set $\{q_{b_i}\}$ of questions in the Boolean subalgebra $\mathcal{B}$ of $\mathcal{M}$, and

$$q_{a_1} \sim q_{b_1} \ ; \ q_{a_2} \sim q_{b_2} \ ; \ q_{a_3} \sim q_{b_3} \ \cdots .$$

Now let p_a be a Boolean polynomial which uses only elements of $\{q_{a_i}\}$ and p_b be the corresponding polynomial involving elements of $\{q_{b_i}\}$. For example, if $p_a = q_{a_1} \vee (q_{a_2} \wedge q_{a_3})^\perp$, then $p_b = q_{b_1} \vee (q_{b_2} \wedge q_{b_3})^\perp$. Can we show that $p_a \sim p_b$?

From this result it would follow that the algebra of equivalence classes of experimental questions (the algebra $\mathcal{M}/\sim$) was a Boolean manifold. (Again, I use Hardegree and Frazer's terminology.) That is, that $\mathcal{M}/\sim$ was a family $\{\mathcal{B}_i : i \in I\}$ of Boolean algebras, such that $\mathcal{B}_i = \langle B_i, \vee_i, \wedge_i, {}^{\perp i}, T_i, F_i \rangle$ such that, for all $i, j \in I$,

(a) $T_i = T_j$ $\qquad\qquad\qquad$ $F_i = F_j$;

(b) there is a k such that $B_i \cap B_j = B_k$;

(c) if $a, b \in B_i \cap B_j$, then

$$a \vee_i b = a \vee_j b \ , \ a \wedge_i b = a \wedge_j b \ , \ a^{\perp i} = a^{\perp j} .$$

Alas, the result in question does not, in general, hold. It certainly does not hold, for instance, if there is only one state, w, in which systems can be prepared. For consider the case when the system is a (fair) die and $q_{a_1}, q_{a_2}, q_{b_1}, q_{b_2}, p_a, p_b$ are the following questions:

q_{a_i} : die shows 6;

q_{a_2} : die shows odd number;

q_{b_1} : die shows 6;

$$q_{b_2} \quad : \quad \text{die shows even number;}$$

$$p_a \quad : \quad q_{a_1} \wedge q_{a_2};$$

$$p_b \quad : \quad q_{b_1} \wedge q_{b_2}.$$

It is an open question what requirements the set of states must satisfy for the required result to hold.

A further postulate yet would be required to show that structure is a partial Boolean algebra, as Kochen and Specker (1965) define the term, i.e., that the further condition is met that, if there are $\mathcal{B}_i$, $\mathcal{B}_j$, $\mathcal{B}_k$ such that $a,b \in \mathcal{B}_i$, $b,c \in \mathcal{B}_j$, $c,a \in \mathcal{B}_k$, then there is a $\mathcal{B}_m$ such that $a,b,c \in \mathcal{B}_m$. An additional postulate which guaranteed that $q_1 \leqslant q_2$ only if there was a $\mathcal{B}_i$ in the family of which the equivalence classes containing q_1 and q_2 were both members would ensure that the partial Boolean algebra was transitive.

Thus the answer to the question we started with may seem disappointingly weak. The structure which Q must possess is that of a Boolean atlas, of which the Boolean algebras of classical mechanics and the transitive partial Boolean algebras of quantum theory are special cases.

Notes

[1] In this paper I assume an elementary knowledge of Boolean algebra and lattice theory. For a slightly fuller account of the classical situation, see Birkhoff and von Neumann (1936).

[2] For more on this correspondence see, e.g., Hughes (1980).

[3] Whether or not he was justified in adding the postulate in the case of quantum theory is moot. See, for instance, Bub (1977) and van Fraassen (1974), especially p. 297.

[4] See, for instance, Jauch and Piron (1969). The discussion on p. 93 of Foundations of Quantum Mechanics makes it quite clear that when Jauch wrote that work he was thinking in terms of alternative (2). Using the taxonomy of quantum logical interpretations suggested by Hughes (1980) we can describe the move as one from a logic of M-statements to a logic of MC-statements.

[5] Actually, by the time he introduces $\vee$, Piron has moved to equivalence classes of questions; my simplification of his presentation doesn't affect the points at issue.

[6] Maczyński actually works in terms of the internal structure of questions, writing, for instance, $(A, \Delta) = (A, R-\Delta)$.

[7] This result may seem surprisingly strong; Maczyński obtains it by an analysis involving the internal structure of questions (see note 6).

References

Birkhoff, G. and von Neumann, J. (1936). "The Logic of Quantum Mechanics." <u>Annals of Mathematics</u> 37: 823-843.

Bridgman, P.W. (1927). <u>The Logic of Modern Physics.</u> New York: Macmillan.

Bub, J. (1977). "Von Neumann's Projection Postulate as a Probability Conditionalization Rule in Quantum Mechanics." <u>Journal of Philosophical Logic</u> 6: 381-390.

Czelakowski, J. (1974). "Logics Based on Partial Boolean σ-Algebras (1)." <u>Studia Logica</u> 34: 371-396.

Finkelstein, D. (1969). "Matter, Space and Logic." In <u>Proceedings of the Boston Colloquium for the Philosophy of Science.</u> (<u>Boston Studies in the Philosophy of Science.</u> Volume V.) Edited by R.S. Cohen and M.W. Wartofsky. Dordrecht: Reidel. Pages 199-215.

Gudder, S.P. (1972). "Partial Algebraic Structures Associated with Orthomodular Posets." <u>Pacific Journal of Mathematics</u> 41: 717-730.

Hardegree, G.M. and Frazer, P.J. (1981). "Charting the Labyrinth of Quantum Logics." <u>Current Issues in Quantum Logic.</u> Edited by E. Beltrametti and B.C. van Fraassen. London: Plenum Press. Pages 53-76.

Hughes, R.I.G. (1980). "Quantum Logic and the Interpretation of Quantum Mechanics." In <u>PSA 1980.</u> Volume One. Edited by P.D. Asquith and R.N. Giere. East Lansing, Michigan: Philosophy of Science Association. Pages 55-67.

Jauch, J.M. (1968). <u>Foundations of Quantum Mechanics.</u> Reading, Mass.: Addison Wesley.

———————— and Piron, C. (1969). "On the Structure of Quantal Propositional Systems." <u>Helvetica Physica Acta</u> 43: 842-848.

Kochen, S. and Specker, E. (1965). "Logical Structures Arising in Quantum Theory." <u>The Theory of Models.</u> (<u>Proceedings of the 1963 International Symposium at Berkeley.</u>) Edited by J.W. Addison, <u>et al.</u> Amsterdam: North Holland. Pages 177-189.

Mackey, G. (1963). <u>The Mathematical Foundations of Quantum Mechanics.</u> New York: Benjamin.

Maczyński, M.J. (1967). "A Remark on Mackey's Axiom System for Quantum Mechanics." <u>Bulletin de l'Académie Polonaise des Sciences,</u> Série des Sciences Mathématiques, Astronomiques et Physiques XV: 583-587.

Piron, C. (1972). "Survey of General Quantum Physics." <u>Foundations of Physics</u> 2: 287-314.

————. (1977). "On the Logic of Quantum Logic." <u>Journal of Philosophical Logic</u> 6: 481-484.

van Fraassen, B.C. (1974). "The Einstein-Podolsky-Rosen Paradox." <u>Synthese</u> 29: 291-309.

The Status and Meaning of the Laws of Inertia

Robert Alan Coleman and Herbert Korte

Department of Mathematics & Statistics
Department of Philosophy
University of Regina
Regina S4S 0A2

1. Introduction

A great deal of literature on the status and meaning of the Laws of
Inertia in spacetime theories has nurtured and given wide currency to
the claim that the laws are conventional in character, that they are
definitions, or circular and without empirical content.

Philosophers who argue for the conventional character of the laws,
do so, either emphasizing epistemological or ontological considerations concerning the structure of spacetime.

Those who argue for their conventional character mainly on epistemic
grounds, point out, that the laws do not supply independent criteria of
what is to count as <u>force-free</u> or <u>natural motion</u>. The only way of
knowing when no forces act on a body is that it moves as a <u>free particle</u>
traveling along the geodesics of spacetime. But how, without already
knowing the geodetic structure of spacetime is one to determine which
particles are free and which are not? And of course to find the geodesics of spacetime one must use free particles.

Those philosophers who argue in support of the conventional character of the laws mainly from ontological considerations concerning the
nature of spacetime structure advance the view that spacetime is
structurally amorphous in the sense that what counts as a standard of
<u>no-acceleration</u> or <u>uniform motion</u> is not dictated by a physically real,
causally efficacious dynamic structure of spacetime. Rather, forces
are defined relative to a conventionally chosen standard of <u>no-acceler-
ation</u> and depend for their existence on that conventional choice.

The problem concerning free or natural motion has recently become a
pressing issue within the particular context of the <u>Constructive</u>

PSA 1982, Volume 1, pp. 257-274

<u>Axiomatics</u> for the General Theory of Relativity (GTR). In a paper
entitled "The Geometry of Free Fall and Light Propagation", Ehlers,
Pirani and Schild (EPS) (1972) propose a set of constructive axioms
which are propositions about the local incidence and differential
topological properties of arbitrary massive particles, freely falling
massive particles and light propagation. The constructive axioms are
propositions about a few general qualitative assumptions concerning
free fall motion and light propagation that can be verified directly
through experience in a way that does not presuppose the full blown
edifice of GTR. From these axioms EPS generate a unique pseudo-
Riemannian metric, unique up to a constant positive factor.

One of the constructive axioms employed by EPS -- the Projective
Axiom -- is a statement of the infinitesimal version of the Law of
Inertia, the Law of <u>free</u> (fall) <u>motion</u> which contains Newton's First
Law as a special case in the absence of gravitation. The problem is
how to introduce a class of preferred motions. How is one to characterize that particular path structure which would be realized by a
free particle, namely a neutral, spherically symmetric non-rotating
test particle and avoid the circularity problem surrounding the notion
of a <u>free particle</u>?

The transition from the kinematic to the dynamic behavior of bodies
in spacetime theories has traditionally been viewed to consist in
singling out a particular class of motions as the standard or <u>natural</u>
<u>motions</u> which are considered as the <u>free motions</u>. The concept of
force is then defined in terms of accelerations relative to these
standards of no-acceleration or <u>free motions</u>. This procedure is meaningful if and only if there exists one and only one standard motion
through each event and timelike direction at that event. Geometrically speaking, the choice of a class of standard motions amounts to
giving the spacetime manifold a <u>path structure</u>. Such a path structure
is geodetic and constitutes a projective structure on the spacetime
manifold.

Since geometrical conventionalism takes physical spacetime to be
factually definite only up to local differential topological descriptions, the projective, conformal, affine and metrical structures are
conventionally assignable features of spacetime. According to geometric conventionalism, what counts as a standard of no-acceleration
or uniform motion is not dictated by a physically real, causally
efficacious dynamic structure of spacetime. Rather, forces are definable
relative to a <u>chosen</u> standard of no-acceleration, or equivalently,
forces are conventionally assignable relative to some conventionally
assignable projective structure of spacetime, depending ontologically
on that conventional choice.

One introduces the concept of a class of standard free motions and
hence a projective structure by postulating the Law of Inertia. Hence,
in terms of the above discussion, conventionalism considers force
descriptions in terms of the Laws of Motion wholly conventional, such

descriptions being constrained only by what we find convenient and elegant to use.

Since EPS do not provide an independent non-circular criterion with which to characterize free (fall) motion, their approach has been charged with circularity by philosophers such as Grünbaum, Salmon, Sklar and Winnie. One has argued essentially that any criterion that determines which bodies are suitable as freely falling test bodies and permits their identification presupposes the specification of geometric structures beyond those implied by the local differential topological structure of spacetime.

To the extent that EPS fail to supply an independent characterization of free (fall) motion in a way that would render their approach non-circular, these criticisms are certainly justified. To the extent, however, that they exploit the longstanding problem surrounding the Laws of Inertia, namely, that these Laws do not supply independent criteria of what is to count as _free motion,_ the circularity charges are symptomatic of a longstanding misunderstanding and confusion concerning the meaning and content of the Inertial Laws. This confusion and misunderstanding is due in part to how the laws have been traditionally formulated.

In particular their meaning and content is especially obscured if one stays within the space _plus_ time formulation. In that context, understanding Newton's First Law poses severe problems with respect to the following questions: What reference or coordinate system should one use to describe the motion of a body? What system of temporal measurement should be used? What system of spatial metric should be adopted? How is one to judge the presence or absence of forces?

One has tried to define a _free particle_ with respect to an _inertial frame_ such that Newton's First Law is realized by a _free particle_ satisfying the equation $d^2x^i/dt^2=0$ in that frame. But how is one to determine what an inertial frame is? If one characterizes an inertial frame as a frame which satisfies the equation $d^2x^i/dt^2=0$ that a _free particle_ would satisfy in that frame, then such a definition is obviously circular. Consequently, one sometimes interprets Newton's First Law to be an existence claim concerning physical inertial frames in which a _free particle_ would satisfy $d^2x^i/dt^2=0$. Of course, a space _plus_ time formulation of Newtonian gravitation theory requires strictly speaking that the above be construed _counter-factually_: If there were inertial frames (i.e., if gravitational forces were absent) free particles would satisfy $d^2x^i/dt^2=0$ in them.

One of our main objectives in this paper is to provide a new formulation of the Law of Inertia which explicitly shows that it is a coordinate or frame independent _empirical_ assertion about the _projective structure_ of spacetime. We shall show that the Law of Inertia is testable in a way that assumes no more structure than the local

differential topology of spacetime; that is, it is testable at a
level at which one need assume no more structure than is necessary for
introducing <u>arbitrary</u> local coordinates. This will answer the question
as to what extent EPS's geodesic method for deriving a unique pseudo-
Riemannian metric constitutes a solution toward the controversy between realism and geometrical conventionalism in favour of geometrical
realism. Since our analysis establishes that we <u>do</u> have epistemic
access to light propagation and free (fall) motion (and hence to the
<u>conformal</u> and <u>projective</u> structure of spacetime respectively) in a way
that does not beset the geodesic method either with logical or derivatively with epistemological circularity, we claim to have shown that
EPS's approach supplemented with our analysis decisively undercuts
geometrical conventionalism. This suggests that the Constructive
Axiomatics so supplemented may be interpreted as constituting a convention-free and--in relevant respects--theory-independent body of
evidence that can adjudicate between spacetime geometries and hence
between spacetime theories that postulate them.

2. The Philosophical Significance of the Causal-Inertial Method

The above mentioned criticisms of EPS's Constructive Axiomatics all
suggest that the causal-inertial method is not convention-free and that
the geodesic hypotheses are in relevant respects theory-dependent
assertions, thereby rendering the geodesic method ineffective in
providing a solution to the controversy between geometrical realism
and conventionalism in favour of realism. The criticisms exploit the
fact that any description of the behavior of probative physical
systems in spacetime that does not exclusively employ the local
differential topology must use additional geometric structures. Since
such post-topological structures are considered to be conventional,
different post-topological geometric assumptions can be made that will
lead to different descriptions. Thus any method for finding the
geometry of spacetime will be unsuccessful, if the corresponding probative physical systems used in the discovery procedure require for
their description geometric assumptions beyond the local differential
topology of spacetime. Therefore the causal-inertial method will of
necessity be circular if it is the case that one requires for the
characterization of the behavior of light rays and freely falling
particles those very post-topological structures, namely the conformal
and projective structures, which are supposed to be revealed through
their behavior. It is clear that a genuine sense of "discovery" of
these structures amounts to being able to describe the behavior of the
corresponding probative systems exclusively in terms of the local
differential topological structure and to be able to distinguish between suitable and unsuitable systems on that basis as well. Moreover,
it must be possible to determine at that level whether or not these
structures are unique.

Therefore any hope that the geodesic method can totally undercut
geometrical conventionalism by providing a factual basis on which an
empirical claim for the physical reality of the conformal and
projective structures can be made in a non-circular and non-conventional

manner, requires that the descriptions of the probative systems used to reveal the conformal and projective structures of spacetime assume only the local differential topological structure.

But suppose we live in a devious world and are related to it in a way that makes it impossible for us to use merely the local differential topological structure for the description of any probative physical systems and that compels us to involve geometrical assumptions beyond those implied by the local differential topological structure. What would follow? In particular what would follow from the above criticisms which seem to suggest that something like this might be the case?

All that would follow is that we would then in fact be unable to identify or single out a class of suitable test objects which would have epistemic bearing on the spacetime geometry in an epistemologically non-circular way.

In particular, our inability to identify or single out a class of suitable test objects in an epistemologically non-circular way whose free motions would exhibit the unique projective structure of spacetime, if spacetime did in fact possess such a structure, means only that the truth of the Projective Axiom concerning free (fall) motion is epistemically undecidable. But any argument from the epistemic inaccessibility of free test particles does not establish that the structures derived from the axioms are ontologically conventional. A property which is objective and factual can at most be conventional in an epistemic but not in an ontological sense. Therefore we cannot know whether the projective structure of spacetime is ontologically conventional unless we first establish whether or not it is an objective feature of spacetime. Therefore, the most that is entailed by epistemological circularity is epistemological conventionality.[1]

Now it is one thing to show that the arguments put forward by conventionalism are not strong enough to establish the ontological conventionality of the geometric structures of spacetime. It is quite another thing to show that the conclusions they do draw are false in our particular world. An effective reply from the realist must therefore be a two-edged one.

On the one hand, the realist must show that the conventionalist's arguments are too weak to sustain the conclusions that he wants.

On the other hand, for the realist to make his case successfully, it is not sufficient to show that the arguments for the conventional assignability of the projective, conformal, affine and metric structures are inadequate, but he must actually succeed in showing the conventionalist's conclusion to be false in our particular world by showing that these geometric structures are real. In particular, he must succeed in showing that the world does not lack a physically real projective structure, that the transition from kinematics to dynamics is non-arbitrary but is in fact dictated by a physically real,

262

causally efficacious unique projective structure.

Since the realist cannot argue for the physical existence of these geometrical structures on a priori grounds, something of an empirical claim has to be made. But in order to incisively undercut geometric conventionalism, the factual basis on which such an empirical claim for the physical reality of the projective, conformal, affine and metrical structures and hence forces rests, must be a basis that is factually definite in a way that makes it immune to the charge that it, too, is conventional. Since the conventionalist views the world to be factually definite and non-conventional only up to local differentiable topological descriptions, it is clear that any decisive challenge to conventionalism requires some demonstration that the projective, conformal (causal), affine and metric structures and hence forces are physically real entities by assuming only the local differential topological structure.

We will now show that the causal-inertial method, when supplemented with a number of theorems that serve as empirical local differential topological criteria for singling out free (fall) motion at a level of testing that requires no more structure than is needed for introducing arbitrary local coordinates, is not beset with either logical or epistemological circularity. The geodesic method so supplemented constitutes a decisive challenge to conventionalism in that it provides a non-conventional factual basis for establishing the physical reality of the projective, conformal, affine and metric structures of spacetime.

3. Local Differential Topological Criteria for the Geodecisity of
   Acceleration and Directing Fields[2]

Consider the set of maps $F(\mathbb{R},M)$ called curves. By curve elements we mean, roughly speaking, Taylor polynomial approximations at $p \in M$ to curves $\gamma \in F(\mathbb{R},M)$ through $p \in M$.

Definition 1: Any two curves γ, $\hat{\gamma} \in F(\mathbb{R},M)$ through a given point $p \in M$ are k-jet equivalent at p, iff i) $\gamma(0)=\hat{\gamma}(0)=p$ and ii) for any chart $(U,x)_p$, the curves $x \circ \gamma$, $x \circ \hat{\gamma}$ through $x(p) \in \mathbb{R}^n$ have the same derivatives at $x(p) \in \mathbb{R}^n$ up to and including order k.

By k-th order curve elements we mean equivalence classes of k-jet equivalent curves. We shall also say that k-jet equivalent curves have contact of order k at $p \in M$. The equivalence class of k-jet equivalent curves at $p \in M$ will be denoted by $j_0^k \gamma$. The equivalence class of 1-jet equivalent curves is denoted by $j_0^1 \gamma$ and corresponds to all curves through p whose first order contact or tangency at p is characterized by a unique tangent vector $v_p \in TM_p$. The set of all k-jet equivalence classes at $p \in M$ is denoted by $J^k(\mathbb{R}_0, M_p)$ and the set of all pairs $(p, j_0^k \gamma)$ is the space $J^k(\mathbb{R}_0, M) = U_{p \in M} J^k(\mathbb{R}_0, M_p)$. Clearly, $J^1(\mathbb{R}_0, M_p)$ and $J^1(\mathbb{R}_0, M)$ are isomorphic to TM_p and TM respectively.[3]

Set $L_s^k(M_p) = J^k(\mathbb{R}_0^s, M_p)$. The manifold $L_s^k(M_p)$ denotes the s^k-speeds at $p \in M$ and the submanifold $H_s^k(M_p) \subset L_s^k(M_p)$ denotes the s^k-frames. Normally, $s \leq n$, and if $s = n$ then we shall simply write $H^k(M_p)$ for the set of n^k-frames. $H_s^k(M_p)$ consists of jets of maximal rank, i.e., $H_s^k(M_p) = L_s^k(M_p) \setminus \{ (p, j_0^k \gamma) \mid j_0^1 \gamma = 0 \}$.

The principal bundle of n^k-frames is the structure $H^k(M) = \langle H^k(M), \pi_{H^k}, M, G_n^k \rangle$. The principal bundle $H^1(M)$ is isomorphic to the bundle of linear frames. The associated fiber bundle of k-order curve elements is the structure
$$H_1^k(M) = \langle H_1^k(M), \pi_{H_1^k}, M, H_1^k(\mathbb{R}_0^n), H^k(M) \rangle, \tag{1}$$
where $H^k(M)$ denotes the principal bundle of n^k-frames.

Consider the curve $\gamma^k : I \to J^k(\mathbb{R}_0, M) = H_1^k(M)$, where $I \subset \mathbb{R}$ and $k \in \{0, 1, 2, \ldots\}$. In terms of local coordinates we have
$$\gamma^k(s) = (\gamma_0^{ki}(s), \gamma_1^{ki}(s), \ldots, \gamma_k^{ki}(s)) \tag{2}$$
where $s \in \mathbb{R}$ and $\gamma_a^{ki}(s) = d^a x^i \circ \gamma(s) / ds^a$ for some $\gamma : I \to M$ which in general depends on $p = \gamma(s) \in M$.

<u>Definition 2</u>: γ^k is a <u>special curve</u> iff $\forall s \in \mathbb{R}$, $\gamma_a^{ki}(s) = \overset{\bullet}{\gamma}_{a-1}^{ki}(s)$; $1 \leq a \leq k$.

<u>Definition 3</u>: A special curve γ^k is called the <u>k-order lift</u> of the curve $\pi_{H_1^k} \circ \gamma^k : \mathbb{R} \to M$. If $\gamma = \pi_{H_1^k} \circ \gamma^k$ then we write $j^k(\gamma) = \gamma^k$ to denote the k-order lift of γ.

Consider again the associated fiber bundle of k-order curve elements (viz. (1)).

<u>Definition 4</u>: An <u>acceleration field</u> is a map $A : H_1^1(M) \to H_1^2(M)$ such that $\pi_{H_2^1}^{H_1^1} \circ A$ is the identity map on $H_1^1(M)$ where $\pi_{H_1^2}^{H_1^1} : H_1^2(M) \to H_1^1(M)$ is the natural projection obtained by dropping the second order term of the Taylor polynomial.

In terms of local coordinates (x^i, γ_1^i) for $H_1^1(M)$ and $(x^i, \gamma_1^i, \gamma_2^i)$ for

$H_1^2(M)$, the acceleration field is given by

$$A(x^i(p),\gamma_1^i)=(x^i(p),\gamma_1^i,A_2^i(x^i(p),\gamma_1^i)) \tag{3}$$

where $A_2^i(x^i(p),\gamma_1^i)$ are n functions of the 2n coordinates $(x^i(p),\gamma_1^i)$.

For any curve $\gamma\in F(\mathbb{R},M)$, the curves $j^1\gamma:\mathbb{R}\to H_1^1(M)$ and $j^2\gamma:\mathbb{R}\to H_1^2(M)$, such that for all $s\in\mathbb{R}$, $j^1\gamma(s)=j_s^1\gamma$ and $j^2\gamma(s)=j_s^2\gamma$, are first and second order lifts of γ. A curve $\gamma:\mathbb{R}\to M$ is a solution of the differential equation defined by (3), iff $A\circ j^1\gamma=j^2\gamma$, which in terms of local coordinates may be written as

$$\frac{d^2\gamma^i}{ds^2}(s)=A_2^i(\gamma^i(s),\frac{d\gamma^i}{ds}(s)). \tag{4}$$

<u>Definition 5</u>: An acceleration field is <u>geodesic</u>, and is denoted by $\Gamma:H_1^1(M)\to H_1^2(M)$, iff $\forall p\in M$ there exists some local system of coordinates $(\bar{x}^i(p),\bar{\gamma}_1^i,\bar{\gamma}_2^i)$ with respect to which at $p\in M$, $\bar{\Gamma}(\bar{x}^i(p),\bar{\gamma}_1^i)=(\bar{x}^i(p),\bar{\gamma}_1^i,0)$ or $\bar{\Gamma}_2^i(\bar{x}^i(p),\bar{\gamma}_1^i)=0$.

<u>Theorem 1</u>: An acceleration field is geodesic iff relative to <u>any</u> local system of coordinates $(x^i(p),\gamma_1^i,\gamma_2^i)$: $\Gamma_2^i(x^i(p),\gamma_1^i)=-\Gamma_{jk}^i(x^i(p))\gamma_1^i\gamma_1^k$.

For each step in the above analysis of curves there is an analogue for paths. Since a path is an equivalence class of curves $\gamma:\mathbb{R}\to M$ (any two of which are related by $\gamma=\gamma\circ\mu$, $\mu:\mathbb{R}\to\mathbb{R}$ a diffeomorphism), it is not surprising that a path element of order k, called a <u>k-direction</u>, is an equivalence class of curve elements of order k.

<u>Definition 6</u>: Any two curves γ, $\hat{\gamma}$ are <u>k-direction equivalent at $p\in M$</u>, iff there exists a diffeomorphism $\mu:\mathbb{R}\to\mathbb{R}$ such that

$$j_s^k\hat{\gamma}=j_s^k(\gamma\circ\mu)=j_{\mu(s)}^k\gamma\circ j_s^k\mu \tag{5}$$

where $\hat{\gamma}(s)=\gamma(\mu(s))=p\in M$. An equivalence class of <u>k-direction equivalent curves at $p\in M$</u> is called a <u>k-direction at $p\in M$</u>.

A k-direction may also be regarded as an equivalence class of paths: Let ξ be a path through $p\in M$. Choose any $\gamma\in[\gamma]=\xi$ such that $\gamma(0)=p$ and $j_0^1\gamma\neq 0$. For every $\hat{\gamma}\in j_0^k\gamma$ there corresponds a path, namely, $\hat{\xi}=\hat{\gamma}_\vdash(\mathbb{R})$. Then the set of all such paths is a k-direction at p and will be denoted by $j_p^k\xi$.

For a given choice of local coordinates (U,x) in M there corresponds to each k-direction $j_p^k\xi$ an equivalence class of Taylor polynomials of degree k. For k=2 the coefficients of any two such Taylor polynomials are according to (5) related by

$$\hat{\gamma}_1^i = (D\mu)\gamma_1^i \tag{6}$$

$$\hat{\gamma}_2^i = (D\mu)^2\gamma_2^i + (D^2\mu)\gamma_1^i$$

where $D\mu$, $D^2\mu$ are the derivatives of μ at $0\varepsilon\mathbb{R}$. The relation $j_0^k\hat{\gamma} = j_0^k\gamma \circ j_0^k\mu$ defines a right action of the group G_1^k on the space $H_1^k(M_p)$ of curve elements of order k at $p\varepsilon M$. The equivalence classes at p defined by this action are called k-directions at $p\varepsilon M$ and the space of these k-directions will be denoted by $\mathbb{D}_1^k(M_p) = H_1^k(M_p)/G_1^k$. The set of all pairs $(p,j_p^k\xi)$ is denoted by $H_1^k(M)/G_1^k = \mathbb{D}_1^k(M)$. Set $\mathbb{D}_1^k(\mathbb{R}_0^n) = H_1^k(\mathbb{R}_0^n)/G_1^k$. Then for k>1, the associated bundle of k-directions is the structure

$$\mathcal{D}_1^k(M) = <\mathbb{D}_1^k(M),\pi_{\mathbb{D}_1^k},M,\mathbb{D}_1^k(\mathbb{R}_0^n),H^k(M)> \tag{7}$$

Since for all $\gamma\varepsilon\xi$, $j_0^1\gamma\neq0$, at least one of the components $\gamma_1^k\neq0$. Also $D\mu\neq0$, since $\mu:\mathbb{R}\to\mathbb{R}$ is a diffeomorphism. Consequently, the manifold $D_1^k(M_p)$ may be covered by n-coordinate charts defined as follows: for $b\varepsilon\{1,2,\ldots,n\}$ choose $D\mu$ so that $\hat{\gamma}_1^b=1$ and choose $D^r\mu$ for each $r\varepsilon\{2,3,\ldots,k\}$ such that $\hat{\gamma}_r^b=0$. Define $\xi_r^\alpha=\hat{\gamma}_r^\alpha$ for $r\varepsilon\{1,2,\ldots,k\}$ and $\alpha\neq b$. For the special case k=2, the coordinates of $j_p^2\xi$ in the chart with b=n are determined by the coordinates of $j_0^2\gamma$ by

$$\xi_1^\alpha = \gamma_1^\alpha/\gamma_1^n = dx^\alpha/dx^n$$

$$\xi_2^\alpha = (\gamma_2^\alpha\gamma_1^n - \gamma_2^n\gamma_1^\alpha)/(\gamma_1^n)^3 = d^2x^\alpha/(dx^n)^2. \tag{8}$$

Note that in spacetime, the coordinates of the k-direction $j_p^k\xi$ are just the first k-derivatives of the space coordinates with respect to the time coordinate along some worldline path. Thus ξ_1^α and ξ_2^α are customarily referred to as <u>three velocity</u> and <u>three acceleration</u> respectively.

Of particular importance for a coordinate free discussion of the equations of motion of material bodies is the notion of a set of distinct <u>directing field structures</u> each of whose members corresponds to a distinct <u>path structure</u> on M and the notion of a subset of such structures called <u>geodesic directing fields</u>, each of whose members corresponds to a distinct <u>projective structure</u> Π on M.

<u>Definition 7</u>: A <u>directing field</u> on M is a map $\Xi: \mathbb{D}^1_1(M) \to \mathbb{D}^2_1(M)$ such that $\pi^{\mathbb{D}^1_1}_{\mathbb{D}^2_1} \circ \Xi$ is the identity map on $\mathbb{D}^1_1(M)$.

A directing field Ξ defines a second order differential equation for paths on M. In terms of local coordinates $(x^i(p), \xi^\alpha_1)$ for $\mathbb{D}^1_1(M)$ and $(x^i(p), \xi^\alpha_1, \xi^\alpha_2)$ for $\mathbb{D}^2_1(M)$ the map Ξ is given by

$$\Xi(x^i(p), \xi^\alpha_1) = (x^i(p), \xi^\alpha_1, \Xi^\alpha_2(x^i(p), \xi^\alpha_1)) \tag{9}$$

where the $\Xi^\alpha_2(x^i(p), \xi^\alpha_1)$ are $(n-1)$ functions of $(2n-1)$ variables. The path ξ on M is a solution to the differential equation defined by Ξ, iff $\Xi \circ j^1\xi = j^2\xi$, which in terms of local coordinates may be written as

$$\frac{d^2 x^\alpha}{(dx^n)^2}(x^n) = \Xi^\alpha_2(x^n, x^\alpha(x^n), \frac{dx^\alpha}{dx^n}(x^n)). \tag{10}$$

The family of all paths ξ corresponding to a given directing field Ξ on M is called a <u>path structure</u> on M.

<u>Definition 8</u>: A directing field is geodesic and is denoted by $\Pi: \mathbb{D}^1_1(M) \to \mathbb{D}^2_1(M)$ iff $\forall p \varepsilon M$ there exist some local coordinate $(\bar{x}^i(p), \bar{\xi}^\alpha_1, \bar{\xi}^\alpha_2)$ corresponding to a chart $(\bar{U}, \bar{x})_p$ such that $\bar{\Pi}(\bar{x}^i(p), \bar{\xi}^\alpha_1) = (\bar{x}^i(p), \bar{\xi}^\alpha_1, 0)$ or $\bar{\Pi}^\alpha_2(\bar{x}^i(p), \bar{\xi}^\alpha_1) = 0$. (Note that every geodesic directing field corresponds to an equivalence class of projectively equivalent symmetric linear connections).

<u>Theorem 2</u>: A directing field is geodesic iff for <u>any</u> local system of coordinates $(x^i(p), \xi^\alpha_1, \xi^\alpha_2)$,

$$\Pi^\alpha_2(x^i(p), \xi^\alpha_1) = \xi^\alpha_1[\Pi^n_{\rho\sigma}(x^i(p))\xi^\rho_1\xi^\sigma_1 + 2\Pi^n_{n\rho}(x^i(p))\xi^\rho_1 + \Pi^n_{nn}(x^i(p))]$$
$$- [\Pi^\alpha_{\rho\sigma}(x^i(p))\xi^\rho_1\xi^\sigma_1 + 2\Pi^\alpha_{n\rho}(x^i(p))\xi^\rho_1 + \Pi^\alpha_{nn}(x^i(p))]. \tag{11}$$

Since $\Pi^i_{ji}(x^i(p)) = 0$, $\Pi^n_{n\rho}(x^i(p))$ and $\Pi^n_{nn}(x^i(p))$ may be eliminated from (11). It is clear from Theorem 2 that if a directing field is geodesic, then $\Pi^\alpha_2(x^i(p)), \xi^\alpha_1)$ is a cubic polynomial in ξ^α_1 in <u>every</u> coordinate chart $(U, x)_p$. The converse is also true.

<u>Theorem 3</u>: If with respect to <u>every</u> coordinate chart $(U,x)_p$ the corresponding map $\Xi_2^{\alpha}(x^i(p),\xi_1^{\alpha})$, which determines the directing field Ξ, is cubic, that is, if

$$\Xi_2^{\alpha}(x^i(p),\xi_1^{\alpha})=A^{\alpha}+B_{\rho}^{\alpha}\xi_1^{\rho}+C_{\rho\sigma}^{\alpha}\xi_1^{\rho}\xi_1^{\sigma}+D_{\rho\sigma\tau}^{\alpha}\xi_1^{\rho}\xi_1^{\sigma}\xi_1^{\tau} \tag{12}$$

where the coefficients A, B, C, D are functions only of $p\varepsilon M$, then Ξ is a geodesic directing field.

4. The Measurement Problem of the Projective Structure

Since the mathematical criteria for geodecisity (Theorems 1, 2, 3) involve local differential topological concepts only, they can be effectively applied as <u>empirical</u> criteria, only if it is possible to measure the acceleration and directing fields by making use of only the local differential topological structure.[4] We shall now briefly characterize the role of the empirical differential topological criterion for the geodecisity of directing fields within the context of the causal-inertial method for discovering the causal, inertial and metric structures of spacetime.

- C1: Suppose it is possible to establish a system of coordinates (such as radar coordinates) in the region of spacetime under investigation.

- C2: Suppose it is possible to <u>track</u> material bodies with respect to this coordinate system. That is, the world line paths of material bodies can be determined by making use of only the local differential topological structure of spacetime required for setting up local coordinates. [5]

- C3: Suppose further that there exists a non-empty set of probative systems the motions of which are governed exclusively by a directing field.

Then the following can be shown to be the case:

- C4: Given C1, C2, C3, the directing field can be measured <u>directly</u> (in the sense that no other fields are required) by using no more structure than is needed for introducing local coordinates.

- C5: Once the directing field is measured the differential topological criterion for geodecisity can be applied to determine whether or not it is also geodesic.[6]

Hence if there exists a set of particles whose motion is exclusively governed by a given directing field, then it is possible to isolate those particles uniquely on the basis of C1 and C2 and hence on the basis of the differential topology. The set of particles thus singled out in this manner constitutes a specific <u>directing field set</u> of objects which are exclusively governed by that specific directing

field. The directing field may now be measured by tracking a
sufficiently large number of particles belonging to this specific
directing field set. Once the directing field is measured the
criterion for geodecisity may be applied to see whether it is also a
geodesic directing field (guiding field). Therefore, if spacetime
has a unique projective structure and if there exist particles which
are exclusively governed by that structure, then we are able to determine that structure empirically at that level of empirical analysis
which assumes no more structure than is required for setting up local
coordinates, namely, the local differential topology of spacetime.

It should be remarked here that the condition that there exists a
class of particles which are exclusively governed by a geodesic directing field, in order to measure that field and to determine its geodecisity by means of them, can be considerably weakened.[7]

5. The Law of Inertia and Forces

It is not difficult to see that the above analysis has direct
bearing on the question concerning the status of the Law of Inertia.
The existence of a unique projective structure $\Pi: \mathbb{D}^1_1(M) \rightarrow \mathbb{D}^2_1(M)$ on
spacetime determines a unique coordinate independent standard of no-
acceleration. This geometric structural field bijectively corresponds
to a unique geodesic directing field or differential equation for paths
in spacetime. The world line solution paths of the field Π constitute
a distinguished class of motions which are the free motions of material
bodies with respect to which forces may be defined.

For _any_ theory of spacetime in which the model of its event set is
a four dimensional differential manifold M, the ontological basis of
the dynamics of material bodies is expressed in the following _empirical_
law:

The Law of Inertia: There exists on spacetime a unique projective
structure Π (or equivalently, a unique geodesic directing field Π).

The concept of _free motion_ may now be characterized on the basis of
the Law of Inertia in the following way:

The Definition of Free Motion: A possible or actual material body is
in a _state of free motion_ during any part of its history, just in case
the corresponding segment of its world line path is a solution path of
the differential equation determined by the unique _projective structure_
of spacetime.

Newton's second law of motion may now be reformulated as follows:

The Law of Motion: With respect to any coordinate system, the path
that describes the world line of a possible or actual material body
satisfies an equation of the form

$$m(\xi_2^\alpha - \Pi_2^\alpha(x^i(p), \xi_1^\alpha)) = F^\alpha(x^i(p), \xi_1^\alpha) , \qquad (13)$$

where m is a scalar constant characteristic of the material body, called its mass, and $F^\alpha(x^i(p), \xi_1^\alpha)$ is the force acting on the body.

In order that the form of equation (13) be independent of the coordinate system chosen, it is necessary that under a coordinate transformation, the forces transform according to

$$\bar{F}^\alpha(\bar{\xi}_1^\alpha) = \frac{\bar{\Lambda}_\rho^\alpha F^\rho(\xi_1^\alpha)(\bar{\Lambda}_n^n + \bar{\Lambda}_\beta^n \xi_1^\beta) - \bar{\Lambda}_\rho^n F^\rho(\xi_1^\alpha)(\bar{\Lambda}_n^\alpha + \bar{\Lambda}_\beta^\alpha \xi_1^\beta)}{(\bar{\Lambda}_n^n + \bar{\Lambda}_\sigma^n \xi_1^\sigma)^3} . \qquad (14)$$

From the fact that the transformation law for forces is linear in the forces and does not contain an inhomogeneous term, it follows that forces are additive and that, if a force is nonzero in any coordinate system, then it is nonzero in all coordinate systems.

The forces $F^\alpha(x^i(p), \xi_1^\alpha)$ may also depend on the variables which describe the orientation of possible internal multipole structures of the material body and the coupling of such structures to various geometric object fields in its microneighborhood. For those particles which have such additional internal structure, it is necessary to supplement (13) with additional equations of motion for the internal variables in order to obtain a complete system of equations adequate for _predicting_ the motion of the particle. However, the Law of Motion (13) _alone_ is adequate for the _analysis_ of an observed motion and for the measurement of the force acting on a body, since the quantities appearing on the left side of (13) (ξ_2^α and $\Pi_2^\alpha(x^i(p), \xi_1^\alpha)$) can be empirically determined by ideal test procedures which presuppose only the local differential topological structure of spacetime.

It is important to emphasize that the Law of Motion (13) makes explicit use of the unique projective structure Π on spacetime. _The Law of Motion, therefore, depends ontologically on the Law of Inertia. Consequently, it is impossible to derive the Law of Inertia from the Law of Motion._

The Law of Inertia and the Law of Motion as formulated above apply to all, relativistic or non-relativistic, curved or flat, dynamic or nondynamic, spacetime theories. The reason for the general character of these laws consists in the fact that they require for their formulation only the differential topological structure of spacetime, a structure which is _common_ to all spacetime theories.

The Law of Inertia is an _empirical_ law. It is _falsifiable_, for if there exist at least two classes of particles each of which is governed by a distinct geodesic directing field (projective structure), then, as discussed above, the particles belonging to these classes may be identified and used to measure in any chosen local neighborhood of spacetime the two distinct projective structures. The discovery in

any local region of spacetime of two distinct projective structures is
sufficient to falsify the Law of Inertia.

If there exists exactly one class of particles governed solely by a
geodesic directing field, then the particles may be identified and
used to measure the unique projective structure of spacetime in any
chosen local neighborhood of spacetime by a procedure which is
coordinate independent.

The fact that the motions of free particles are uniquely determined
by the projective structure of spacetime makes them particularly suitable as instruments for probing this structure because their motions
reveal the unique projective structure of spacetime in a particularly
simple manner. However, as explicitly formulated above in the Law of
Motion, the unique projective structure of spacetime constrains in
part the motions of all material bodies. Hence, it should be expected
that a sufficiently sophisticated analysis would permit the use of more
complicated material bodies for the measurement of the projective
structure of spacetime.[8]

The measurability of the projective structure of spacetime at the
local differential topological level constitutes an empirical non-
conventional basis which establishes that the projective structure of
spacetime is a real physical field.

It is clear that forces are measurable at a level that uses only
the local differential topological structure of spacetime. For, only
this structure is required to measure the projective structure Π,
to measure the paths ξ of arbitrary material bodies and to compute
the lifts $j^1\xi$ and $j^2\xi$. Aside from the inertial mass m which is just
a constant, all quantities on the left side of the Law of Motion (13)
are measurable. Consequently, the force acting on a material body
is measurable. The measurability of forces just described establishes
that forces, conceived of as causal dynamical agents are real physical
entities.

The main projective axiom in EPS constitutes an application of the
infinitesimal version of the Law of Inertia. It is clear from the
above, that this axiom can be tested at a level which makes use of
only the local differential structure required to set up local
coordinates.

Notes

[1]We distinguish between epistemological and ontological conventionality in the following sense: While the world may or may not lack
definite objective features that would be sufficient to uniquely
determine the extension of theoretical terms and thus provide an
empirical basis for choosing from among several possible alternatives,
we either do not know or cannot know of those features. Thus

epistemological conventionality leaves open whether the extension of
some theoretical term is or is not in fact underdetermined. Epistemological conventionality does not therefore preclude a realistic
interpretation of the theoretical terms in question. But we either
do not know or cannot know whether or not the world has the relevant
objective features that would support such an interpretation.

Ontological conventionalism makes the factual claim that the world
lacks certain objective features that would be sufficient to uniquely
determine the extension of certain relevant theoretical terms. Ontological conventionalism certainly entails epistemological conventionalism. The absence of certain relevant objective features of the
world trivially entails our corresponding lack of knowledge of them.
But the converse need not be true. Lack of knowledge of certain
relevant objective properties of the world does not entail their
absence.

[2]The theorems presented here without proofs are proved in Coleman,
Korte (1980). Theorem 2 was first proved in Ehlers, Köhler (1977)
using the T(M) and T(T(M)) formalism. We prove it more simply by
using the more appropriate jet formalism of Ehresmann (1951-52). For
a new approach to spacetime G-Structures see Coleman, Korte (1981a).

[3]Some illustrative material of these notions may be found in Coleman,
Korte (1981b).

[4]See Coleman, Korte (1981b).

[5]See Coleman, Korte (1981b).

[6]See Coleman, Korte (1980, 1981b).

[7]It is shown in Coleman, Korte (1981b) that even if particles
governed solely by the projective structure of spacetime do not exist
it is still possible to measure the projective structure of spacetime
using more general probative systems, namely, charged monopoles. It
is also shown there that charged monopoles may be used to measure not
only the projective structure of spacetime but also the ratios of the
components of the electromagnetic field as well as the conformal
structure of spacetime.

[8]A not too complicated example of such an analysis is provided in
Coleman, Korte (1981b) for charged monopoles.

References

Coleman, R.A. and Korte, H. (1980). "Jet Bundles and Path Structures." _Journal of Mathematical Physics_ 21: 1340-1351. Erratum- 23: 345 (1982).

------------------------------. (1981a). "Spacetime G-Structures and their Prolongations." _Journal of Mathematical Physics_ 22: 2598-2611.

------------------------------. (1981b). "A Realist Field Ontology of the Causal-Inertial Structure." Unpublished University of Regina preprint.

Earman, J. and Friedmann, M. (1973). "The Meaning and Status of Newton's Law of Inertia and the Nature of Gravitational Forces." _Philosophy of Science_ 40: 329-359.

Ehlers, J. _et al._ (1972). "The Geometry of Free Fall and Light Propagation." In _General Relativity, Papers in Honour of J.L. Synge._ Edited by L. O'Raifeartaigh. Oxford: Clarendon Press. Pages 63-84.

----------. (1973). "Survey of General Relativity Theory." In _Relativity, Astrophysics and Cosmology._ Edited by W. Israel. Dordrecht: Reidel. Pages 1-125.

---------- and Schild, A. (1973). "Geometry in a Manifold with Projective Structure." _Communications in Mathematical Physics_ 32: 119-146.

---------- and Kohler, E. (1977). "Path Structures on Manifolds." _Journal of Mathematical Physics_ 18: 2014-2018.

Ehresmann, C. (1951-52). "Les Prolongements d'une Variete Differentiable, I-V." _Comptes Rendus Hebdomadaires des Séances de l'Académie des Sciences_ 233: 598-600, 777-779, 1081-1083. 234: 1028-1030, 1424-1425.

Einstein, A. (1919). What is the Theory of Relativity?" _The London Times_ November 28, 1919. (As reprinted in _Essays in Science._ New York: Philosophical Library, 1934. Pages 53-60.)

------------. (1970). "Remarks Concerning the Essays Brought Together in this Co-operative Volume." In _Albert Einstein Philosopher-Scientist._ Vol. II. Edited by P.A. Schilpp. LaSalle, Illinois: Open Court Press. Pages 665-688.

Ellis, B. (1965). "The Origin and Nature of Newton's Law of Motion." In _Beyond the Edge of Certainty._ (University of Pittsburgh Series in the Philosophy of Science._ Volume 2.) Edited by R.G. Colodny. Englewood Cliffs, New Jersey: Prentice Hall. Pages 29-68.

Grünbaum, A. (1973). "General Relativity, Geometrodynamics and Ontology." In <u>Philosophical Problems in Space and Time,</u> 2nd enlarged edition. <u>(Boston Studies in the Philosophy of Science,</u> Volume XII. Edited by R.S. Cohen and Marx Wartofsky.) Dordrecht: Reidel. Pages 728-803.

------------. (1977). "Absolute and Relational Theories." In <u>Foundations of Space-Time Theories,</u> <u>(Minnesota Studies in the Philosophy of Science,</u> Volume VIII.) Edited by J. Earman, <u>et al.</u> Minneapolis: University of Minnesota Press. Pages 303-373.

Hanson, N.R. (1965). "Newton's First Law: A Philosopher's Door into Natural Philosophy." In <u>Beyond the Edge of Certainty,</u> <u>(University of Pittsburgh Series in the Philosophy of Science,</u> Volume 2.) Edited by R.G. Colodny. Englewood Cliffs, New Jersey: Prentice Hall. Pages 6-28.

Hooker, C.A. (1974). "Defense of a Non-Conventionalist Interpretation of Classical Mechanics." In <u>Logical and Epistemological Studies in Contemporary Physics,</u> <u>(Boston Studies in the Philosophy of Science,</u> Volume XIII.) Edited by Robert S. Cohen and Marx W. Wartofsky. Dordrecht: Reidel. Pages 123-191.

Hunt, I.E. and Suchting, W.A. (1969). "Force and Natural Motion." <u>Philosophy of Science</u> 36: 233-251.

Kobayashi, S. and Nomizu, K. (1963). <u>Foundations of Differential Geometry,</u> Volume I. New York: John Wiley & Sons.

Munkres, J.R. (1963). <u>Elementary Differential Topology,</u> Princeton, New Jersey: Princeton University Press.

Nijenhuis, A. (1972). "Natural Bundles and Their General Properties." In <u>Differential Geometry: In Honour of Kentaro Yano,</u> Edited by S. Kobayashi, <u>et al.</u> Tokyo: Kinokuniya Book-Store Co. Pages 317-334.

Pommaret, J.F. (1978). <u>Systems of Partial Differential Equations and Lie Pseudogroups,</u> New York: Gordon and Breach.

Porteous, I.R. (1969). <u>Topological Geometry,</u> London: Van Nostrand Reinhold.

Reichenbach, H. (1969). <u>Axiomatization of the Theory of Relativity,</u> Berkeley: University of California Press.

Salmon, W. (1977). "The Curvature of Physical Space." In <u>Foundations of Space-Time Theories,</u> <u>(Minnesota Studies in the Philosophy of Science,</u> Volume VIII.) Edited by J. Earman, <u>et al.</u> Minneapolis: University of Minnesota Press. Pages 281-302.

Sklar, L. (1977). "Facts, Conventions and Assumptions." In
     _Foundations of Space-Time Theories. (Minnesota Studies in the
     Philosophy of Science,_ Volume VIII.) Edited by J. Earman, _et al._
     Minneapolis: University of Minnesota Press. Pages 206-274.

Trautman, A. (1966). "The General Theory of Relativity." _Uspekhi
     Fizicheskikh Nauk_ 89: 3-37. (English translation: _Soviet
     Physics Uspekhi_ 9: 319-339.)

Weyl, H. (1922a). _Raum, Zeit, Materie._ 4th ed. Berlin-Heidelberg-New
     York: Springer-Verlag. (As reprinted as _Space-Time-Matter._
     (trans.) H.L. Brose. New York: Dover, 1952.)

--------. (1922b). "Zur Infinitesimalgeometrie: Einordnung der
     projektiven und konformen Auffassung." In _Nachrichten von der
     Königlichen Gesellschaft der Wissenschaften zu Göttingen,
     Mathematisch-physicalische Klasse aus dem Jahre 1921._ Pages
     99-112. (As reprinted in _Gesammelte Abhandlungen,_ Volume II.
     Berlin-Heidelberg-New York: Springer-Verlag, 1968. Pages
     195-207.)

--------. (1923). _Mathematische Analyse des Raum Problems._ Berlin:
     J. Springer. (As reprinted in _Das Kontiuum und Andere
     Monographien._ H. Weyl, _et al._ New York: Chelsea, 1960.)

--------. (1927). _Philosophie der Mathematik und Naturwissenschaft.
     (Handbuch der Philosophie.)_ München und Berlin: R. Oldenbourg.
     (As reprinted as the revised and augmented English edition:
     Philosophy of Mathematics and Natural Science. (trans.) O.
     Helmer. Princeton, New Jersey: Princeton University Press,
     1949.)

Winnie, J.A. (1977). "The Causal Theory of Space-Time." In
     _Foundations of Space-Time Theories. (Minnesota Studies in the
     Philosophy of Science,_ Volume VIII.) Edited by J. Earman, _et al._
     Minneapolis: University of Minnesota Press. Pages 134-205.

Part VII

Tachyons, Temporal Becoming and Thermodynamics

Tachyon Signals, Causal Paradoxes,
and the Relativity of Simultaneity[1]

Steven F. Savitt

The University of British Columbia

1. Tachyon Signals

The basic equations of the Special Theory of Relativity (STR)
imply that no massive particle travelling at a velocity less than
the velocity of light in vacuo (call that 'c') can be accelerated
to a velocity either equal to or greater than c. For instance, the
total energy E of a particle is given by the formula

$$(1) \quad E = m(1-v^2)^{-\frac{1}{2}}$$

where m is the rest mass of the particle and v its velocity in some
given co-ordinate system (CS). One can choose units so that c = 1;
hence for ordinary particles ($|v| < 1$) as v approaches 1, E grows
without bound. That is, as a particle accelerates, its total energy
increases and ever-increasing amounts of energy are required to
accelerate it. So no ordinary particle (tardyon, bradyon) can be
accelerated continuously from velocity v < 1 to velocity v = 1, much
less to velocity v > 1.

During the 1960's, however, it was conjectured that it would be
consistent with STR for there to exist particles which had throughout their entire lifetimes velocity greater than c.[2] In formula
(1), for instance, when v > 1 the denominator of the fraction is the
square root of a negative number (i.e., it is an imaginary number).
If the rest mass of a particle with v > 1 were also imaginary, then
the i's in the numerator and denominator would cancel, and E would
be just a real number. Since a particle with v > 1 cannot be at
rest in any CS moving with sub-luminal velocity with respect to us,
its rest mass is a theoretically defined rather than an experimen-
tally measurable quantity. There is no difficulty or inconsistency
in considering it to be an imaginary quantity.

Gerald Feinberg (1967) called particles with velocity v > 1 "tachyons". There is very little experimental evidence indicating that tachyons exist, but there is a flourishing theoretical literature concerning their properties.[3] As we shall soon see, there is much to interest philosophers in these arguments concerning tachyons since fundamental features of time and causation seem to be challenged by their existence.[4]

Before plunging into these arguments, we need one preliminary clarification. The existence of "things that move faster than light" is not controversial.[5] Imagine a powerful searchlight rotating rapidly at the centre of a large sphere so that its beam sweeps out a flat plane. For a sufficiently large value of the radius of the sphere, the searchlight beam will sweep along the inner side of the sphere with a velocity greater than c. The key point is that if A and B are two points on the path swept out by the beam, no causal influence is propagated from A to B at velocity v > 1. Unless otherwise noted, then, I mean by "tachyons" particles which move with velocity v > 1 and which do propagate causal influence with that same velocity.[6]

Furthermore, I shall assume that tachyons can be used to convey information from one space-time point to another with velocity v > c. Like any speculation concerning tachyons, this assumption may turn out to be false; but I see now no reason to suppose that it is false, and I shall use this assumption freely in what follows.

2. Causal Paradoxes

Much of the controversy concerning the existence of tachyons results from the observation that the assumption of their existence (in conjunction with some standard principles of STR) seems to lead to paradoxical conclusions. Let us turn then, using diagram 1, to one version of these causal paradoxes.[7]

The lines labelled 'A' and 'T' in diagram 1 represent the world-lines of two observers, whom I shall call 'Alice' and 'Ted'. The line labelled 'x' represents one spatial axis (we confine ourselves to one spatial axis for convenience), and the axis perpendicular to that is the time axis. The x- and ct-axes define a CS S in which Alice and Ted are at rest. Again for convenience, let us say that these two observers are separated by 10 spatial units.

Now let us introduce another pair of observers, Bob and Carol, represented by the lines B and C, and suppose that they are both moving in S in the positive x-direction with velocity .8. They then define a new CS S' in which they are at rest, and I shall suppose that in S' Bob and Carol are separated by 10 spatial units.

When the world-lines of observers cross, we say that they coincide. The world-lines of Alice and Bob, for example, coincide at a point which is labelled E_4 in the diagram. Let E_4 be the origin of

Diagram 1

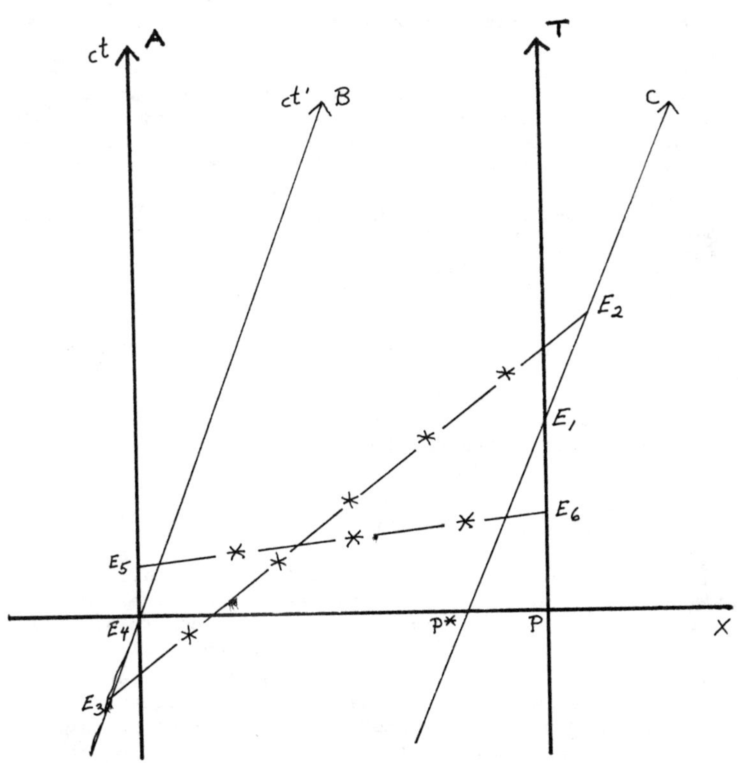

Table of Co-ordinates for Points E_1 to E_6

	S	S'
E_1	(10,5)	[10,-5]
E_2	(14,10)	[10,-2]
E_3	(-4/3,-5/3)	[0,-1]
E_4	(0,0)	[0.0]
E_5	(0,1)	[-4/3,5/3]
E_6	(10,2)	[14,-10]

both CSs. That is, we suppose that in S E_4 has the (x, t) coordinates (0,0) and in S' it also has [x', t'] coordinates [0,0]. (Although it is redundant, I shall distinguish coordinates in S from those in S' both by indicating the CS explicitly on all occasions and by writing the former with parentheses and the latter with brackets.)

We note that there are standard equations in STR (the Lorentz Transformations) which enable one to determine the coordinates of a point in one CS if one knows the coordinates of that point in another CS and the relative velocity of the two CSs. For instance, since the relative velocity v of S' with respect to S is known, if we are given the coordinates [x',t'] of a point in S', we can find the coordinates (x,t) that point in S by using the following equations:

$$(L1) \qquad x = (x' + vt') \; (1-v^2)^{-\frac{1}{2}},$$

$$t = (t' + vx') \; (1-v^2)^{-\frac{1}{2}}.$$

Conversely one can find the coordinates of a point in S' if one knows the (x,t) coordinates in S and the relative velocities of S and S'. The relevant transformations are:

$$(L2) \qquad x' = (x-vt) \; (1-v^2)^{-\frac{1}{2}},$$

$$t' = (t-vx) \; (1-v^2)^{-\frac{1}{2}}.$$

Notice that the point P, Ted's coincidence with the x-axis, has coordinates (10,0) in S, but in S' its second (or time) coordinate is <u>not</u> 0. (Use the Lorentz transformations to show this.) Thus from Alice's point of view (i.e., in S) P is simultaneous with E_4, but from Bob's point of view it is not. From Bob's point of view (i.e., in S') the point P' with coordinates [10,0] is simultaneous with E_4; but its time coordinate in S is not 0, so in S it is not simultaneous with E_4. These facts represent no paradox; they merely illustrate a fundamental feature of STR, the relativity of simultaneity, the principle that observers moving relative to each other will determine different pairs of points (or events) to be simultaneous by their own standard clocks. Although I have not proved it, it is useful to note that lines parallel to the x-axis contain points simultaneous in S whereas lines parallel to the line connecting E_4 to P' contain points simultaneous in S'.

Now, finally, let us tell a little story about our four observers, starting at E_1. The time of E_1 in S' is -5 (see the table of coordinates accompanying Diagram 1), and at E_1 Ted meets Carol and tells her that he loves her. At E_2 at time -2 in S' Carol emits a tachyon signal telling Bob that she is leaving him for Ted. At E_3 (time = -1 in S') Bob receives Carol's message. At E_4, which has time = 0 in both CSs, Bob tells Alice that Ted is leaving her for

Carol. Alice has never trusted Ted. At E_5, which occurs at t = 1 in S, she emits a tachyon signal which reaches Ted at E_6 (t = 2 in S) and detonates a tiny bomb that Alice has planted in the heel of Ted's shoe. Ted is blown to bits; hence E_1 does not occur since E_6 occurs on Ted's worldline <u>before</u> E_1. But if E_1 does not occur, then neither do E_2, E_3, E_4, E_5 and E_6, and it was the occurrence of E_6 that caused E_1 not to occur; so E_1 does occur. That is, E_1 occurs if and only if it does not occur. Here we have a genuine paradox.

3. The Reinterpretation Principle: We are not amused

The astute reader may have noticed a curious feature of the causal paradox described in the preceding section. As we told the story, shifting from one CS to the other, each tachyon signal moves forward in time. In S' the signal connecting E_2 to E_3, for instance, leaves E_2 at $t_2' = -2$ and arrives at E_3 at $t_2' = -1$, but in S this signal travels backwards in time, leaving E_2 at $t_2 = 10$ and arriving at E_3 at $t_3 = -5/3$.

This is not an isolated oddity. The Lorentz transformations can be expressed in the following form for time intervals:

$$\Delta t' = (\Delta t - (\Delta x)v)(1 - v^2)^{-\frac{1}{2}}$$

$$= \Delta t (\Delta t/\Delta t - (\Delta x/\Delta t)v)(1 - v^2)^{-\frac{1}{2}}$$

$$(2) \qquad = \Delta t (1 - Vv)(1 - v^2)^{-\frac{1}{2}}$$

where v is the velocity of S' relative to S and V is the velocity in S. We can see from formula (2) that if in S a tachyon signal moves forward in time ($\Delta t > 0$), but it has velocity V sufficiently large that Vv > 1, then that signal must move backwards in time in S' ($\Delta t' < 0$), since the quantity $(1 - v^2)^{-\frac{1}{2}}$, for $-1 < v < 1$, is considered to have physical significance only when it is positive.

This curious feature of tachyons, in conjunction with an even more curious one, provides the basis for the most important response by the Friends of Tachyons to the causal paradoxes. We can see from the following formula

$$(3) \qquad E' = E (1 - Vv)(1 - v^2)^{-\frac{1}{2}},$$

where E,E' are the energies of a particle in S, S' respectively, that a particle in S with positive energy could have in some other CS S' negative energy if its velocity V were sufficiently large. Formulas (2) and (3) are written in a form that makes it clear that a tachyon will appear in S' to have negative energy precisely when in S' it appears to be travelling backwards in time. This coincidence gave rise to the ideas that:

282

(a) A tachyon travelling backwards in time with negative energy
    can consistently be reinterpreted as a tachyon travelling
    forward in time with positive energy;
(b) This reinterpretation principle (RIP) dissolves the causal
    paradoxes.[8]

Let us suppose that tachyons can, like photons, be emitted and
absorbed by atoms. Then diagram 2 illustrates RIP. (After
Feinberg, 1970). The observer in S notes the following sequence of
events. At time t_0 atom A is at rest in its ground state and atom B
is at rest in an excited state. At t_1 atom B emits a tachyon (wavy
line), recoils (dashed line), and drops to its ground state. Atom A
is unchanged. At t_2 A absorbs the tachyon, jumps to an excited
state, and recoils.

Now let us describe using RIP these events as seen in some CS S'
in which $\Delta t = t_2 - t_1 < 0$ (that is, in which the event that occurs
at t_2 in S occurs at a time t_2' in S' and the event that occurs at
t_1 in S occurs at t_1' in S' and $t_2' < t_1'$). At t_0' atom A must be
in its ground state and atom B in its excited state; both must be
moving (dashed lines). At t_2' (the time when, in S, Atom A absorbed
the tachyon) Atom A emits a (positive energy) tachyon, loses some of
its kinetic energy, but jumps to an excited state. At t_1' (the time
when, in S, Atom B emitted the tachyon) Atom B absorbs the tachyon,
gains kinetic energy, and drops from its excited to its ground
state. Each sequence of events is consistent; neither involves a
negative energy tachyon moving backwards in time.

John Earman seems to doubt the very intelligibility of this
switching of the roles of absorbers and emitters. Let me quote him:

> In classical SRT, the momentum and energy of a particle form a
> four vector, and as such, the momentum-energy of a particle has
> a meaning independently of emission and absorption and energy
> transfer processes in general. If the four vector nature of
> momentum-energy is maintained for tachyons, it follows that the
> energy component of this vector will be negative for certain
> inertial observers. If this poses a difficulty for tachyons...
> how does any amount of talk about emission and absorption of
> energy transfer solve it? (1972, p.230)

Talk of emission and absorption is, I would suppose, meant to be
merely illustrative of the class of interactions between tachyons
and bradyons. The RIP could have been illustrated just as well in
terms of interactions with measuring instruments and their readings. The RIP is a suggestion as to how these readings would be
used to assign values to the components of the four vector, just as
other formulas of SRT tell us how to assign times to events based on
clock readings. These magnitudes do have independent meaning, but
they are covariant. The RIP would change the sign of the value

Diagram 2

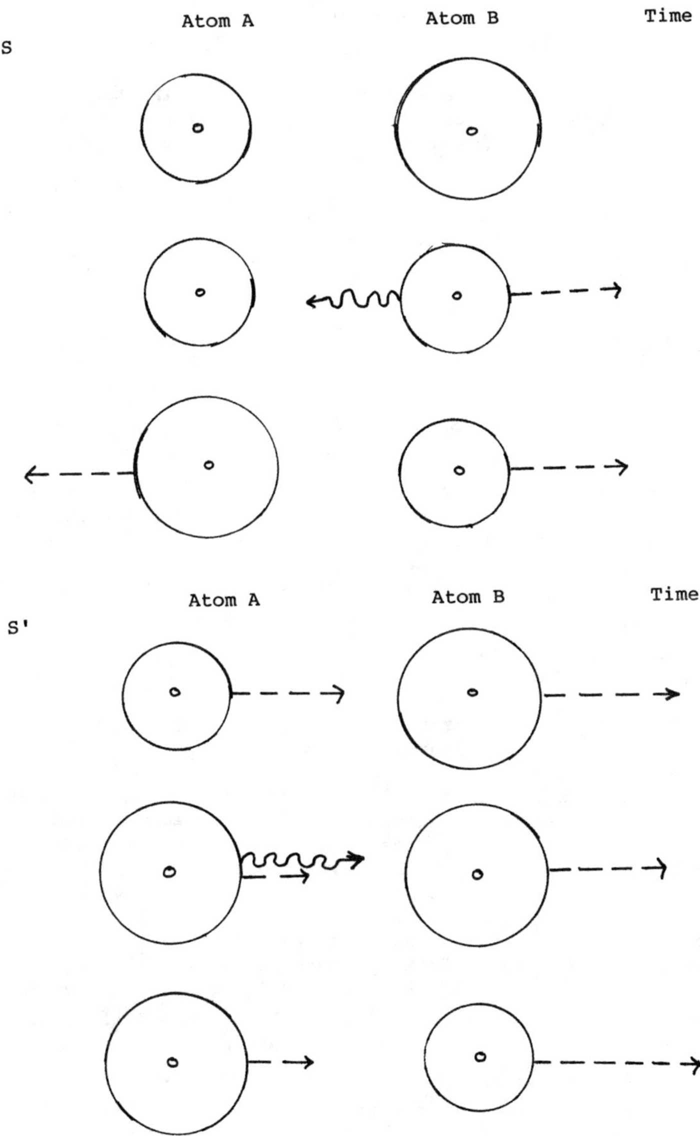

assigned to E in certain cases by changing the sign of a time interval.[9] Objecting that in those cases E must be negative seems to me no more forceful than objecting that for two events E_1: (x_1, t_1) and E_2: (x_2, t_2), if $t_1 - t_2 < 0$ (is negative) in S, then $t_1' - t_2' < 0$ in all other S' as well!

Earman's uneasiness may be caused by an ambiguity in statements of RIP that is reflected in (a) above. Does RIP say that there are indeed negative-energy particles but that they may always be consistently thought of as (reinterpreted as) positive-energy particles or that there are in fact no negative energy particles at all? The first reading would raise just the questions that Earman raises; the second reading escapes them, I think, so long as values can be assigned to the four vector that are consistent with assumed laws and conservation principles.[10]

Although Earman may have a deeper point in mind than I have recognized, let us suppose that RIP is at least _prima facie_ acceptable. How is it supposed to dissolve the causal paradoxes?

To see how, let us first notice a peculiar consequence of RIP. In the story I told in Section 2 a tachyon is emitted at E_2 by Carol and received at E_3 by Bob, moving forward in time in S'. Assuming that tachyons transmit causal influence, the absorption of the tachyon at E_3 is an effect of its emission at E_2. But since in S that tachyonic transmission seems to move backwards in time, RIP has us switch emission/absorption roles and thus in S E_3 is the cause of E_2 rather than its effect. Assignment of causes and effects is thus not Lorentz invariant; to adopt RIP is to abandon the following principle:

> (P2) Causal order is objective; i.e., it is the same for all observers.   (Earman 1972, p.226).

The Friends of Tachyons then claim that by using RIP a consistent causal story can be told in any given CS and thus that no paradox can arise. Root and Trefil, for instance, are credited (by the Friends of Tachyons, at least) with dissolving Bohm's paradox in "An Amusing Paradox Concerning Tachyons." (1970).

Transposed into my notation, their analysis is this. Consider first the whole set of events from the point of view of Alice and Ted. As before, they see a signal sent from E_5 to E_6 (when Ted is -- as they used to say -- neutralized). But, Root and Trefil claim, Alice and Ted also see a tachyonic message emitted at E_3 but now _not_ absorbed at E_2 (since E_1 does not occur and hence E_2 does not). The message, they claim, simply travels past Carol to be absorbed ultimately at some distant point in the universe. No paradox here.

"The amusing point", they write, "appears when we ask how this
sequence looks to [Bob and Carol] in the moving frame." (p. 414)
They see a tachyon signal sent from E_6 (at which time Ted disappears) to E_5. "Then", they write, "at precisely the moment when
[Carol] would have sent the tachyon from [E_2] a 'cosmic-ray tachyon'...would pass [Carol] and be absorbed at [E_3]. Thus it appears
as if a chance event...has acted in just such a way as to preserve
causality." (p. 414)

We are not amused by this point. Root and Trefil have not dis-
solved the paradox; they have just changed the story. E_1 and E_2
have been omitted and E_3 has been given a miraculous new sufficient
condition in the form of cosmic ray tachyons. But then we can
change our story. After being stood-up by Bob at E_1, Carol sets up
her cosmic ray tachyon deflector for a test and -- By golly! -- it
works with a bunch of tachyons that happens along at t_2. Then
what? If E_3 does not occur, why should E_4 or E_5? But if not E_5,
how can E_6 occur?

There is, however, a profound point lurking in the neighborhood
of all this foolishness, and it starts with an observation of Gerald
Feinberg's. In his 1967 paper Feinberg (pp. 1102-1103) claims

(i) that a tachyon signal moving backwards in time between two
observers A,B will not be regarded as carrying a message from A to B
but rather as a spontaneous emission from B to A, and
(ii) that observation (i) dissolves the causal paradoxes.

Now (ii) must be incorrect, for in the Bohm paradox each signal
goes forward in time from sender to receiver in their own rest
frame. But the profound point is that (i) must also be incorrect,
as was argued by Newton (1967).

Recall diagram 2, our explanation of RIP. Suppose that we have a
large number of tachyoctive, excited B-type atoms collected into a
volume -- a sufficiently large number that on the macro-level a
steady stream of tachyons is emitted. We funnel the stream through
a shutter which we control. We have a computer that randomly generates 0's and 1's at given intervals; and whenever a 0 is generated
we open the shutter for a short time. When the shutter is opened, a
burst of tachyons escapes to be absorbed by a collection of A-type
atoms some distance away.

That, at least, is how an observer 0 at rest (or moving slowly)
with respect to all this apparatus would describe it. But what says
an observer 0' moving so quickly with respect to this apparatus that
the time-order of events is reversed? Using RIP, he would have to
claim that spontaneous emissions occurred at the A-collection of
atoms just in time for them to travel through the shutter during the
brief random intervals when it is open, thence to be absorbed at the

B-collection. But Newton claims, and I think he is right, that the correlations between the random shutter openings and the emissions (as 0' sees them) "rules out" that the emissions from the A collection occurred spontaneously. It would indeed be astounding if the A collection should spontaneously emit bursts of tachyons which arrive precisely at the time the shutter, which is controlled by the random number generator, opens.

But if that option is indeed incredible, then all that 0' can conclude on the basis of the correlations he observes and his knowledge that the shutter is controlled by the random number generator is that the events at the shutter cause the events at the A-collection even though the events at the A-collection occur <u>before</u> the events at the shutter. According to Newton, then, even if tachyons exist, one must preserve the principle (P2) but at the cost of abandoning

(P1)   Effect never precedes cause.

And if we abandon (P1) we would have little reason to accept Feinberg's claim (i) above.

4.   The Relativity of Simultaneity

Thus were tachyons to exist, we would seem to be faced with a dilemma. It looks at first as if we must abandon either (P1) or (P2). Adopting RIP (and thus giving up (P2)) gets us little since, if the argument of Section 3 is correct, RIP will not dissolve the causal paradoxes. Should we then give up (P1)?

To throw some light on this question, let me turn to an argument presented in Feinberg's 1967 <u>Physical Review</u> article. "If faster than light particles existed", he wrote, "it might appear natural to use them to synchronize the clocks of observers in relative motion." (p. 1091). Indeed, the most natural assumption concerning tachyons is that their maximum velocity (Vmax) is infinite (a word more about this below). So between points on the world-lines of any two inertial observers, if these points are spacelike separated, it should be possible to send a tachyonic signal with infinite velocity -- that is, the transmission time must = 0. Why could this not be a time signal?

But if it can be a time signal, the results are dramatic. In diagram 1, for instance, we noted that in S the points E_4 and P were simultaneous, but that in S' the points E_4 and P' were simultaneous (transforming co-ordinates, of course, by the Lorentz transformations). Now we suppose that at E_4 Alice sends out a signal with $V = \infty$ in S (i.e., its worldline is the x-axis) announcing that time = 0. Carol receives this signal at the point (let's call it P*) at which her world-line crosses the x-axis. When she does receive the signal her clock should, according to the Lorentz transformations,

read -11 1/3. Carol might regard the signal as travelling backwards
in time; but if she does not do so -- that is, if she accepts (P1)
-- and she accepts Newton's arguments for (P2), then Carol has no
choice but to set her clock to 0 at least.

Notice that when Carol re-sets her clock the possibility of
causal paradox is eliminated. That is, if all inertial observers
set their clocks as Carol does, to agree with the time signals sent
out by Alice, there can be no tachyonic transmission backwards in
time, no negative energy tachyons. Does this at a bold stroke solve
the problems raised by the existence of tachyons for which the
friends of tachyons invoked RIP? Well -- yes, it does <u>that.</u>

But it creates its own problems in turn. For events in the CS of
Alice and Carol, as Feinberg observes, "would be related not by
Lorentz transformations, but by a new group of transformations, and
part of the justification for the requirement of Lorentz invariance
would be lost... . [I]n general, the laws of physics will not be invariant under the transformations obtained in this way." (p. 1091).
Feinberg clearly takes this argument to be a <u>reductio ad absurdum,</u>
and rightly so. Its conclusion could not be <u>true unless STR</u> were
radically false, since the principle of relativity -- the principle
that the laws of physics must be the same for all observers -- is at
the very core of STR, and STR is a highly confirmed theory. But
what, then, is Feinberg's argument a <u>reductio</u> of?

Of the idea that tachyon signals can transmit temporal information? If so, some reason is needed to convince us of this. I can
see no reason at all why tachyon signals, if there can be such signals at all, could not be used to set clocks.

The logical situation, then, looks to me like this. The Friends
of Tachyons use STR, though typically in an extended form, as the
basis for theorizing about tachyons. The Friends of Tachyons generally accept (P1). Friends of Tachyons, of course, take seriously
the possibility that

   (T) Tachyons exist

is true. The point of my (Feinberg's?) argument is that T, P1 $\vdash$
$\neg$STR. My conclusion is that the Friends of Tachyons have a problem And the problem can't be solved simply by giving up (P1) and
allowing retrocausation, for then

   (a) the causal paradoxes loom up again since we used only T,
       STR to derive Bohm's paradox, and
   (b) even if the causal paradoxes could somehow be blocked, the
       Tachyonite must begin again to make sense of negative energy
       particles.

There are other ways one might try to block my argument. Perhaps
there is a finite upper bound on the velocity of tachyon signals,

precluding instantaneous transmission. Is this maximum velocity (Vmax) invariant (like the velocity of light) or not?

Feinberg claims that it cannot be invariant.

> The invariance of the velocity of light from observer to observer depends not only on the use of light rays to synchronize clocks, but also on the empirical fact that for any observer the velocity of light is independent of its energy, i.e., of the velocity of the source of light. Since this cannot be made the case for tachyons, their velocity will also vary from observer to observer. (pp.1091-1092)

Nor could Vmax vary. Then there would have to be some CS or CSs which had the maximum Vmax, and it/they would be distinguished from all the others. Again, the principle of relativity would not hold.

Another possibility is that, although there are no infinite velocity (or "transcendent") tachyons, there is no finite upper bound on the velocity of tachyon signals in any CS. In this case Wesley Salmon (1975, pp. 118-122) has shown how such super-luminal signals could be used to define absolute simultaneity. After a brief consideration of the causal paradoxes raised by tachyon signals, he concludes:

> It seems to me that we are faced with the following dilemma: either tachyons cannot be used to send messages (and if not, why not?) or they can be used to establish absolute synchrony and absolute simultaneity, thereby eliminating the relativity of simultaneity which is so fundamental to the entire special theory.[11]

5. A Final Question.

Thus the position that I am arguing for, that T and STR are incompatible, is not novel. What may be novel is my explicit use of the following dilemma in arguing for it:

(a) $\neg$ P1, T, STR $\vdash$ Causal Paradoxes (section 3)
(b)   P1, T, $\vdash$ $\neg$ STR   (section 4)

This dilemma is an interesting one to Friends of Tachyons because, as Sudarshan put it, "...[W]e hardly think seriously of a physical theory which is not in accordance with the principles of the theory of special relativity." (1970, p. 129).

This dilemma is an interesting one to philosophers, too, but for other reasons. To see this, let's begin with a simple question: how could the existence of a tachyon be established? The simple answer is that one finds a particle at one place at one time and at another place at another time and calculates that its average veloc-

ity is greater than c. But how is the particle to be detected at a
place at a time; that is, how do tachyons interact with the ordinary
bradyonic matter out of which any detection device we use must be
constructed?

As far as I know, all theories developing the properties of tachyons are extensions of STR. If the thesis of this paper is correct,
unless some way is found to render the causal paradoxes innocuous,
then the existence of tachyons is incompatible with the truth of
STR. Finding a tachyon would, therefore, falsify the theory on the
basis of which the existence of the tachyon was established. Is
this a self-defeating procedure, or are there historical parallels
to show that science can progress in this apparently broken-backed
fashion? In other words, if the incompatibility thesis is correct,
what is the logic of the relation between experiment and theory in
the search for tachyons? I think that this question has not been
raised before, but it is worth serious consideration.

Notes

[1] I have had help from Paul Fitzgerald, R.I.G. Hughes, Arthur
Millman, Howard Pfeffer, and Tom Settle, to whom I extend my thanks.

[2] See Bilaniuk et al. (1962).

[3] A review of the theoretical and experimental literature up to
1974 can be found in Barashenkov (1974). See also the papers in
Recami (1978).

[4] Despite the fundamental importance of the topic, the philosophical literature on tachyons seems to consist of just three papers:
Fitzgerald (1971), Earman (1972), and Maund (1979).

[5] See Taylor and Wheeler (1966): 71-72.

[6] In other words, I take the proposition (T) "Tachyons exist" to be
logically equivalent to the proposition (C) "Events separated by a
space-like interval can be causally connected".

[7] This version of the causal paradoxes first appeared in Bohm
(1965). Tom Settle called my attention to it.

[8] A useful discussion of these suggestions, as well as most of the
basic issues concerning tachyons, can be found in Bilaniuk and
Sudarshan (1969) and in the replies in Bilaniuk et al. (1969).

[9] The expression for the energy component of the four vector is $E =
M(dt/d\tau)$ (Taylor and Wheeler, p. 112). The effect of RIP is to
change the sign of dt or of $(dt/d\tau)$.

[10] In Bilaniuk and Sudarshan (1969) RIP is unequivocal: "A 'nega-tive-energy' particle that has been absorbed first and emitted later is nothing else but a positive-energy particle emitted first and absorbed later... ." (p. 47).

[11] (1975, p. 122). Salmon ignores the RIP in his discussion of tachyons. His argument for the incompatibility of T and STR explic-itly uses (P1) (pp. 120-121).

References

Barashenkov, V.S. (1974). "Tachyons: Particles Moving with Velocities Greater than the Speed of Light." _Uspekhi Fizicheskikh Nauk_ 114: 113-149. (English translation: _Soviet Physics Uspekhi_ (1975) 17: 774-782.)

Bilaniuk, O.M.P. _et al._ (1962). "'Meta' Relativity." _American Journal of Physics_ 30: 718-723.

---------------- and Sudarshan, E.C.G. (1969). "Particles Beyond the Light Barrier." _Physics Today_ 22(5): 43-51.

---------------- _et al._ (1969). "More about Tachyons." _Physics Today_ 22(12): 47-52.

Bohm, David. (1965). _The Special Theory of Relativity._ New York: W.A. Benjamin.

Earman, John. (1972). "Implications of Causal Propagation Outside the Null Cone." _Australasian Journal of Philosophy_ 50: 222-237.

Feinberg, G. (1967). "Possibility of Faster-Than-Light Particles." _Physical Review_ 159: 1089-1105.

------------. (1970). "Particles that Go Faster Than Light." _Scientific American_ 222(2): 69-77.

Fitzgerald, Paul. (1971). "Tachyons, Backwards Causation, and Freedom." In _PSA 1970. (Boston Studies in the Philosophy of Science._ Volume VIII.) Edited by R.C. Buck and R.S. Cohen. Dordrecht: Reidel. Pages 415-436.

Maund, J.B. (1979). "Tachyons and Causal Paradoxes." _Foundations of Physics_ 9: 557-574.

Newton, Roger G. (1967). "Causality Effects of Particles that Travel Faster than Light." _Physical Review_ 162: 1274.

----------------. (1970). "Particles that Travel Faster than Light?" _Science_ 167: 1569-1574.

Recami, Erasmo (ed.). (1978). _Tachyons, Monopoles, and Related Subjects._ Amsterdam: North Holland.

Rolnick, William B. (1969). "Implications of Causality for Faster-than-Light Matter." _Physical Review_ 183: 1105-1108.

Root, R.G. and Trefil, J.S. (1970). "An Amusing Paradox Involving Tachyons." _Letters al Nuovo Cimento_ 3: 412-414.

Salmon, Wesley. (1975). _Space, Time, and Motion._ Encino, California and Belmont, California: Dickenson Publishing Co., Inc.

292

Sudarshan, E.C.G. (1970). "The Theory of Particles Traveling Faster
     Than Light I." In <u>Symposia on Theoretical Physics and
     Mathematics,</u> Volume 10. Edited by Alladi Ramakishnan. New York:
     Plenum Press. Pages 129-151.

Taylor, Edwin F. and Wheeler, John Archibald. (1966). <u>Spacetime
     Physics.</u> San Francisco: W.H. Freeman and Company.

Temporality, Secondary Qualities, and the Location of Sensations

Paul Fitzgerald

Baruch College/ City University of New York

Several philosophers in recent years, e.g., Baker (1974-5, 1979), Grünbaum (1971), Rescher (1973) and Salmon (1974) have argued that something they call "temporal becoming" is mind-dependent. They see the issue as analogous to the traditional one about the mind-dependence of secondary qualities. Do directly seen colors, felt coldness, and salty taste exist in the mind-independent world? Or only within experience? We know how the dialectic dances around that question....

"Roses are red, violets are blue, in the same way that some of our afterimages and sense-data are."

"No, roses and violets don't really have colors. Roses only have the property of selectively reflecting light in the 7200 Å region. That makes them *look* red to us. But they aren't really red. Science shows that."

"You're both wrong. Of course roses are red. But to be red *just is* to be the sort of thing which would *look* red to normal observers under standard conditions. Similarly, to be soluble is simply to have the dispositional property of dissolving-if-placed-in-a-solvent."

How about phenomenal spatiality and phenomenal temporality? Discussion of the mind-dependence of temporality has tended to *assume* that the classical secondary qualities are mind-dependent, and then argue that we should likewise accept the mind-dependence of temporal becoming. It has also tended to assume that the close temporal analogue of directly experienced phenomenal qualities of color, taste, and so on is *nowness* (cf. Baker 1979, Salmon 1974, and Grünbaum 1971). Nowness is supposed to be a property or "A-determination" (Gale 1968) expressible by such irreducibly indexical predicates as 'is occurring now' or 'exists at present'.

This second assumption is mistaken, as is argued by Fitzgerald (1980).

PSA 1982 Volume 1, pp. 293-303

The really close analogue of phenomenal sensory qualities is the phenomenal temporality or *elapsiveness* of which we are directly aware in our experience. This has no closer connection with an indexical *nowness* property than the phenomenal spatial features of our afterimages have with any alleged property of *hereness*. This claim will not be argued here. See Fitzgerald (1980).

Here I want to argue that defenders of the mind-dependence of temporal becoming are right at least to this extent. The classical problem of secondary qualities does arise for temporality (and spatiality as well). The visual appearances which the physical world presents to us are not located outside of our bodies. Our somatic sensations are not located in the parts of our bodies where we ostensibly feel them. Your toothache isn't in your tooth. If these sensations are anywhere in physical space they are in the brain. But even if they have physical location in the brain (whether literally, or in some extended sense), they also have *phenomenal* spatial and temporal features. "Features" refers here not to repeatable universals but to non-repeatable *abstract* or *Stoutian particulars*, to use the term coined by D.M. Armstrong (1978). Those features might well be numerically distinct from the physical spatio-temporal features of physical brain states. They might be *qualitatively* different as well. This is the analogue of the old question about whether the colors of our visual sense-data resemble anything in the physical world.

Apply this to time. I have a sense-datum which lasts for awhile. Is its durational or elapsive character numerically distinct from the physical duration of any physical event or chunk of spacetime? Is it qualitatively similar to any such physical duration or proper time? These questions are genuine ones, I hope to show. Defenders of the claim that "temporal becoming" is mind-dependent may be right in thinking that phenomenal temporality is mind-dependent. They are wrong only in thinking that it has some peculiarly close connection with indexicality, or that reflecting on "nowness" and the ways of indexicals could take us far toward solving the real analogue for time of the secondary qualities problem.

Many would deny that there is a genuine "secondary qualities problem" for spatiality and temporality. Some deny there's a problem because, as direct realists, they deny the split between sensory appearance and physical reality which raises the problem. Some deny it on the grounds that sensations are brain states and have no non-physical phenomenal features. I want to argue, against both these groups, that there is a genuine problem of the relation(s) between phenomenal spatiality and temporality, and physical spatio-temporality. Direct realism is wrong. Both the visual appearances of material objects and our own somatic sensations have no location in physical space, except possibly in the brain of the perceiver. And if sensations aren't in physical space, how can they literally be in physical time, given the way relativity theory welds space and time together? Even if sense-data are somehow located in the brain, they have *phenomenal* spatial and temporal features which *may* be distinct and different in kind from any *physical* spatio-temporal features. Ugly old dualism threatens, at least dualism of *aspects*. There is in fact a non-physical phenomenal realm whose connections with physical space and time are problematic. This strains the commonsense belief that sensations are located in physical time in the same straightforward way that physical events are. I don't like this dualism at all. But I find all the attempts to avoid it specious and unacceptable.

Part of the case for this dualism consists in showing that somatic sensations are not located in the parts of our bodies where we ostensibly *feel* them. Your toothache isn't in your tooth. The case of somatic sensations is important because of all phenomenal entities they seem commonsensically to have the strongest claim to be located in physical space, and outside the brain. Here it will be argued that neither they nor physical sense-data are outside the brain. And even if they are in some sense *in* the brain, they still have phenomenal spatial and temporal features whose connection with physical spacetime is problematic.

Some say that it is a matter of "logical grammar", not to be gainsaid by philosophical argument, that somatic sensations are where we ostensibly feel them. Of course a toothache is in a tooth. A few would even say that when I sincerely believe that my toothache is in my right bicuspid, or the searing is in my left arm, *it must be so*. But how could my toothache be in my right bicuspid if the dentist has extracted it and it's been ground to powder by a garbage disposal unit? The searing pain can't be in my left arm if the arm's been amputated and is now miles away. For that reason defenders of the "conceptual truth" claim might be tempted to add that when we say our aches are in our teeth or limbs we are not claiming *literal* spatial location for them at all.

I would welcome this non-literalist interpretation, if it could be established. It would place me in the happy position of not being in conflict with what ordinary people think and say. I could argue my case that somatic sensations are not *literally* located in physical space. Then I would add that when you say, in an everyday way, that there's a pain in your tooth, you are not saying anything that disagrees with my claim. For you are not attributing literal location in physical space to the toothache. And that literal locatedness is all I meant to deny.

Alas, I see no way to establish this non-literalist interpretation. How can we show that people are not attributing literal spatial location to their sensations when they locate them in parts of their bodies? To be sure, you don't expect the dentist to *see* or *feel* your toothache when he probes your tooth. And the oculist won't *see* the spots-in-front-of-your-eyes when he gazes at the space before your eyes. But these facts don't show that you are not attributing literal location in physical space to the aches and spots. Maybe you just haven't thought through the implications of what you're saying. Or maybe you don't expect literally located sensations to be publicly observable, because they are not publicly observable kinds of things even though they do have literal physical locations. Or maybe you are implicitly attributing to them a kind of literal spatial location but a *relational* location, as when we say that the glare is *on* the blackboard *from* over here. To see the glare you have to occupy the place *from* which it exists on the blackboard. Other people are not located in the place (presumably that occupied by your brain) *from which* they are there before your eyes. That is why they can't see the spots. Because of all this, I find it hard to establish that people are not attributing to their sensations literal locations in physical space. Even if they are not, we can go on to ask whether sensations do have such locations.

Many philosophers would reject our question about whether aches, afterimages, and the like are located in physical space by saying that it presupposes an illegitimate act-object analysis of sensations. This analysis involves an alleged distinction between the act (state) of sensing or feeling, and the X which is immediately sensed or felt, and which is also called a *sensation* or *feeling*. (Indeed, that distinction is presupposed here). It is then argued against me that this act-object analysis is a mistake. We should adopt instead an *adverbial analysis*. "I have a toothache in my right bicuspid." should not be analyzed as predicating a four-place relation of a person, a time, a tooth, and an ache. On the adverbial analysis the statement is viewed as analogous to "She broke the vase by accident.". This does not predicate a four-place relation binding a person, a past time, a vase, and something called "accident" *by which* she broke the vase. Rather, it is equivalent to "She broke the vase accidentally.". Similarly, the statement about the toothache is equivalent to "I feel *(toothache-in-right-bicuspid)-ly*.", where the italicized phrase is an adverbial modifier of the verb "feel". And instead of wondering whether visual sense-data exist in physical space, have backsides, consist of unextended geometrical points, and are three-dimensional, we are encouraged to say simply that we *are appeared to redly,* or *roundly,* thus avoiding reification of appearances and phony philosophical puzzles. The classical secondary qualities problem does not arise. Neither do its analogues for temporal becoming or spatial extension. Adverbial theories were championed by G.Dawes Hicks (1917), C.J. Ducasse (1951), R. Chisholm (1957,1977), W. Sellars (1968,1975), J.Cornman (1975) and others. They have been attacked by F. Jackson (1977), R. Grossmann (1976), R. Clark (1975) and P. Butchvarov (1980). Ducasse applied an adverbial theory to the spatio-temporal features of experience explicitly, in saying that we "may or must speak of being aware not only *bluely* but also *briefly* (or perhaps lengthily), *extensionally* (or perhaps punctually), *here-ly* (or perhaps there-ly)... ." (Ducasse 1951, quoted by Butchvarov 1980, p. 269).

It seems to me that the adverbial theory will not work. Phenomenological inspection reveals that the act-object analysis *is* appropriate for visual, somatic, and other sensations. The problems which the adverbial theory seeks to avoid should not be brushed under the rug by artifices like adverbial paraphrase. They should be faced squarely and solved.

Consider. It is perfectly natural to say things like "I just had two afterimages, the smaller one a pale green, and located to the right of the larger, deeper green one.". This on the face of it seems to be saying that there were two Xs, each colored green, spatially located (in some sense), and having spatial properties and relations. We could infer from the quoted statement that something pale green stood in a spatial relation to something else, which was both darker and larger than the first thing. These inferences make sense if an act-object analysis of visual experience is correct, and we are here talking about the objects of that experience. To capture the inferences we would want to use the whole apparatus of pronominal cross-reference which is captured formally by quantification over individual variables: "$(\exists x)(\exists y)(x \neq y \,\&\, Rxy....)$".

On the adverbial analysis, we would presumably have something like this. "I was just appeared to *(two afterimages, the smaller one a pale green, and located to the right of the larger, deeper green one)*-ly." The best that the adverbial theorist can plausibly hope for is that the references to afterimages will be sealed up within some kind of brackets, the whole to be regarded as an adverb (formally, a predicate-forming operator on predicates). Inside the adverb we have quantifications, in story-talk about a world of non-real Xs such as afterimages and toothaches. This story-talk cannot be sullied by quantifiers from without. It must be hermetically sealed off from talk about the real.

But should it be thus sealed off? Don't we want to be able to say things like "There are three letters on the eye chart, but without my glasses I see only two blurs when looking at it. So there is one more letter than there are blurs."? This requires quantifiers to trespass into the sealed precincts of the adverbial construction. (Of course, we also want to make such numerical comparisons between *fictional* entities and real ones. I am not here arguing that sense-data are *real*, as opposed to fictional or merely possible, because we need to quantify over them. My argument is just that *adverbial* "eliminations" of them won't work, because we need to quantify over them.)

Besides, what is gained by the adverbialization maneuver, once even purely internal quantification is admitted within the adverbial operator? The same old questions about sense-data arise, only now they appear in the form of questions about *which* intra-bracket inferences are permissible. From the fact that I was just appeared to redly and roundly, we can infer that I was just appeared to coloredly and shapedly. Can we infer that I was just appeared to three-dimensionally? Was I appeared to *something-with-a-backside*-edly? Answering that last question is tantamount to answering the old conundrum about whether or not sense-data have backsides. Many old problems about sense-data will simply trot out again, dressed now as questions about the logical relations among appearing predicates or operators. Nothing is gained by the adverbial appearing theory. We have to quantify over appearances or sense-data in any complete story about the world, and the appearing theory won't really help us avoid the problems this raises.

Similarly, some have tried to conjure away visual impressions, afterimages, hallucination images, and other kinds of sense-data by means of *behaviorist* or else *topic-neutral analyses*. I take it that Jackson (1977) and others, such as Bradley (1963), Noren (1970), Campbell (1971) and Schaffer (1968) have shown the inadequacies of such attempts.

Suppose then we adopt an act-object analysis of sensation. Still, why can't somatic sense-data be in those regions of physical space where we spontaneously locate them! Isn't your toothache *in your tooth*? Or, if the tooth is gone, isn't the toothache in the place by your gum where the tooth was when you had one? I would like to argue against this suggestion by calling attention to some general similarities between visual sense-data and other kinds, such as somatic. An important similarity is that both kinds are proximately caused by states of the perceiver's brain. So it is plausible to hold that if somatic sense-data were located *in* the regions of the body where they seem to be, then the visual sensory appearances of things would be located outside the body, where *they* seem to be. But visual sense-data are *not* located outside of the perceiver's body. (We'll see why not in a minute.) Therefore somatic sense-data are not in the places where they seem to be, either.

Imagine you were one of the Chinese astronomers who in 1054 A.D. saw the supernova in the Crab nebula. The supernova explosion had taken place 6000 years earlier. It reveals itself to you by a bright, glittery spot in your visual field. If you hold a rose-tinted, transparent glass in front of your eyes the spot changes hue, taking on a rosy cast. That rose-tinted appearance of the long-gone explosion is a visual sense-datum. Is it located outside your body, in physical space?

I think not. The proximate cause of the sense-datum consists of neural events in your brain (perhaps in the visual centers of the occipital lobes, perhaps not). Do these neural events cause the sensa-datal appearance of the supernova to pop into existence outside you in physical space? If so, *where* in space, exactly? I will argue that it is impossible to assign a location in any reasonable way, and that *that* is a good reason for saying the sense-datum is not located outside you. But let's check all this.

Is the sense-datal appearance of the supernova located where the supernova occurred and emitted the light that is just reaching you? Sounds good. But what does it mean? What we count as the same place lasting through time depends on what reference frame we freely decide to pick. There is no such thing as *the* place, lasting through time, at which the explosion occurred. Even within pre-Einsteinian physics there would be an infinity of non-overlapping place-times simultaneous with your here-and-now brain events, each of which can be regarded as a later stage in the history of the place at which the supernova occurred, 6000 years ago. Are we to suppose that the brain picks out a particular one of these place-times and causes the sense-datum to leap into life at that particular one? What is the brain mechanism whereby this selective feat of causality at a distance is brought about? Why does the brain "choose" the particular place-time which it does? And how does it "rule out" ineligible candidates, ones which could not possibly be later stages of any place where the supernova occurred? We have no reason to believe there is any such mechanism, and no way of discovering the location of the sense-datum out in space. To say that it is at a specific place out in space is to posit a state of affairs which is empirically undetectable in principle. That is the sort of thing which led Einstein to deny that there is real rest and motion with respect to the ether. There is not. And there is no such thing as the place in space outside the perceiver where his sense-datum exists as he is sensing it. See Diagram I.

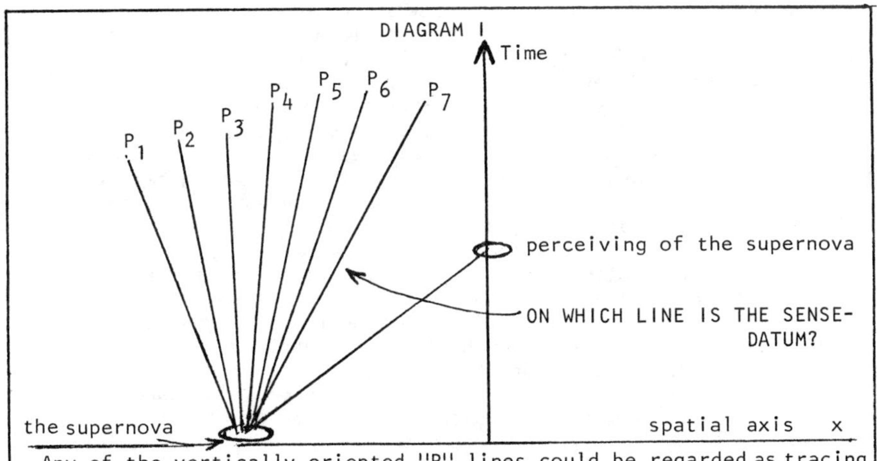

Any of the vertically oriented "P" lines could be regarded as tracing the later history of "*the* place" where the supernova occurred. At which of these places is the perceiver's sense-datum of it located?

There is another point. According to relativity, there is no such thing as *the* time when anything happens, either. Simultaneity is frame-relative. So we have an additional problem for the claim that the perceiver's brain events cause a sense-datum to come into existence out in space *simultaneously* with them. Simultaneously, relative to what reference frame? And why that particular one? How does the brain "know" or "decide" which *frame-time* to choose as the time at which the supernova sense-datum is to be brought into existence, way far away? Once again, we have a question which is empirically undecidable in principle.

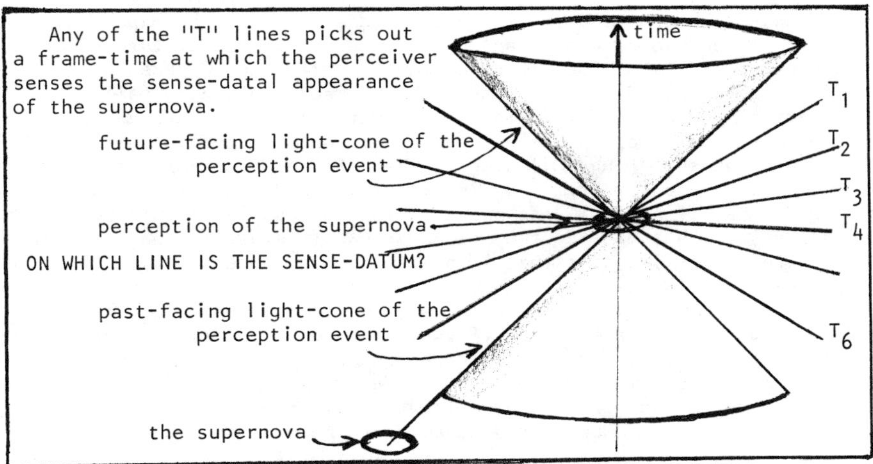

Relativity to reference frames is what causes the problem here. There
is a way to avoid that. We might try the idea that when you look at the
supernova, your supernoval sense-datum is caused by your brain to come into existence back at the place-time where the supernova occurred 6000
years before. Of course, that place-time is in the absolute past of the
brain events which "generate" your sense-datum. So we get a cause which
is later than its effect. Bizarre. But less trouble than it might seem.
For we can hold, plausibly, that the sense-datum existing back there at
the place-time of the supernova can have no physical effects back there
and then. So we have no worry about temporally closed causal chains.

The trouble lies elsewhere. How does the perceiver's brain "know" the
place-time at which the supernova presently seen actually occurred? The
light waves hitting the observer's retinas carry no signs betraying the
identity of the place-time from which they were dispatched. And the
brain has no fine-tuned mechanism for causing the sense-datum to come into existence at the exact place-time 6000 years or so ago where the light
waves started out from the explosion. If the sense-datum is not located
at that (somewhat fuzzily demarcated) place-time, so far away and long ago,
then *when* and *where* is it located? Whatever (doubtless arbitrary) answer
is given, the same answer should be given in the case of a simulated Crab
supernova, staged by cleverly arranged lighting on the roof of a planetar-
ium, fifty feet above the perceiver. For the physical stimuli impinging
on the observer in *that* case may be exactly similar to those which hit
the Chinese observers who saw the real thing. Is each sense-datum fifty
feet from the brain which causes it? Six thousand light-years away?

Clearly, neither of these. There is no plausible way to defend the
claim that *any* visual sense-data are located in spacetime outside the body
of the person who senses them. In short, the *spots-in-front-of-your-eyes*
are not in front of your eyes. Nor, by the way, are they *in* your eyes.
Think of having two eyes and a single afterimage. Which is the lucky eye
in which the image lurks?

I suggest that what holds for visual sense-data can be expected to hold
for tactual and somatic ones as well. Take *phantom pain,* throbbing "in"
a non-existent limb. Does it throb out there in empty space, where the
sufferer's leg would be, if he still had a leg? Do others fail to feel
it there simply because it exists there not *simpliciter* but *perspectivally,*
that is, *in* that empty place but only *from* the sufferer's brain, where its
proximate cause is found? I suggest that if the pain exists anywhere in
space outside the sufferer's brain, it should exist out there where his
leg would be, if he still had one. It is implausible to hold that if the
sufferer still has his leg then his brain would cause the pain to exist
out there in the leg, but that if his leg is no longer there, an identical
brain state will cause the pain to exist somewhere else instead, such as
in the stump. Further, differences between tactual-somatic and visual
sense-data depend not on the difference between the natures of the nerve
messages transmitted from different sense organs, but only on the fact that
they terminate in different places in the brain. A bit of re-wiring would
translate the visual stimuli from the Crab supernova into somatic sense-
data. A stubbed toe would show up as stars before the eyes. This makes

me believe that if visual sense-data are not outside the brain, then somatic ones aren't either.

Why not *in* the brain, then? An identity or a double-aspect theory (which accepted an act-object analysis of sensation) might well locate our sense-data in our brains, where their proximate physical causes or correlates are. This could always be done in a rough way, assuming that the required physical brain correlates do exist. A sense-datum would be located where its physical correlate is. But unless further conditions were satisfied, this would be a brute verbal convention, defining a physical location for sense-data only in a sense *analogous to* that in which physical things and events themselves have location in spacetime. To see this, suppose for example that the physical correlate of a single, circular afterimage were a spatially scattered brain state. We could adopt the convention of saying that the single (arcwise connected) afterimage had as location the mereological sum of scattered places in the brain where its physical correlate occurred. But this would not permit us to locate the various subparts of the image within corresponding subparts of the scattered physical location. And it would not permit us to identify the particular *roundness* of the afterimage with any physical shape-feature of its physical correlate. (The roundness of which I speak is not the universal but the abstract particular; what D.M. Armstrong (1978) calls a "Stoutian particular".) If the geometry of the phenomenal visual space differed from that of physical apace, for example, in not having uncountably many points between any pair of locations, then our convention of locating sense-data in the brain would be somewhat arbitrary. They would have location there only in a weak or analogous sense.

On the other hand, if there were a strong isomorphism between the phenomenal spatial features of our sense-data and the physical spatial features of their brain correlates, then the sense-data would be *strongly located* in the brain. I have in mind a situation in which, e.g., phenomenal parts of sense-data could plausibly be paired off with physical parts of their brain correlates. In that event, we might argue that the phenomenal spatial features of the sense-data (say, the roundness of the afterimage) are numerically identical to the physical spatial features of the brain correlates (e.g., the roundness of the place in the cortex in which the physical correlate occurs). Partly analogous remarks hold for the phenomenal *durational* features of our sense-data and the physical temporal features of their correlates, if any.

Since it is not known at present even that there are physical brain correlates of our sense-data, much less what their features are, we don't know at present whether sense-data are strongly, or only weakly located in the brain. Since there is no location in physical time without location in physical spacetime, the way in which sense-data are spatially located in the brain bears on the ontological status of their temporal features. We can at present only set up the questions and alternative answers as clearly as possible. Further empirical input is needed to give us the answers. Perhaps acts of sensing are literally brain states, strongly located in physical spacetime, but having a non-physical simultaneity relation with their sense-datal objects, which thereby get weak location there. Just a speculation. Let us have some more.

References

Armstrong, David M. (1978). _Universals and Scientific Realism._ 2 Vols. London: Cambridge University Press.

Baker, Lynne Rudder. (1974-5). "Temporal Becoming: The Argument from Physics." _The Philosophical Forum_ 6: 218-236.

——————————————. (1979). "On the Mind-Dependence of Temporal Becoming." _Philosophy and Phenomenological Research_ 39: 341-357.

Bradley, M.C. (1963). "Sensations, Brain Processes, and Colours." _Australasian Journal of Philosophy_ 41: 385-393.

Butchvarov, Panayot. (1980). "Adverbial Theories of Consciousness." In _Studies in Epistemology._ (_Midwest Studies in Philosophy,_ Volume 5.) Edited by Peter A. French, _et al._ Minneapolis, Minn.: University of Minnesota Press. Pages 261-280.

Campbell, Keith. (1971). _Body and Mind._ London: Macmillan.

Chisholm, Roderick. (1957). _Perceiving, A Philosophical Study._ Ithaca, N.Y.: Cornell University Press.

——————————————. (1977). _Theory of Knowledge._ 2nd ed. Englewood Cliffs, N.J.: Prentice Hall.

Clark, Romane. (1975). "The Sensuous Content of Perception." In _Action, Knowledge, and Reality._ Edited by Hector-Neri Castañeda. Indianapolis, Ind.: Bobbs-Merrill. Pages 109-127.

Cornman, James W. (1975). _Perception, Common Sense, and Science._ New Haven: Yale University Press.

Dawes Hicks, G. (1917). "Are the Materials of Sense Affections of the Mind?" _Proceedings of the Aristotelian Society_ 17: 434-445. (Symposium with G.E. Moore, W.E. Johnson, J.A. Smith, and James Ward.)

Ducasse, Curt J. (1951) _Nature, Mind, and Death._ LaSalle, Ill.: Open Court Press.

Fitzgerald, Paul. (1978). Review of Mundle (1971). _Philosophy of Science_ 45: 165-169.

——————————————. (1979). Review of Jackson (1977). _International Philosophical Quarterly_ 19: 103-113.

——————————————. (1980). "Is Temporality Mind-Dependent?" In _PSA 1980._ Volume 1. Edited by Peter D. Asquith and Ronald N. Giere. East Lansing, Mich.: Philosophy of Science Association. Pages 283-291.

Gale, Richard. (1968). _The Language of Time._ New York: Humanities
    Press.

Grossmann, Reinhardt. (1976). Review of Cornman (1975). _International
    Studies in Philosophy_ 8: 210-213.

Grünbaum, Adolf. (1967). _Modern Science and Zeno's Paradoxes._
    Middletown, Conn.: Wesleyan University Press.

----------------. (1971). "The Meaning of Time." In _Basic Issues in
    the Philosophy of Time._ Edited by E. Freeman and W. Sellars.
    LaSalle, Ill.: Open Court Press. Pages 195-228.

Jackson, Frank. (1977). _Perception: A Representative Theory._ New
    York: Cambridge University Press.

Mundle, C.W.K. (1971). _Perception: Facts and Theories._ New York:
    Oxford University Press.

Noren, S.J. (1970). "Smart's Materialism, The Identity Theory, and
    Translation." _Australasian Journal of Philosophy_ 48: 54-66.

Rescher, Nicholas. (1973). _Conceptual Idealism._ Oxford: Basil
    Blackwell.

Salmon, Wesley. (1974). "Memory and Perception in _Human Knowledge._"
    In _Bertrand Russell's Philosophy._ Edited by George Nakhnikian.
    London: Gerald Duckworth & Co. Pages 139-167.

Schaffer, Jerome. (1968). _Philosophy of Mind._ Englewood Cliffs, N.J.:
    Prentice Hall.

Sellars, Wilfrid. (1968). _Science and Metaphysics._ New York:
    Humanities Press.

----------------. (1975). "The Adverbial Theory of the Objects of
    Sensation." _Metaphilosophy_ 6: 144-160.

Smart, J.J.C. (1963). _Philosophy and Scientific Realism._ New York:
    Humanities Press.

The Apparent Inconsistency of Moulines'
Treatment of Equilibrium Thermodynamics

John H. Harris

University of Otago

1. Introduction

   One of the main advantages that has been claimed for the Sneedian
"structuralist" approach is that with it one can actually represent
real scientific theories. (I consider a "structuralist" approach to
be one in which the objects of study are structures in the model-
theoretic sense, with at least some such structures thought of as
representing physical situations or processes.) Using just Suppes'
structuralist approach, one can already give a reasonable representation of many features of scientific theories. Examples of such are
the representations of classical particle mechanics (CPM) in McKinsey,
Sugar & Suppes (1953), relativistic particle mechanics in Rubin &
Suppes (1954), and rigid body mechanics in Adams (1959). Then with
Sneed's (1971) modifications of Suppes' approach, in particular with
Sneed's introduction of the notion of a constraint between structures,
the representational strength of a structuralist approach seems to
increase dramatically. Kuhn in his (1976, p.182) describes the purpose of Sneedian constraints as insuring that "the values assumed by
theoretical functions in one application must, for example, be compatible with those assumed in others." He goes on to say the following.
"Together with the correlated notion of applications, that of constraints constitutes what I take to be the central conceptual innovation of Sneed's formalism." (Kuhn 1976, p.182).

   Though Sneed includes constraints with what he calls the mathematical formalism of a scientific theory, he never shows that they play a
key role in the presentation or theoretical development of such
formalism (as versus playing a key role in how the formalism is related
to applications of the theory.) It is in the work of Moulines (1975),
and so far mainly in this paper, that one finds convincing evidence
that Sneedian constraints play an essential role even in the purely
theoretical development of the mathematical formalism of at least some
scientific theories, in particular, for simple equilibrium thermodynamics (S.E.T.). In §2 I will attempt to show that Moulines' treat-

---

PSA 1982, Volume 1, pp. 304-311

ment of S.E.T., call it S.E.T.$_M$, is either inconsistent or unintelligible, with the problem being due to the way Moulines _represents_ constraints. In §3 I will ouline an alternative approach to representing constraints which is both formally intelligible and consistent.

2. The Apparent Inconsistency of S.E.T.$_M$

S.E.T.$_M$ is rather technical, as it rightly should be. But the problem with S.E.T.$_M$ is a very basic one; it is even an obvious one once you see it. So fortunately we will not need to review and use very much of the complicated Sneedian formalism.

A Sneedian (representation of a) scientific theory is an ordered pair $\langle K,I \rangle$, where K is a class called the _core_ of the theory and supposedly represents the mathematical formalism of the theory, while I is roughly speaking the set of (representations of) successful applications of the formalism. But Moulines never works with I and just focuses on the core part K.

A Sneedian core K is a 5-tuple $\langle M_p, M_{pp}, r, M, C \rangle$ where the nature of the various components are described in any good treatment of the Sneedian approach. (For two of the latest cf. Stegmüller (1979) or Niiniluoto (1979).) Moulines (1975, p. 122) -- from now on all page references will be to that paper - presents the Sneedian core for S.E.T. as $K^T = \langle \widehat{ET}, \widehat{SS}, r_T, \widehat{SET}, \mathcal{C}^T \rangle$ (where the 'T' is, I presume, to be suggestive of 'thermodynamics'). The components $\widehat{SS}$, $\widehat{ET}$ and $\widehat{SET}$ are well-defined (apparently proper) classes, being the extensions of three earlier defined set-theoretic predicates: SS(x); ET(x) - x is an equilibrium thermodynamics system; SET(x) - x is a simple ET system (cf., D1-D3, pp.111-112). The third component, r_T, is quite superfluous, being definable in terms of $\widehat{ET}$ and $\widehat{SS}$. The last component, $\mathcal{C}^T$, the "Sneedian constraint class", is loosely defined on p.122 in terms of eight constraint conditions, represented by C1-C8, which Moulines introduced earlier. (pp. 119-122)

Though the problem with S.E.T.$_M$ has to do with constraints, it has nothing to do with $\mathcal{C}^T$ for Moulines never makes use of $\mathcal{C}^T$. In fact I don't see how anyone could and I wonder whether Sneed even thinks that anyone could make use of a constraint class (as versus specific constraints) when _using_ (as versus just trying to abstractly characterize) the formalism of a specific scientific theory.

The problem with S.E.T.$_M$ also has nothing to do with Moulines' choice of constraints. The ones he lists seem quite reasonable when he describes them verbally. The problem with S.E.T.$_M$ is caused by the combination of the way Moulines _formally represents_ constraints and the way he uses these representations.

His representations of constraints are the rather formal set-theoretic formulas C1-C8 mentioned earlier. (We will soon list and consider two of these.) And he really does make use of these formulas. In particular he uses C1-C8 in giving rather formal looking proofs of two very significant results that a physicist or physical chemist would expect to hold in any good formulation of S.E.T., namely the positivity of thermodynamic temperature (Th.4, p.123) and Clausius' principle (Th.5, p.124). For example in his proof of Th.4 he says that step (6) is justified "by [steps] (2),(3),(5),C8"; that step (7) is justified "by (2),(5),C4", etc. Now in any proof the various steps are justified by axioms, previous steps and/or rules of inference. Since constraints don't seem to be rules of inference, they or at least their representations C1-C8 must be treated as axioms, you would think.

There is a problem with Sneedians using C1-C8 as axioms: it seems to lead to S.E.T.$_M$ being inconsistent. And in such cases we don't even need C1-C8 collectively to get an inconsistency. Any one of these formulas, if conjoined as an axiom with set theory seems to lead to an inconsistent theory.

The easiest way to get the idea of what seems to be wrong with adding any of the formulas C1-C8 as axioms is to give an example not in S.E.T.$_M$ but in a simpler mathematical theory where everyone is familiar with the notions. For instance, rather than considering $ET(x)$, let us instead consider the predicate 'x is a group'. In particular, say we define x is a group iff $x = \langle G, \circ \rangle$ for some non-empty set G and some binary operation $\circ : G \times G \to G$ satisfying the usual group axioms. Let $\mathcal{G}$ be the class of all groups. And instead of considering one of Moulines' constraints on $\widehat{ET}$-structures, let us consider a constraint on $\mathcal{G}$-structures, i.e., on groups. In particular, consider the constraint that group operations associated with groups of the type we want to consider should agree where their domains overlap. In the spirit of Moulines, such a constraint would be expressed by say

$$\text{CG} \quad (\wedge x,y,G,G',\circ,\circ')(\langle G,\circ \rangle \in \mathcal{G} \wedge \langle G',\circ' \rangle \in \mathcal{G} \wedge x \in G \cap G' \wedge y \in G \cap G'$$
$$\to \quad x \circ y = x \circ' y).$$

But the formula CG is incompatible with the formalism of group theory that we have introduced because we can find, within set theory, examples of groups not satisfying CG, hence in set theory we can prove

$$\overline{\text{CG}} \quad (\vee x,y,G,G',\circ,\circ')(\langle G,\circ \rangle \in \mathcal{G} \wedge \langle G',\circ' \rangle \in \mathcal{G} \wedge x \in G \cap G' \wedge y \in G \cap G'$$
$$\wedge \quad x \circ y \neq x \circ' y)$$

which contradicts CG. Thus conjoining the constraint CG with the theory of groups gives a formally inconsistent system.

The above way of generating an inconsistent theory is a rather glaring one. Let us now look at Moulines' work and see if he has indeed made a similar mistake. To set the stage we first need to introduce some of Moulines' special notation. He defines $\widehat{ET}$ to be the class of all 8-tuples $x = \langle Z,V,I,N,P,E,U,S \rangle$ satisfying certain conditions such as $\emptyset \neq E \subseteq Z$, $U:Z \to \mathbb{R}$ and $S:Z \to \mathbb{R}$. And he defines $\widehat{SET}$-structures to be $\widehat{ET}$-structures satisfying certain additional conditions, what you might call the fundamental laws of S.E.T. (In the $\widehat{ET}$- and $\widehat{SET}$-structures we are interested in, the first (Z), sixth (E), seventh (U) and eighth (S) components will represent a continuous process, the set of equilibrium states of that process, the energy function, and the entropy function on the states that make up that process, respectively.) He (p. 118) also introduces a suggestive indexing notation such that, for example, S_Z denotes the entropy function and E_Z, the set of equilibrium states associated with process Z.

We now have sufficient notation to present two of Moulines' constraints exactly as he did. They are:

C1 $\bigwedge z,Z,Z'(z \in E_Z \wedge z \in Z' \to z \in E_{Z'})$;

C5 $\bigwedge z,Z,Z'(z \in Z \cap Z' \to U_Z(z) = U_{Z'}(z) \wedge S_Z(z) = S_{Z'}(z))$.

Before we consider how to interpret these formulas, let us note that Moulines' subscripting notation is actually unacceptable. After all, given two $\widehat{ET}$-structures with the same first components, there is no guarantee that they will have the same second through eighth components. But this is what his notation 'E_Z','U_Z','S_Z', etc., presupposes.

In fact what would legitimize such notation are the very constraints C1, C5 and some others which he hasn't given. To avoid confusion, we will use a better notation. Define eight functions Z,V,I,N,P,E,U,S on $\widehat{ET}$ as follows. For any $x \in \widehat{ET}$, let Z_x be the first component of x,... and S_x be the eighth component of x (itself a function from Z_x into the reals). In this new notation Moulines' C1 and C5 become

C1' $\bigwedge z,x,y (z \in E_x \wedge z \in Z_y \to z \in E_y)$;

C5' $\bigwedge z,x,y (z \in Z_x \cap Z_y \to U_x(z) = U_y(z) \wedge S_x(z) = S_y(z))$.

Now to discuss the intuitive constraints which the formulas C1 and C5 (or C1' and C5') are supposed to represent. As Moulines puts it (p. 119) C1 is to represent the idea that "the property of being an equilibrium state is independent of the system to which the state belongs." And according to what he says (p. 120) we should think of C5 (or our C5') as representing - he says "being" - "...the identity constraint for energy and entropy analogous to the identity constraint for mass in mechanics stated by Sneed." So in words, what C5 (or C5') is <u>meant</u> to say is that the energy and entropy functions associated with two $\widehat{ET}$-structures <u>of the type we want to consider</u> should agree where their domains overlap. But his notation doesn't seem to cope with the "of the type we want to consider" aspect, which is the very aspect constraints are supposed to cope with.

Let us look at the formulas C1' and C5' (or C1 and C5). Now the very first thing a logician or even a mathematician is going to ask when he looks at the '$\bigwedge z,x,y$' part (or the '$\bigwedge x,Z,Z'$' part) of the formulas is "What are the ranges of those quantified variables?" The only two classes of x's already defined by Moulines for which the Z_x, E_x, U_x and S_x appearing inside Cl' and C5' make sense are $\widehat{ET}$ and $\widehat{SET}$. So as a first guess one would assume that Moulines means x and y to range over one of these classes, take your pick. But it is then not difficult to construct examples of $\widehat{ET}$-, even $\widehat{SET}$-structures disobeying Cl' and C5', hence it is easy to prove within "$\widehat{SET}$-theory"

$\overline{Cl}'$ $\quad \bigvee z,x,y \ (z \in E_x \wedge z \in Z_y \wedge z \notin E_y)$ $\quad$ and

$\overline{C5}'$ $\quad \bigvee z,x,y \ (z \in Z_x \cap Z_y \wedge (U_x(z) \neq U_y(z) \vee S_x(z) \neq S_y(z)))$

where x and y are interpreted as ranging over $\widehat{ET}$ or $\widehat{SET}$, whichever one you chose. Clearly $\overline{Cl}'$ and $\overline{C5}'$ are incompatible with Cl' and C5' respectively, being the negations of Cl' and C5'; hence $\widehat{SET}$-theory conjoined with Cl' or C5' is inconsistent, hence S.E.T.$_M$ seems to be inconsistent, as I claimed.

One of the weak points in our argument was the need to guess at what Moulines would have meant the range of x and y to be in Cl' and C5' when we quantified over them in the form '$(\bigwedge z,x,y)$'. We guessed it to be $\widehat{ET}$ or $\widehat{SET}$. Maybe he meant some other class. Let us rewrite Cl' and C5' to allow for this possibility. For any $X \subseteq \widehat{ET}$, consider

Cl'_X $\quad (\bigwedge z)(\bigwedge x,y \in X)(z \in E_x \wedge z \in Z_y \rightarrow z \in E_y)$;

$C5'_X$ $\quad (\bigwedge z)(\bigwedge x,y \in X)(z \in Z_x \cap Z_y \rightarrow U_x(z) = U_y(z) \wedge S_x(z) = S_y(z))$.

Now for any class X that we might <u>define</u>, it is clear that using just set theory we will probably be able to prove Cl'_X , in which case there is no point in stating it as a constraint, or prove the negation of Cl'_X, in which case adding the constraint Cl'_X as an axiom would lead to an inconsistent formalism; and similarly for $C5'_X$. (I say 'probably' because there would be those rare examples of a defined X where Cl'_X is undecidable within any present standard version of set-theory.) So generalizing our previous argument about the inconsistency of S.E.T.$_M$ to the case when Cl' or C5' are replaced by Cl'_X or $C5'_X$ for any definable X you want to choose doesn't improve the situation: if we gain anything by conjoining such constraint axioms, we gain an inconsistent system.

The other weak point in our argument, the one I imagine Sneedians would jump on considering their extreme anti-linguistic bent, is our assumption that Cl and C5 are axioms or could even be represented by axioms. But since Cl-C8 are used to justify various steps of some of Moulines' proofs, if they aren't axioms, then they must be (functioning as) rules of inference, at least according to the present day canons of what constitutes a proof. But if they are rules of inference, then they don't seem to be of any type studied so far by logicians, and apparently they wouldn't be finitary rules of inference. To put Moulines' treatment of S.E.T. on a firm foundation, do we really need

to go to such extremes as leaving the realm of finitary proof theory?
And Sneedians can't pull out that over-worked escape clause that they
are using acceptable informal set theory, so everything is alright.
Informal it may be, but it sure doesn't seem to be acceptable.

In summary, by the canons of what mathematicians and especially
logicians accept these days, S.E.T.$_M$ seems to be either inconsistent
or unintelligible.

Maybe Sneedians have a simple solution to the problems raised above.
I don't know. But in the next section I will present a very simple
non-Sneedian solution to these problems.

3. A Consistent Approach to Representing Constraints

Whether one prefers a "statement" or a "non-statement" view of
scientific theories, one surely must admit that scientists do use the
mathematical formalism of a scientific theory to carry out deductive
style arguments , sloppy and full of hidden premises though they may be.
It is very clear on Suppes' "non-statement structuralist" view how one
should represent and think of such derivations. One defines certain
set-theoretical predicates such as SET(x) and maybe the corresponding
classes of set-theoretic structures such as $\widehat{\mathrm{SET}}$. One can then derive
within set-theory (augmented only by various definitional axioms)
various statements about such structures. Since such set-theoretical
statements are analytically true, following as they do from just set
theory, everything is still in keeping with a "non(-contingent)-
statement" view.

When it comes to representing an actual scientific theory on a
Sneedian "non(-contingent)-statement" view, one has a problem as we
have seen: How do we represent the constraints-part of the mathematical
formalism in a way that could actually be used in derivations carried
out by scientists and philosophers of science interested in foundations? Moulines' approach to this problem in the case of S.E.T. seems
to fail as we have shown in §2. Another approach, following the method
of representing scientific theories used in Harris (1978) or (1983),
seems to solve this representational problem in a particularly simple
way, as we will now see.

All those who favour a "non-statement" view, either of the type
proposed by Suppes or Sneed, labour under what seems to me to be the
needless constraint of requiring the <u>total</u> separation of the mathematical formalism of a scientific theory from any indication of its
interpretation. Let us see how easy things become if one drops this
constraint. In the case of representing S.E.T. via statements about
$\widehat{\mathrm{SET}}$-structures, why not think of there being a class, I^T, of "intended"
$\widehat{\mathrm{SET}}$-structures which S.E.T. is meant to be about. We can't <u>define</u> I^T
just as we can't define Sneed's class I in a Sneedian theory $\langle K,I \rangle$.
In fact we might not even have a good heuristic idea of what is in I^T.

But that can't stop us from introducing 'I^T' as a class symbol in the language of set-theory to represent I^T, as vague as it is, and proposing some axioms about I^T - i.e., about the use of 'I^T' - which will narrow down the classes which 'I^T' could represent. One such axiom would be $\ulcorner I^T \subseteq \widehat{SET} \urcorner$. (Thus I^T wouldn't be the same as Sneed's class I since in the case of S.E.T. Sneed would require $I \subseteq \widehat{SS}$, or in the latest version I know of, $I \subseteq \mathcal{P}(\widehat{SS})$ - cf., Stegmüller (1979, p.90); and $\widehat{SS}$-structures are 5-tuples whereas $\widehat{SET}$-structures are 8-tuples.) Other axioms would be to the effect that structures in I^T are related by Moulines' constraints. For example, we could use $C1'_X$ and $C5'_X$ with $X = I^T$, call them C1" and C5". It would be trivial to go through Moulines' paper and convert C2 to a corresponding C2", etc. And note: the Sneedian constraints are now being fruitfully represented by axioms. With just such a simple change as suggested above it seems that we could salvage all of Moulines' work.

Note how our method of showing S.E.T.$_M$ inconsistent is no longer available. We can still find $\widehat{ET}$-, even $\widehat{SET}$-structures x,y such that
$$(*) \qquad (z \in E_x \wedge z \in Z_y \wedge z \notin E_y) \text{ for some } z.$$
But we obviously can't show such structures are in I^T because we haven't pinned I^T down that much. In fact C1" claims that I^T doesn't contain any $\widehat{ET}$-structures for which (*) holds.

All of the above is compatible with a "non-statement" view if one assumes that axioms about I^T such as $\ulcorner I^T \subseteq \widehat{SET} \urcorner$, C1" and C5" are meaning postulates on I^T (or on the use of 'I^T'), hence are analytically true. In this case anything derived from just these axioms would also be analytically true, hence would represent a non-contingent statement. My suggestion also seems to be compatible with Niiniluoto's idea - see his (1979, p.16) - of modifying a Sneedian theory to a triple $\langle K,I,J \rangle$ where in the case of S.E.T. the set J of "theoretical applications of the core K" would be I^T. But my suggested approach to modifying S.E.T.$_M$ is also compatible with a statement view when some of the axioms on I^T are considered to be proper (contingent) axioms. In either case we definitely get a contingent theory when we also add some axiom to the effect that a particular Sneedian non-theoretical type structure - some particular $\widehat{SS}$-structure in the case of S.E.T. - can be extended to an I^T-structure.

References

Adams, E.W. (1959). "The Foundations of Rigid Body Mechanics and the Derivation of its Laws from Those of Particle Mechanics." In _The Axiomatic Method._ Edited by L. Henkin, _et al._ Amsterdam: North Holland. Pages 250-265.

Harris, J.H. (1978). "A Semantical Alternative to the Sneed-Stegmüller-Kuhn Conception of Scientific Theories." In _The Logic and Epistemology of Scientific Change_ (_Acta Philosophica Fennica_ 30.) Edited by I. Niiniluoto and R. Tuomela. Amsterdam: North Holland. Pages 184-204.

------------. (1983). "Existential Claims in Physical Theories." Forthcoming in _Boston Studies in the Philosophy of Science._

Kuhn, T.S. (1976). "Theory-Change as Structure Change: Comments on the Sneed Formalism." _Erkenntnis_ 10: 179-199.

McKinsey, J.C.C., Sugar, A.C., and Suppes, P. (1953). "Axiomatic Foundations of Classical Particle Mechanics." _Journal of Rational Mechanics and Analysis_ 2: 253-272.

Moulines, C.U. (1975). "A Logical Reconstruction of Simple Equilibrium Thermodynamics." _Erkenntnis_ 9: 101-130.

Niiniluoto, I. (1979). "The Growth of Theories: Comments on the Structuralist Approach." In _Proceedings of the Second International Congress for History and Philosophy of Science._ Edited by E. Agassi and J. Hintikka. Dordrecht: Reidel. Pages 3-46.

Rubin, H. and Suppes, P. (1954). "Transformations of Systems of Relativistic Particle Mechanics." _Pacific Journal of Mathematics_ 4: 563-601.

Sneed, J. (1971). _The Logical Structure of Mathematical Physics._ Dordrecht: Reidel.

Stegmüller, W. (1979). _The Structuralist View of Theories._ Berlin: Springer-Verlag.

Part VIII

Levels of Explanation in Biology

<u>The Levels of Selection</u>

Robert Brandon

Duke University

It is a mistake to suppose that the units of selection controversy in biology centers around a single question. In this paper I will take Wimsatt's recent work (1980 and 1981) as defining the question 'What are the units of selection?'. I will show that there is another important question, what I will call the <u>levels of selection</u> question, separable from the first but easily confused with it. Finally, I will try to show why the levels of selection question is important.

First I must make a terminological point. In this paper I will adopt the distinction between fitness and adaptedness. I will not defend the distinction here since that has been done elsewhere (Brandon 1978 and 1981), but I will briefly indicate what the distinction is. 'Fitness' in this usage refers to actual reproductive success. 'Adaptedness' refers to an expected fitness value, in the mathematical sense of expected value.[1] The role of relative adaptedness in evolutionary theory is to explain differential fitness <u>via</u> the principle of natural selection. The distinction is not as simple as the above might indicate; some complications are relevant to the topic at hand, but exploring them goes beyond the scope of this paper.

I am going to distinguish levels of selection from units of selection. This distinction has not been made before but the seeds of it may be found in Mayr (1963). There he marshaled two arguments against the view that adaptedness can and should be attributed to genes. First, he pointed out that the fitness, i.e., reproductive success, of a gene depends as much on the surrounding genes (the genetic environment) as it does on the ecological environment. Thus, he argued, "no gene has a fixed selective value, the same gene may confer high fitness on one genetic background and be virtually lethal on another." (1963, p. 296). This he termed the <u>genetic theory of relativity</u>. (Also see Mayr 1954.) Second, he pointed out that "natural selection favors (or discriminates against) phenotypes, not genes or genotypes." (1963, p. 184).

PSA 1982, Volume 1, pp. 315-323

316

One may, as I believe Mayr did, take both points as arguments
against attributing adaptedness to genes, but they may also be
taken as two distinguishable arguments for two closely related but
importantly different conclusions. The first point, the genetic
relativity point, addresses the units of selection question; the
second the levels of selection question.

Lewontin (1974) developed Mayr's genetic relativity point in a
rigorous and convincing manner. In brief, he argued as follows:
In standard cases of organismic selection (i.e., selection among
organisms within an interbreeding population) genes do reproduce
at different rates and some alleles increase in frequency relative
to others. Thus genes can be assigned fitness values (values which
are just summaries of what has happened at the genic level). But
because of interactive effects among genes and also gene linkage
these fitness values are mere statistical abstractions; they are
of little or no predictive value. Thus, the argument goes, genes
are not the units of organismic selection since they are not the
proper units of fitness. The proper unit of organismic selection is,
in general, the genome.

Lewontin's work provides the foundation for Wimsatt's definition
of units of selection which is as follows:

> A <u>unit of selection</u> is any entity for which there is heritable
> <u>context-independent</u> variance in fitness among entities at that
> level which does not appear as heritable context-independent
> variance in fitness (and thus, for which the variance in fitness
> is <u>context-dependent</u>) at any lower level of organization.
> (Wimsatt 1981, p. 144)[2].

So in standard cases of organismic selection where there are
epistatic interactions among genes, genes are not the units of
selection due to the fact that the variance in fitness is not context-
independent at the genic level. Thus the units of selection in such
cases are some higher level entities, in the limit, the whole genome.

There is another important point one might make concerning
organismic selection, a point corresponding to Mayr's dictum that
selection acts directly on phenotypes, not genes or genotypes.
That is, that what I have characterized as organismic selection is
<u>in fact</u> selection at the organismic level. How and why might one
argue for this point?

Most biologists would give lip-service to Mayr's dictum yet many
have not taken it seriously. Thus reductionistically inclined
biologists will admit that the process I've called organismic
selection has as its mechanism the differential reproduction of
organisms, but will claim that organismic selection <u>really</u> acts
directly on genes. Using the Salmon-Reichenbach (Salmon 1971)
notion of <u>screening-off</u> it can be shown that this claim is simply
wrong.

The basic idea behind the notion of screening-off is this:
If $\underline{A}$ renders $\underline{B}$ statistically irrelevant with respect to outcome $\underline{E}$
but not _vice versa_, then $\underline{A}$ is a better causal explainer of $\underline{E}$ than is
$\underline{B}$. In symbols, $\underline{A}$ screens off $\underline{B}$ from $\underline{E}$ if and only if:

$$P(\underline{E},\ \underline{A}\cdot\underline{B}) = P(\underline{E},\underline{A}) \neq P(\underline{E},\underline{B})$$

(Read '$P(\underline{E},\underline{A}\cdot\underline{B})$' as the probability of $\underline{E}$ given $\underline{A}$ and $\underline{B}$.) The
screening-off relation is an asymmetric relation defined in terms
of statistical relevance that is meant to capture causal relevancies.
Salmon has argued that causes screen off symptoms from effects. For
example, a sudden drop in atmospheric pressure screens off a drop
in barometer reading from the occurrence of a storm since,
$P(\underline{s},\underline{a}\cdot\underline{b}) = P(\underline{s},\underline{a}) \neq P(\underline{s},\underline{b})$. (Where '$\underline{s}$', and '$\underline{a}$' and '$\underline{b}$' stand for the
obvious.) Similarly he has argued that common causes screen off
common effects from one another. And most importantly for present
purposes, more immediate causes screen off more remote causes.[3]

When concerned with the differential reproduction of organisms
(which is the mechanism of organismic selection) phenotypes screen
off genotypes (and so genes as well). That is, $P(\underline{n},\underline{g}\cdot\underline{p}) = P(\underline{n},\underline{p}) \neq$
$P(\underline{n},\underline{g})$. (Where '$\underline{n}$' stands for 'having $\underline{n}$ offspring', '$\underline{g}$' for 'having
genotype $\underline{g}$' and '$\underline{p}$' for 'having phenotype $\underline{p}$'.) _Gedanken_ experiments
suffice to establish the relevant equality and inequality. Basically
the idea is that tampering with the phenotype without changing the
genotype can affect reproductive success (as my castrated cat
testifies), while tampering with the genotype without changing
the phenotype cannot affect reproductive success.[4] What is true
of the relation between phenotype and genotype obviously holds of
the relation between phenotype and gene as well.

I have argued that in episodes of organismic selection phenotypes
screen off both genotypes and genes from the reproductive success
of organisms. This, I think, is a precise and systematic way to
explicate the principle that organismic selection acts on phenotypes,
not directly on genes or genotypes. Importantly, the screening-off
relation between phenotype and genotype _vis-a-vis_ organismic
reproductive success is asymmetric. This is in spite of the fact
that causality is, we may suppose, transitive. Schematically we
may say that having genotype $\underline{g}$ causes phenotype $\underline{p}$ which causes
reproductive success $\underline{n}$. But that does not imply that the relation
between phenotype and genotype is symmetric; as we have seen it is
not. Thus screening-off provides us the means for answering the
question: 'At what level does the causal machinery of organismic
selection really act?'

I have argued that the level of organismic selection is the
organismic phenotype. Notice that this conclusion is different
from the Lewontin-Wimsatt conclusion that the unit of organismic
selection is, in general, the genome. To show that two arguments
are different it should suffice to show that they have different
conclusions; but to be safe let me point out that the premises of

the two arguments are different as well. The Lewontin-Wimsatt
argument depends on certain genetic facts, most important among them,
that epistatic interactions do occur (epistasis) and that not all
genes segregate independently (gene linkage). If the facts were
different, their conclusion would be different. If the "bean-
bag" picture of genetics were true, that is, if genes segregated
independently and if the one-gene-one-trait hypothesis were true
(and further if there were no interactive effects among traits at
the phenotypic level), then genes would be the units of organismic
selection. In Wimsatt's terms the variance in fitness at the
genic level would be context-independent. None of those facts are
relevant to my screening-off argument. Let the "bean-bag" picture
be true, phenotypes would still screen off genotypes from the
reproductive success of organisms and so selection would still be at
the level of the organismic phenotype.

Have I argued that all selection must be at the level of the
organismic phenotype? No. Selection may occur at many levels.
Shortly I will offer a general definition of levels of selection,
but first let us consider selection at the group level.

Natural selection is the differential reproduction of biological
entities which is <u>due to</u> the differential adaptedness of those
entities to a common environment (Brandon 1978, 1981). Group
selection then is the differential reproduction of biological groups
which is due to the differential adaptedness of those groups to a
common environment. Thus a necessary condition for the occurrence
of group selection is that there be differential reproduction
(propagation) among groups. But this necessary condition is not
sufficient. For the differential reproduction of groups to be
group selection, i.e., selection at the group level,[5] there must
be some group property (which abstractly characterized is the group
adaptedness) which screens off all other properties from group
reproductive success.

It is by no means necessary that such a property exists. For
instance, if the fitness of a group is simply a linear function
of the sum of the fitness values of its members, then group
"adaptedness" does not screen off all nongroup properties from
group reproductive success.[6] In particular it does not screen off
the aggregate of the individuals' adaptedness values. The
probability of such a group producing $\underline{n}$ propagule groups given the
"adaptedness" value of the group and the adaptedness values of each
of the members of the group does equal the probability of the
group producing $\underline{n}$ propagules given just the adaptedness value of
the group. In symbols, $P(\underline{n}, A_G \cdot [a_1 \cdot a_2 \cdot \ldots \cdot a_k]) = P(\underline{n}, A_G)$.
(Where A_G is the adaptedness value of the group and a_i is the
adaptedness value of the ith member of the group.) However,
it is not the case that $P(\underline{n}, A_G) \neq P(\underline{n}, a_1 \cdot a_2 \cdot \ldots \cdot a_k)$. Thus the
group adaptedness does not screen off the aggregate of the group
members' adaptedness values.

Notice that the aggregate doesn't screen off the group adaptedness
value either; the relation between the two is symmetric. That
symmetry relative to group reproductive success gives sense to the
claim that the group "adaptedness" value is "nothing more than" the
aggregate of the individual adaptedness values.[7]

In summary, group selection occurs if and only if: (1) There
is differential reproduction of groups; and (2) Group adaptedness
values screen off all other properties (of entities at any level)
from group reproductive success (see footnote 8 for qualification).
One way to restate clause (2) is this: Differential group
reproduction is best explained in terms of group level properties
(in terms of differential group adaptedness to a common environment).
Still another way to restate (2), a way that would have appeared
question-begging prior to what has been said concerning screening-
off, is this: The causal process of selection acts directly on
groups.

What has been said about group selection is easily generalizable
into the following definition:

> Selection occurs at a given level if and only if:
> (1) There is differential reproduction among the entities at
>   that level; and
> (2) The adaptedness values of these entities screen off the
>   adaptedness values of entities at every other level from
>   reproductive values at the given level.[8]

Let us return to the case mentioned above where there was
differential group reproduction but no group selection. Recall that
the group "adaptedness" values did not screen off the aggregate
of the individual adaptedness values from group reproductive success.
Thus, according to our definition, the case was not a case of group
selection. But also recall that the individual adaptedness values did
not screen off the group adaptedness values from group reproductive
success. Should one conclude from this that selection is not
occurring at the individual (organismic) level either? No. That
individual adaptedness values do not screen off group adaptedness
values from group reproductive success does not entail that individual
values do not screen off group values from individual (organismic)
reproductive success. The later is what is required by our definition
for selection to be at the level of the individual organism.

I should also point out that our definition allows, as it should,
selection to act simultaneously at various levels. For example,
in the well known case of the t-allele in the house mouse Mus
musculus (see Lewontin and Dunn 1960) selection acts at three levels.
I will not discuss this case in detail since Sober (1981) has
recently done so. But briefly, in this case there is differential
reproduction at three levels: the chromosomal, the organismic and
the group. In the formation of sex cells chromosomes containing the

t-allele are disproportionately produced relative to their homologous competitors lacking the t-allele. At this level a chromosomal property screens off all others from the differential reproduction of this pair of chromosomes. At the organismic level males homozygous for the t-allele are sterile. This organismic property, sterility, (organismic adaptedness of 0) screens off all others from this lack of organismic reproductive success. Finally at the group level, groups whose males are all homozygous for the t-allele go extinct. This is not analogous to the case of differential group reproduction which was not group selection since here group reproductive success cannot be satisfactorily explained in terms of the aggregate of the group members' adaptedness values. That would be the case if we were dealing with groups all of whose members were sterile. But here only the males are sterile. That fact explains the group extinction only when conjoined with other facts about the group structure, such as, females mate only with males from their own group. In sum, according to the screening-off criterion the above example is indeed one where selection occurs at three distinct levels.

(I should at least mention that cases of simultaneous multi-level selection raise complications for the notion of adaptedness. But dealing with these complications will have to await another paper.)

I have shown that my definition of levels of selection is not equivalent to Wimsatt's definition of units of selection. In cases of organismic selection, i.e., selection at the level of the organismic phenotype, the units are genomes, gene-complexes or even genes, depending on the amount of epistasis and gene linkage. Yet it is curious that at many other levels of selection, e.g., chromosomal, kin-group, group, the units and levels analyses agree.[9] (I must leave it to the reader to confirm this assertion for himself.)

Why is this? Some units (e.g., groups) interact directly with their environment. Differential reproduction of these entities is best explained in terms of differences in their (direct) interactions with the environment. Other units (e.g., genes or even whole genomes) interact with their environment mediately through their phenotypes. The screening-off argument shows that their differential reproduction is best explained in terms of properties of their mediating phenotypes. For directly interacting units the units and levels analyses agree. For mediately interacting units the two analyses disagree, with the level of selection being the level of the mediator.[10] I am not convinced that this answer is wholly satisfactory. I am convinced it is headed in the right direction.

At the beginning of this paper I said that there are at least two important questions that have heretofore been conflated by people working on what has been called the units of selection problem. I have offered a bipartite defense of this claim: (1) I have given a definition of levels of selection and have shown that it is not equivalent to what is the best extant definition of units of selection; and (2) I have tried to motivate the levels of selection question.

Basically I have argued that if one is concerned with explaining the causal process of natural selection then one is interested in the levels of selection question. Put differently, if one is concerned with the question 'At what level(s) does the causal machinery of natural selection really act?', one is concerned with levels of selection. The above question is fairly described as an ontological question. We reify the things that we decide are _really_ acted on by selection. For example, genes have been, mistakenly I believe, reified so that selfishness is attributed to them. The question is of obvious philosophical, as well as biological, interest.

My argument that there are two important questions in this area has an obvious gap. I have not explained why the units of selection question is important. I have pointed out that some of the considerations that have previously been used to motivate that question properly apply only to the analysis of levels of selection. It remains for others to motivate the units of selection question.

Notes

[1]This point is also made by Mills and Beatty (1979) but they stick with the term 'fitness'.

[2]The terminology used here is potentially misleading. To say that variance in fitness is context-independent is not to say that the actual fitness values are not dependent on the actual context, i.e., environment. Obviously fitness values are context-dependent in that sense. But the actual _variance_ in fitness may or may not be attributable to differences between entities in a given environment. That part of the variance that can be so accounted for is context-independent variance. Sober (1981, especially footnote 7) fell victim to this confusion. The reason context-independence is important is that only the context-independent variance is heritable. For a discussion of the same point in different terms see Brandon (1981).

[3]Although the first two claims are debatable, I know of no plausible counterexample to the claim that more immediate causes screen off more remote causes from their effects.

[4]The latter claim is controversial. One might argue, as David Hull and an anonymous referee have suggested, that it is possible to tamper with the sex cells (e.g., by irradiation) and affect re-productive success without changing the phenotype. I view this as a conceptual question and would take my claim as a fixed point in drawing the phenotype-genotype and levels of selection distinctions.

322

[5]Henceforth when I speak of group selection I will just mean
selection at the group level; likewise for chromosomal selection,
organismic selection, kin-group selection, etc.

[6]Here adaptedness values are attributed to groups only tentatively
while the hypothesis of group selection is still alive. See Sober
(1981) for discussion of an example of differential group reproduction which is not group selection.

[7]Salmon's notion of screening-off has been used before in the
literature on the units of selection. In discussing a case
relevantly similar to the above Wimsatt (1980, 1981) claimed that
the group adaptedness value is screened off by the individual
values. This, as we have seen, is not the case since the relevant
inequality does not hold.

[8]A stronger, I think unnecessarily stronger, version of (2) reads
as follows: (2) The adaptedness values of these entities screen off
all other properties of entities at any level from the reproduction
values at the given level. I have eschewed this stronger clause
since in this paper I cannot demarcate legitimate properties from
the potentially problematic crazy pseudoproperties philosophers
are fond of dreaming up.

[9]I will not discuss Sober's (1981) definition of group selection.
Wimsatt (1981) argues that when suitably generalized Sober's
definition is equivalent to his definition of units of selection. I
think that Wimsatt is wrong and that Sober's aim is closer to mine,
but I am not sure of this.

[10]In Hull's (1980) terminology, the units and levels analyses
agree if and only if the units are <u>interactors</u>.

References

Asquith, P.D. and Giere, R.N. (eds.). (1981). _PSA 1980,_ Volume 2. East Lansing, Michigan: Philosophy of Science Association.

Brandon, R.N. (1978). "Adaptation and Evolutionary Theory." _Studies in History and Philosophy of Science_ 9: 181-206.

------------. (1981). "A Structural Description of Evolutionary Theory." In Asquith and Giere (1981). Pages 427-439.

Hull, D. (1980). "Individuality and Selection." _Annual Review of Ecology and Systematics_ 11: 311-332.

Lewontin, R.C. (1974). _The Genetic Basis of Evolutionary Change._ New York: Columbia University Press.

-------------- and Dunn, R. (1960). "The Evolutionary Dynamics of a Polymorphism in the House Mouse." _Genetics_ 45: 705-722.

Mayr, E. (1954). "Change of Genetic Environment and Evolution." In _Evolution as a Process._ Edited by J. Huxley, _et al._ London: Allen and Unwin. Pages 157-180.

--------. (1963). _Animal Species and Evolution._ Cambridge: Harvard University Press.

Mills, S. and Beatty, J. (1979). "The Propensity Interpretation of Fitness." _Philosophy of Science_ 46: 263-286.

Salmon, W.C. (1971). _Statistical Explanation and Statistical Relevance._ Pittsburgh: University of Pittsburgh Press.

Sober, E. (1981). "Holism, Individualism and the Units of Selection." In Asquith and Giere (1981). Pages 93-121.

Wimsatt, W.C. (1980). "Reductionistic Research Strategies and Their Biases in the Units of Selection Controversy." In _Scientific Discovery, Volume II: Historical and Scientific Case Studies._ Edited by Thomas Nickles. Dordrecht: Reidel. Pages 213-259.

--------------. (1981). "The Units of Selection and the Structure of the Multi-level Genome." In Asquith and Giere (1981). Pages 122-183.

Grades of Organization and the Units of Selection Controversy[1]

Robert C. Richardson

Department of Philosophy, University of Cincinnati
History and Philosophy of Science, University of Pittsburgh

1. Organization and Group Adaptation

R. C. Lewontin (1970) proposed three criteria as jointly sufficient for a unit of evolutionary change and immediately went on to
argue that selection could operate simultaneously at a variety of
levels of organization. For natural selection to operate on a unit,
there must be _variation_ in the objects in question: natural selection "chooses" between alternatives, so there must be different alternatives. This variation must affect _fitness_: the variation available must create different probabilities of reproductive success.
Third, this variation in fitness must be _heritable_: there must be a
correlation between parents and offspring for the traits yielding
variation in fitness. In brief, for there to be evolution by natural selection, there must be _heritable variation in fitness_.

Among the concerns that might be raised over Lewontin's formula,
one is particularly important. It was clearly brought out by
Williams (1966) and is emphasized by Elliott Sober (1981). As
Williams explains, we should distinguish sharply between a population of adapted individuals and an adapted population: if each
individual in a population is adapted, then since each individual
will enjoy reproductive success (measured, e.g., in contribution to
further generations), the population considered on the whole will also be successful (measured, e.g., in terms of the lapse of time or
generations to extinction). Thus, population success is not necessarily indicative of a higher level "biotic adaptation", in which
there are mechanisms designed to promote the success of biota, or
groups. It could as well be due to adaptation at the individual
level or lower. Williams' constraint on units of selection thus
acknowledges a preference for lower levels: "Any feature of the
system that promotes group survival and cannot be explained as an
organic [i.e., lower level] adaptation can be called a biotic adaptation." (1966, p. 108). Williams' standard may be thought of as

---

PSA 1982, Volume 1, pp. 324-340

adding a principle of parsimony to the group selection issue. So
understood, one may consider the alternative explanations for various adaptations as concerned to find the operative level for selection processes. When,e.g., an explanation is offered for warning
cries in birds in terms of mutual benefit, then the explanation in
terms of individual advantage standardly is taken to screen off and
render impotent any explanation in terms of higher level units; e.g.,
in terms of benefit to the flock (cf., Trivers 1971, esp. pp. 43-45).

Williams does argue that the effects of gene interaction and environmental dependence on individual genes can, in principle, be
redescribed in such a way that adaptation can "be attributed to the
effect of selection acting independently at each locus." (1966, p. 57).
This argument has been subjected to searching criticism at the hands
of William Wimsatt (1980 and 1981) and Elliott Sober (1981, 1982a,
and 1982b), driving for the conclusion that this argument provides
for little more than "genetic bookkeeping" in which the complexities
due to interactions at higher levels are redescribed; what it fails
to show us is that a focus on individual loci will yield significant
explanations or reveal the operative level for selection processes.

Williams is not wholly oblivious to such concerns. The caution
that "biotic adaptations" may be no more than artifacts is directed
especially toward groups that are fundamentally aggregates (cf.,
Hamilton 1971): fleetness of a herd of deer is not a genuine biotic
adaptation, selected for because of differential survival of groups;
it is, rather, an epiphenomenon arising because fleetness is adaptive for individuals which constitute the group. Specialization,
Williams acknowledges, is the crucial parameter. Herds of deer, he
suggests, might minimize predation if they adopted an "organized program of bear avoidance" in which some served as sentinels and others
as decoys. "Such individual specialization in a collective function,"
he concludes, "would justify recognizing the herd as an adaptively
organized entity. Unlike individual fleetness, such group-related
adaptation would require something more than the natural selection
of alternative alleles as an explanation." (Williams 1966, p. 17).
To acknowledge that organization and specialization are important is
of marginal help unless we understand how they are important. Are
there any plausible standards for demarcating systems that will yield
to an atomistic vision? What complex systems are such that their behavior can be explained, at least in principle, in terms of theories
proper to lower levels? And what complexity forebodes explanatory
wholes?

2. Grades of Organization

Complexity in its most interesting forms is hierarchical. A hierarchical system, as Herbert Simon explains, is "a system that is
composed of interrelated subsystems, each of the latter being, in turn,
hierarchic in structure until we reach some lowest level of elementary subsystem." (1962, p. 196). Hierarchical systems come in a variety of forms. We shall be concerned with hierarchical systems which

are dynamic, evolving, systems in which there are variable constraints on constituents of the systems. These are systems in
which there is significant change in state, and this change in
state tends to promote "quasi-independence" (Simon 1981, p. 12):
the goal or the equilibrium state is attained or maintained independently of environmental variation. The variation in state is
a consequence of variable constraints on constituent subsystems.
In this sense, the variability is essential to understanding system
function. (For an account of function harmonious with an emphasis
on dynamic hierarchies, see Wimsatt (1972).) The basic contrast
is with structural hierarchies, in which change of state is incidental to systemic integrity (or inimical to it) and constraints
on constituents are marginally variable, e.g., a crystalline
structure is hierarchic but not dynamic (cf., Pattee 1970). Dynamic
hierarchies in turn assume a number of forms (see figure 1).

INTEGRATED SYSTEMS	Functional subordination, functional diversification, and functional specialization.
COMPOSITE SYSTEMS	**Interlocking Systems**: constituent function is organizationally determined.
	Component Systems: intrinsic determination of component function.
EQUIPOTENTIAL SYSTEMS	Simple Organized Systems: inter-substitutable elements, organizational relevance to system function.
	Aggregative Systems: no organization relevant to system function, and inter-substitutable elements.

Figure 1. Diagrammatic representation of varied grades of organization in dynamic hierarchies. This portrayal ignores intermediate stages between types of systems.

The simplest dynamic systems are equipotential systems. These are
hierarchical systems in which the constituents of the systems are the
same in kind. More precisely, if under a given decomposition a system
S has as constituents $\{c_1,\ldots,c_n\}$ with a specified organization, the
functions of S are not affected by substitutions of constituents provided the organization of S is preserved, then under that decomposition S is equipotential. We may take a system as equipotential in
an extended sense if its constituents are of a relatively small number of kinds, and constituents within each kind are inter-substitu-
able. We could construct a measure of the degree of equipotentiality for a system with constituents $\{c_1,\ldots,c_n\}$ of kinds $\{k_1,\ldots,k_m\}$

under a specified decomposition as the ratio of m to n. Equipotential systems, too, are a diverse lot. As a limiting case, we have <u>aggregative systems</u>, in which systemic properties are statistical functions of the constituent properties (cf., Levins 1970, p. 76); or in which whatever organization is present <u>within</u> the system is not a primary determinant of the relevant systemic properties. The flow of fluids, or the movement of a herd, might approximate such a case. At the other end of the spectrum, we have <u>simple organized systems</u> in which systemic properties are determined, at least in part, by the organization of the system, or the "boundary conditions" for the constituents. A bureaucratic organization is common example: the tendency toward "bureaucratic inertia" in which the organization "runs itself" is an instance of organizational determinants.

Within this continuum, there is as yet no demand to embrace an explanatory holism. The simplest sort of system, in which constituents are under no internal constraints proves unpredictable in practice at the lower level and only statistically predictable at higher levels; and when the aggregate behavior is predictable, it is because of external constraints, as when one can predict the movement of a herd because of terrain. Imposing (internal) constraints on constituent functioning may not make the system easier to deal with: in systems with large numbers of constituents, we trade significant degrees of freedom in constituents for an equally overwhelming interactive complexity. Yet so long as the difficulty is <u>only</u> that there is a large number of constituents with a high frequency of interaction, non-predictability does not seem to be a matter of principle. Organizational constraints thus change very little of the picture: prediction may be impractical in the extreme, but, in principle, quite tenable.

A more complex form of hierarchical system is a <u>composite system</u>. In such systems, the organization of the system again determines, at least in part, the systemic properties; however, the constituents are not inter-substitutable. If S has as constituents $\{c_1,\ldots,c_n\}$ with a specified organization under a given decomposition, then S is strictly composite if no c_i and c_j in $\{c_1,\ldots,c_n\}$ are inter-substitutable, and the organization is partly determinative of S's behavior. We could readily regard equipotential systems and composite systems as lying within a two-dimensional space.[2] The two relevant dimensions of variation would be in degree of intersubstitutability and degree of aggregativity. Pure aggregates maximize both parameters and strictly composite systems minimize both. Simple organized systems maximize only the former and minimize the latter.

Composite systems vary along another continuum. At one end we have <u>component systems</u>.[3] In such systems, the functional properties of the constituents can be determined in isolation from other components: the organization imposed on $\{c_1,\ldots,c_n\}$ is determinative of the function of S but not of the functions of constituents. It may, of course, be true that each constituent must have a specific function for S to function properly; in component systems, though,

<u>the constituent function is intrinsically determined</u>. (This is not
to say its internal state, or the variation in its internal state, is
intrinsically determined; neither does it imply that its sequence of
outputs is intrinsically determined.) In <u>Interlocking systems</u>, the
organization is determinative of constituent functioning as well as
of system functioning. That is, in an interlocking system, constituent subsystems can realize their functions only within the system: it may be, e.g., that subsystems are mutually correcting or
that they have significant feedback relations.

The distinction between component systems and interlocking systems is akin, but not equivalent, to Simon's notion of near decomposability: "(1) in a nearly decomposable system, the short-run behavior of each of the component subsystems is approximately independent of the short-run behavior of the other components; (2) in
the long run, the behavior of any one of the components depends in
only an aggregate way on the behavior of the other components."
(Simon 1962, p. 210). In <u>nearly decomposable systems</u> the causal interactions <u>within</u> subsystems are much stronger than interactions
<u>between</u> subsystems (cf., Simon 1962, p. 209). Wimsatt (1974, p. 72) sug-
gests we characterize such systems in terms of a parameter ε_c which
is a "measure of the relative magnitudes of intra- and inter-systemic interactions for these subsystems". This may be taken to be
the ratio of the interaction strengths. ε_c then also may be taken
as an estimate of the likely error in predictions based on models
that assume inter-systemic interactions are negligible. Composite
systems, whether component or interlocking, allow any such system
to have significant inter-systemic interactions (ε_c may be quite
large). In a simple composite system in which components are se-
quentially organized such that output from $\underline{c_i}$ is the sole input to
$\underline{c_{i+1}}$ and the output (behavior) of $\underline{c_{i+1}}$ is a function of input, the
interaction (ε_c) between subsystems $\underline{c_i}$ and $\underline{c_{i+1}}$ might be gauged to
be quite high. By way of contrast, if components are organized in
parallel, minimizing any feedback between units, interaction (ε_c)
would be low. Either counts as a composite system, provided the
interactions responsible for determining subsystem function are
dominated by intra-systemic interactions. For illustrative purposes, we may think of realizing a function as realizing a machine
table. There may be intra- and extra-systemic factors responsible
for $\underline{c_i}$ realizing a function f_i. There will also be intra- and extra-
systemic factors responsible for the actual behavior of $\underline{c_i}$. Factors
responsible for the determination of actual behavior may be negligible
factors as contributing to f_i. Thus, we might construct a measure
$\varepsilon^*_{\underline{c_i}}$ of the relative magnitudes of intra- and extra-systemic interactions responsible for the realization of f_i in c_i. $\varepsilon^*_{\underline{c_i}}$ may be
small even when $\varepsilon_{\underline{c_i}}$ is large. This would be the relevant measure
for the continuum between component and interlocking systems, defin-
ing a third dimension of variability: aggregates would maximize intersubstitutability, aggregativity, and this modified decomposability,

$(1/\varepsilon^*_c)$; interlocking systems lie opposite them in three dimensional space, minimizing all three dimensions; and component systems minimize equipotentiality and aggregativity, while maximizing decomposability $(1/\varepsilon^*_c)$. (See figure 2.)

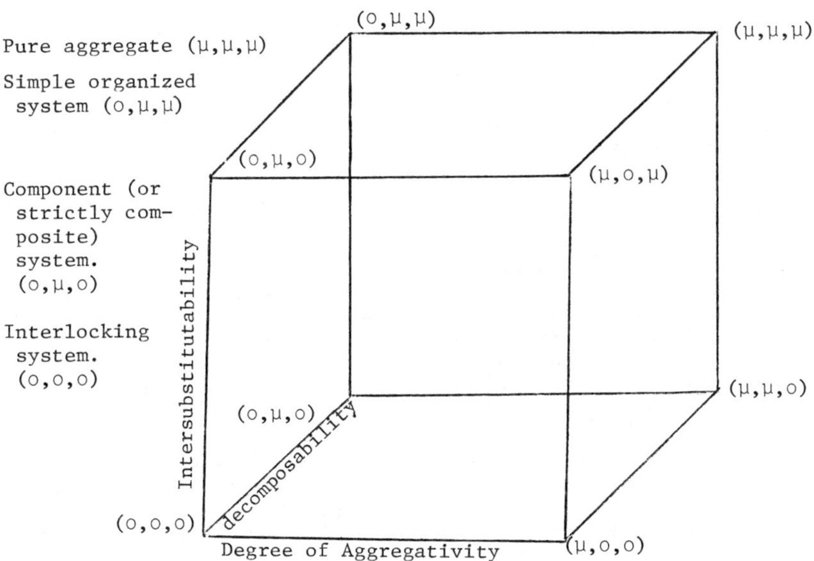

Figure 2. A three dimensional representation of the relationships between composite and equipotential systems.

Some of the most interesting and bewildering problems arise with systems that are interlocking, minimizing all three dimensions. Richard Levins comments: "This is a system in which the component subsystems have evolved together, and are not even obviously separable; in which it may be conceptually difficult to decide what are the really relevant component subsystems. Thus, for example, we might consider that a simpler multicellular organism is composed of cells, and yet the cells may be more profitably regarded, under other circumstances, as simply spatial subdivisions, partly isolated, of an organism." (1970, p. 77). The paradigm of an interlocking system is one in which elements are co-adapted: evolution has wrought changes in individual functioning such that the elements rely on each other even to the point that some functions are mediated by the co-evolved partner. Many flowering plants are incapable of

fertilization and reproduction without the intervention of bees.
(Notice that such fertilization is not an evolved trait of bees, and
not the "function" of any organ or behavior of theirs in any viable
sense, but a side effect.) When such dependence is mutual, organisms
may lose even physical independence. Thus, mitochondria are specu-
lated to have once been independent organisms, though they are now
parts of a cell.

With interlocking systems, many theorists will acknowledge the existence of higher level units. The crucial factor is the possibility that, within a group of organisms, there may be specialization of
function that cannot but be regarded as cooperative; the joint effects
can, and frequently do, have decisive effects on individual fitness.
Williams' acknowledgment of diversification and specialization as indicative of higher level units of selection lies in this category.
In a similar vein, after appealing to Williams' artifact argument,
Elliott Sober (1981) proposes this standard: "Group selection acts
on a set of groups if, and only if, there is a force impinging on
those groups which makes it the case that for each group, there is
some property of that group which determines one component of the
fitness of every member of the group." (p. 107). He has in mind the possibility, simply put, that differences in group membership might affect
individual fitness of group members because of a property of the
group determining member fitness and evolved as a response to some
force acting on the group.

It is possible that group selection is responsible for the evolution of such systems. It is even likely that group selection contributes to the spread of such systems. Nonetheless, the proposal
advocated by Sober and suggested by Williams is too liberal. The organization present among troops of anubus baboons fulfills their
standard readily (for details, see Wilson 1975, ch. 26), but does not
appear to require appeal to group selection for its explanation. On
the savanna, the marching order places dominant males and females with
infants at the center, juveniles flanking them, and other adults at
the front and rear. Confronted with a potential predator, dominant
males move to meet it. The group does display diversification, much
of the sort Williams imagines deer might have; the organization fa-
cilitates effective responses to predators; and, surely, this organization does affect the fitness of individuals. Yet it is at
least plausible to view such "group adaptations" as artifacts: the
strict dominance hierarchies, e.g., appear to increase the investment
of dominant males in the group; their fitness accordingly depends both
upon defending both individuals in the group and upon maintaining the
integrity of the group. Anubus baboons do not even have an especially
ornate social structure. Consider, by contrast, the hamadryas baboon.
Whereas the social unit of <u>Papio anubus</u> consists of females, offspring,
and multiple males, <u>Papio hamadryas</u> has a much more complex organization. The <u>troop</u> is a sleeping unit which consists of a number of
<u>bands</u>. The latter are confederations of one-male units (consisting
of a single male together with the harem) which cooperate in foraging.
The size of troops increases with lack of shelter, and thus

presumably has some defensive functions. Again we have organizations
that meet the standard of Sober and Williams. Yet, again, there is
no necessity to appeal to group selection or to view such "group
adaptations" as more than artifacts; for the advantages accruing to
individuals from cooperation offer a plausible explanation for such
behavior (cf., Trivers 1971).

Composite systems, whether component or interlocking, do not appear
to demand treatment as units, though, again, they do permit it. Setting examples aside, the reason for this is quite compelling. Aggregative, component, and interlocking systems may be viewed as forming an evolutionary hierarchy. Simon has shown that a hierarchical
system which is aggregative is much more likely to evolve than one
that is not, and that if aggregation proceeds by forming stable intermediate aggregations, this likelihood is further increased (cf.
Simon 1962, pp. 200 ff.). In fact, he shows that the time required
to "assemble" a system with a given number of units is proportional
to the number of levels in the system. That is, the rate of evolution is increased if there is decomposability at several levels. As
Wimsatt points out (1974, Part III), once such a system is formed,
divergence and coadaptation will destroy decomposability. The first
stage in such evolution would presumably be the cooperative dependence of component systems, and further fine-tuning could lead to interlocking systems. Wimsatt makes the point that, by breaking systems down into component systems, efficiency and reliability are significantly increased. Several less complicated parts may jointly
serve the function of one more complicated unit. In the evolutionary scenario set by Simon, diversification and coadjustment would
allow a composite system to realize functions which would not evolve,
or not so quickly, by aggregation. The evolutionary scenario is
symbiotic: each element evolves in such a way as to improve its via-
bility, and, in some cases, this requires diversification and inter-
twining of functions. But just as symbiotic relations need not be indicative of higher level units, so too systems manifesting interdependence of function need not thereby be rendered irreducible.

There is yet a higher grade in the organizational hierarchy: integrated systems. Integrated systems are hierarchically organized
systems which manifest diversification and within which there is functional subordination. H. H. Pattee comments: ". . . the essential
characteristic of hierarchical organization, [is] that the collective
constraints which affect the individual elements always appear to
produce some integrated function of the collection. In other words,
out of the innumerable collective interactions of sub-units which
constrain the motions [or functions] of individual sub-units, we
recognize only those in which we see some coherent activity."
(1970, p. 125) The notion of an integrated system is best explained
by appeal to the notion of a functional hierarchy. One of the long-
standing difficulties for functional analysis has been the discrimina-
tion of functional and incidental effects. The only approach which
leaves any hope of a motivated solution involves relativizing the
functions of subsystems to the functions of supersystems of which

they are parts. Wimsatt (1972) has proposed a normal form for function statements which requires just such relativization. Even if the brain did serve to cool the blood, that would _not_ be enough to assign that effect as a function. In order that an effect be functional or have a specific function, it must be the case either (a) that having function is advantageous or (b) that having that function contributes to a function that is advantageous. Thus, let us take the function of a system (e.g., pumping blood is the function of the heart) as relative to the function of a more encompassing system (e.g., replenishing oxygen supplies is a function of the circulatory system). In a hierarchically organized system, this will mean that a proper decomposition of the system will yield components at each level of analysis whose functions are subordinated to the functions of the units of which they are components. In this sense, we will have a functional hierarchy, under a given set of decompositions, when the organization of a system is such that its components at each level are functionally subordinated (generally mediated by intervening levels) to the functions of a single system. Hence, in saying integrated systems are systems within which there is functional subordination, the point can be more properly put as the requirement that functional systems are functional hierarchies. Since integrated systems will also generally be interlocking systems, they also will have components with diversified functions and the functions of components will be context dependent. Schematically, if $\underline{S}$ has as constituents $\{\underline{c}_1,\ldots,\underline{c}_n\}$, then $\underline{S}$ is an _integrated system_, relative to this decomposition, provided: (1) there is a low degree of intersubstitutability between $\{\underline{c}_1,\ldots,\underline{c}_n\}$, and, therefore, the components manifest functional specialization and diversification; (2) the organization of $\{\underline{c}_1,\ldots,\underline{c}_n\}$ is at least partly determinative of the functioning of $\underline{S}$; (3) the components of $\underline{S}$ are functionally interdependent and, thus, $\varepsilon^*_{\underline{c}_i}$ is high, on the average, for $\{\underline{c}_1,\ldots,\underline{c}_n\}$; and (4) the function of each $\underline{c}_i$ $\varepsilon\{\underline{c}_1,\ldots,\underline{c}_n\}$ is subordinate to the function of $\underline{S}$, or to a supersystem which is a constituent of $\underline{S}$ and whose function is subordinate to that of $\underline{S}$, or finally to the function of a supersystem of which $\underline{S}$ is a constituent and to which the function of $\underline{S}$ is subordinate.

Social symbioses serve as a useful model for understanding composite systems, and, by contrast, can highlight the significance of the fourth condition; for though symbiotic relations vary considerably in form and complexity, they are not appropriately embedded in a functional hierarchy, at least not until they become obligate. Symbiosis can take three general forms: commensualism, parasitism, and mutualism. With commensualism, benefit and dependence are unidirectional, but not detrimental to either species involved. In parasitism, the relation is again unidirectional, but detrimental to the host species. These symbiotic relations may be viewed as intermediate between component and interlocking systems: the organism deriving the benefit is often highly specialized, and incapable of survival outside the relationship. The cases are readily viewed as instances of exploitation, in which one organism is used, whether to its detriment or not,

to facilitate the fitness of another: the host organism counts as
little more than a resource. Adaptation at the individual level is
clearly sufficient to explain such relationships,

In mutualistic symbioses, we have a working model for interlocking
systems, exemplifying diversification, organization, and interdependence (cf., Wilson 1975, ch. 17; and Wilson 1971, ch. 10 and ch. 20).
It often fails nonetheless, to constitute an appropriate hierarchy.
In mutualistic interactions, species will have coevolved in such a
way that they are mutually dependent and reinforcing. Commonly, the
benefit one species derives is incidental to the other, much as in
commensal relations: in the course of producing a desired (functional) result, there will be numerous side effects which other species
can utilize. There are numerous arthropods that live as scavengers
on the refuse piles of army ants. Producing refuse piles is, of
course, not crucial for the ants, though it is an inevitable side
effect of mass foraging. It is particularly clear in such cases that
a functional hierarchy is lacking.

A more detailed example of social mutualism involving ants will reinforce the point. In cases of trophobiosis, aphids or one of a number of other insects excrete a substance called "honeydew", which,
though rich in nutrients is, evidently, merely a waste product. The
aphids are taken in by ant colonies, and benefit in gaining virtual
immunity from predation and parasitism. The production of honeydew
is generally taken to be a by-product of aphid functioning. The
aphids feed on plant sap and extract only part of the nutrients, leaving the residue which ants have come to use. Likewise, protection of
aphids is an inevitable result of defending the nest. The functions
are interdependent, but not hierarchically embedded. In more elaborate instances, the same observations are appropriate, In some cases,
the aphids have lost the defensive capabilities common among free
species, becoming, for all intents and purposes, obligate symbionts.
They correspondingly acquire special adaptations to facilitate trophobiosis, primarily in the form of hairs to retain honeydew droplets.
Host ants undergo modification as well. Workers will seek out aphids,
maintaining the eggs throughout the winter and even transporting newly
hatched nymphs to appropriate food sources in the spring. Queens of
at least one ant species even carry coccids (a trophobiont) in their
mandibles during nuptual flights.

Even where there is mutual adaptation and interdependence, there is
still a lack of functional subordination <u>within</u> any one system. In
fact, in some cases, whether we regard an activity as functional
rather than incidental and, in other cases, what we regard the function of an activity (or system) to be may change depending on which
symbiont we consider. Thus, though honeydew is an aphid byproduct,
the production of honeydew is the function of aphid "farming". Whatever egg maintenance may do for aphids, for the ants it is food stor-
age. In these cases, as before, it is possible that there are selective forces at the group level favoring mutualist systems, But in the
absence of evidence concerning specific influences, we focus on the

334

result; and the type of organization manifested in interlocking systems does not appear to demand explanation in terms of higher level
units.

3. Group Selection and Organization

What I shall propose should by now be no surprise. <u>Integrated
Systems are potential units of selection and potentially units of
evolutionary change</u>. When a system is integrated, manifesting interdependence as well as functional subordination, we have a system whose
adaptations may require explanation at higher levels. Systems with a
lower grade of organization permit systemic properties to be treated,
even if only in principle, as artifacts of selection at lower levels.

Evolutionary considerations are intimately related to judgments
concerning the mode of organization a system manifests. Lack of
evolutionary significance for a trait is assumed to be indicative of
a lack of function as well. Conversely, in estimating the functional significance of a trait, we again turn to its evolution. In explaining the evolution of a trait, if it is necessary to relativize
its function to the functions of more inclusive units of which it is
a part (or if it is a unit to which component functions are subordinated), then it is part of a functional hierarchy. The more encompassing units in the hierarchy will have properties subject to an
evolutionary explanation, properties relevant to "survival" of the
organism. If these were artifacts of adaptation at lower levels, it
would <u>not</u> have been necessary to appeal to them in evaluating component functions. The hierarchy would end at lower levels.

That integrated systems are potentially units of evolutionary
change is grounded finally in the fact that if a system is an integrated system, then there are superordinate functions to the system.
<u>If the superordinate functions assigned to the system are indeed of
adaptive significance, then the system is a unit of selection</u>. Thus,
evolutionary reflections may uncover higher grades of organization
precisely because they reveal units on which selection acts. In other
words, an integrated system will be a unit of selection, not only potentially but actually, if there are forces impinging on the system
affecting the reproductive potential of the system and which are contingent on superordinate functions of the system. The existence of
these selective influences will guarantee variation in fitness. The
integration of the system guarantees any resulting adaptation is not
artifactual. <u>A unit of selection is a unit of evolutionary change
provided that, in the context, the relevant properties are heritable</u>.

The import of these restrictions is clearer in the context of the
group selection controversy. The typical representative of group selection is V. C. Wynne-Edwards. The fulcrum for his analysis is the
existence of intra-populational controls on population levels. He
contended that the existence of controls on natural population levels
was best explained by the hypothesis that individuals "voluntarily"
reduced their own reproductive levels "for the good of the species".

Thus, Wynne-Edwards writes: "the greatest benefits of sociality
arise from its capacity to override the advantage of the individual members in the interests of the survival of the group as a whole."
(1963, p. 181).

Dawkins offers two fundamental replies to the group selectionist
portrait in addition to expressing doubts about the relative efficacy
of group selection. The first is this: "The quick answer of the
'individual selectionist' to the argument just put might go something like this. Even in the group of altruists, there will almost
certainly be a dissenting minority who refuse to make any sacrifice.
If there is just one selfish rebel, prepared to exploit the altruism of the rest, then he, by definition, is more likely than they
are to survive . . . . After several generations of this natural
selection, the 'altruistic group' will be overrun by selfish individuals." (1976, p. 8). The point is clearly an informal statement
of the view that a pure altruistic strategy is not evolutionarily
stable. In an altruistic population, mutant cheaters flourish. The
second reply is related. A mixed hawk/dove strategy is stable, and
will maximize individual fitness. Yet, Dawkins comments, "If <u>only</u>
everybody would agree to be a dove, every single individual would
benefit. By simple group selection, any group in which all individuals mutually agreed to be doves would be far more successful than
a rival group sitting at the ESS [evolutionary stable strategy] evolutionarily stable ratio." (1976, p. 77). That is, if individuals were
willing to make "sacrifices" for the "good of the group", then the
strategy we would expect to encounter would be a pure altruistic
strategy. Dawkins evidently thinks this is not what we find.

This sort of treatment confuses the appropriate levels of organization, and accordingly mislocates the place of group selection.
Dawkins' arguments both treat group, or <u>inter</u>demic, selection as if
it were meant to explain the <u>intra</u>demic distribution of traits. Yet
for the group selectionist, both the forces and the facts are to be
described at the higher level. It is fundamental to the group selectionist case that there be forces acting for the differential selection and elimination of groups. It is a simple enough matter to
see how this could occur even using the standard categories. Let us
suppose that we have a species with exactly two variant forms: Suckers, who are pure altruists, and Cheaters, who are specialized to exploit the Suckers. Whenever a Cheater invades a Sucker population,
the population is quickly overrun by Cheaters. Yet, when the service
provided by Suckers is crucial enough to the survival of individuals,
this will mean the population will decline. When a population (or
subpopulation) is overrun by Cheaters, it eventually disappears. The
parts of the population that are most likely to send out colonists to
occupy the vacated parts of the environment are the Sucker populations. The Cheater populations are always on the decline. Thus, provided

> (1) The services of the pure altruists are crucial to the
>   survival of both the altruists and non-altruists;

336

> (2) There is sufficient group structure to guarantee that
> the probability of within-group mating is more likely
> than is inter-group mating; and
>
> (3) Colonies tend to reflect the composition of the group
> propogating the colony;

it would be possible for a pure altruistic variant to spread in the
population as a whole even though the strategy is not an evolutionarily stable strategy. When the services the Suckers provide is
more costly to the Suckers, the speed with which a mutant Cheater
strain will spread is increased. When the services rendered are
more beneficial to recipients, the rate of population decline as
Suckers are overwhelmed by Cheaters increases correspondingly. Both
of these will increase colonization rate (by increasing the rate at
which new empty habitats are produced) and the probability that
founders will be altruists (by attenuating the life of Cheater colonies). Thus, the more essential the services provided by the altruists are, the more likely the altruistic variant is to dominate
the population.

A number of comments on the example are in order if its full import is to be grasped. _First_, there must be significant variation
in the composition of the various elements of the subpopulation.
This requires that the various subpopulations be relatively isolated with significantly lowered interpopulational mating (cf.,
Wade 1978). As Wynne-Edwards comments, "One of the most important
premises of intergroup selection is that animal populations are
typically self-perpetuating, tending to be strongly localized and
persistent on the same ground." (1963, p. 184). _Second_, there must
be forces acting to bring about differential extinction of groups
which are contingent on the variation in group composition or group
organization. It is not sufficient that group membership should
have an effect on individual fitness, just as it is not sufficient
for the existence of units above the level of the gene that fitness
be frequency dependent. What is necessary is that the forces act for
the differential survival of groups. It is thus not true that a pure
altruistic group would in all cases be more able to compete with
other groups than would a group having semi-altruistic members.
Dawkins neglects between-group competition, rather than arguing a-
gainst it.

What is crucial for group selection is that the forces impinging
on the groups create differential survival of groups and be contingent on some organizational properties of the groups. The stronger
these forces are in comparison to forces which favor differential survival of individuals within the groups, the more significant group
selection will be. What I wish to suggest is that in integrated systems group selection is likely to assume a strong influence. Are
there in fact such cases? E. O. Wilson provides, with full knowledge,
a clear example of group selection fitting this paradigm. It is
typical for a number of species of wasps to have a short colony life

cycle. Even in the tropics, where weather cannot be a significant factor, a 3-4 month cycle is maintained in some cases. In Amazonian colonies of <u>Mischocyttarus</u> there is a well-defined life cycle, although the cycles of different colonies are not synchronized. Wilson explains:

> The decline appears to be due at least in part to an increasingly heavy investment in males and larger females ("queens") which contribute little to the welfare of the colony and which in any case often disperse after only a short residence in the nest. What would be the advantage of such early programmed colony senility? The answer is very likely that mortality due to external causes is so heavy, and the average life span of a colony correspondingly so short, that it is of selective advantage for colonies to reach maturity quickly and to invest totally in reproductives. (1971, p. 14).

Embracing such an explanation of colony life cycles commits us to acknowledging, with Wilson, that the colony is the unit of selection. This case is not lacking its problems. There is no evidence of prior variation and, hence, the "programmed colony senility" may not be an adaptation at all. Wilson's explanation does, though, give us an accounting of the forces relevant to the differential selection of such colony traits. Patterns of reproduction, at both the colony and individual level, would be selected for on the basis of such factors as predation, acting on entire colonies.

This sort of intricate interdependence is not unique to social insects. Similar cases would include lichens (an association of algae and fungi) and the relationship between termites and the gut bacteria that facilitate digestion. We have dwelt at length on the several dimensions of complexity to be found in organized systems. Yet this last sort of case is also but an extreme variant of mutualist symbiosis in which the relationship is an obligate one. Thus, if we here find warrant for postulating higher level units of selection, we, <u>ipso facto</u>, have grounds for suspecting they are to be found at lower grades of organization, and are far more ubiquitous than an atomistic vision would bid us assume.

Notes

[1] I am indebted in many points to discussions with William Bechtel and William Wimsatt, and particularly to Richard Burian whose comments saved me from a number of mistakes. This work was supported by a grant from the National Endowment for Humanities and support from the Taft Committee at the University of Cincinnati.

[2]In a hierarchical system, we might well have a composite system with constituents that are equipotential. Information processing systems approximate such a case, since the input and output devices are quite heterogenous relative to the computer, but the computer's components are intersubstitutable. A hierarchical system may be strictly composite relative to one decomposition and only approximately composite relative to another--and, perhaps, even equipotential relative to a third.

[3]Levins' "composed systems" include component systems and all non-aggregative equipotential systems. (Cf., Levins (1970)).

[4]The situation is exacerbated by the difficulty of obtaining independent empirical estimates of the magnitude of forces at individual and group levels; accordingly, distinguishing the effects at the two levels is impossible when the direction of their influence is the same.

References

Dawkins, Richard. (1976). *The Selfish Gene.* Oxford: Oxford University Press.

Hamilton, W.D. (1971). "Geometry for the Selfish Herd." *Journal of Theoretical Biology* 31: 295-311.

Levins, Richard. (1970). "Complex Systems." In *Towards a Theoretical Biology.* Edited by C.H. Waddington. Edinburgh: Edinburgh University Press. Pages 73-88.

Lewontin, R.C. (1970). "The Units of Selection." *Annual Review of Ecology and Systematics* 1: 1-18.

Pattee, H.H. (1970). "The Problem of Biological Hierarchy." In *Towards a Theoretical Biology.* Edited by C.H. Waddington. Edinburgh: Edinburgh University Press. Pages 117-136.

Simon, Herbert. (1962). "The Architecture of Complexity." *Proceedings of the American Philosophical Society* 106: 467-482. (As reprinted in Simon (1981). Pages 192-229.)

----------------. (1981). *The Sciences of the Artificial.* 2nd enlarged ed. Cambridge: MIT Press.

Sober, Elliott. (1981). "Holism, Individualism, and the Units of Selection." In *PSA 1980.* Volume Two. Edited by P.D. Asquith and R.N. Giere. East Lansing, Michigan: Philosophy of Science Association. Pages 93-121.

----------------. (1982a). "Selection of and Selection for." Comments on papers by G.C. Williams and Michael Wade, University of Chicago Conference on Persistent Issues in Evolutionary Theory, March 3-5.

----------------. (1982b). "Reifying the Selfish Gene." Delivered at a Colloquium on Philosophical and Methodological Issues in Evolutionary Biology, University of Cincinnati, April 9-10.

Trivers, Robert L. (1971). "The Evolution of Reciprocal Altruism." *The Quarterly Review of Biology* 46: 35-57.

Wade, M. (1978). "A Critical Review of the Models of Group Selection." *Quarterly Review of Biology* 53: 101-114.

Williams, G.C. (1966). *Adaptation and Natural Selection.* Princeton: Princeton University Press.

Wilson, E.O. (1971). *The Insect Societies.* Cambridge: Harvard University Press.

------------. (1975). *Sociobiology: The New Synthesis.* Cambridge: Harvard University Press.

Wimsatt, William. (1972). "Teleology and the Logical Structure of Function Statements." _Studies in History and Philosophy of Science_ 3: 1-80.

----------------. (1974). "Complexity and Organization." In _PSA 1972. (Boston Studies in The Philosophy of Science, Volume XX.)_ Edited by K. Schaffner and R.S. Cohen. Dordrecht: Reidel. Pages 67-86.

----------------. (1980). "Reductionistic Research Strategies and Their Biases in the Units of Selection Controversy." In _Scientific Discovery, Volume 2: Historical and Scientific Case Studies._ Edited by T. Nickles. Dordrecht: Reidel. Pages 213-259.

----------------. (1981). "Units of Selections and the Structure of the Multi-Level Genome." In _PSA 1980,_ Volume Two. Edited by P.D. Asquith and R.N. Giere. East Lansing, Michigan: Philosophy of Science Association. Pages 122-183.

Wynne-Edwards, V.C. (1963). "Intergroup Selection in the Evolution of Social Systems." _Nature_ 200: 623-626. (As reprinted in Caplan, A. (ed.). _The Sociobiology Debate._ New York: Harper & Row, 1978. Pages 181-190.)

The Insights and Oversights of Molecular Genetics:
The Place of the Evolutionary Perspective[1]

John Beatty

Arizona State University

Nothing in biology makes sense except in the light of evolution.
(Dobzhansky 1973)

1. Introduction

   Along with the notion that DNA is at the bottom of things biological,
goes the notion that molecular genetics is on top. That is, along with
the notion that DNA is the informational basis of life, goes the notion
that the deepest explanations in biology are molecular-genetic. But even
if the former notion is correct, the latter is not. More specifically,
the molecular-genetic perspective alone is inappropriate for explaining
those biological generalities that call out instead for an evolutionary
account. Moreover, the molecular-genetic accounts that are brought to
bear upon biological generalities are often themselves subject to evolutionary scrutiny in the final analysis. I cannot entirely rule out the
possibility of an explanation of a biological generality that is, in the
final analysis, molecular-genetic and nonevolutionary. I am, in this regard, in somewhat the position of the natural historian. "In natural
history," as Gould and Lewontin put it, "all possible things happen sometimes; you generally do not support your favoured phenomenon by declaring
rivals impossible in theory. Rather, you acknowledge the rival, but cir-
cumscribe its domain of action so narrowly that it cannot have any importance in the affairs of nature. Then, you often congratulate yourself
for being such an ecumenical chap." (1979, p. 585). Sometimes that is the
best one can do in philosophy as well.

   I will argue a general case about the insights and oversights of
molecular genetics by arguing two specific cases: the first concerns
the bearing of molecular genetics on Mendelian genetics, and the second
concerns the bearing of molecular genetics on the replicability of the
genetic material. As for the first case, I will argue that Mendel's
law of segregation cannot be explained wholly in terms of molecular

---

PSA 1980, Volume 1, pp. 341-355

342

genetics--the law demands evolutionary scrutiny as well. As for the
second case, I will argue that an account of the replicability of the
genetic material in terms of molecular genetics is not entirely independent of evolutionary considerations, in the sense that it raises
further evolutionary questions. The limitations of the molecular-
genetic approach in these cases point to the limitations of that
approach in general.[2]

2. Mendelian and Molecular Genetics

Consider Mendel's "law". In the simplest terms, Mendel's law
describes the statistical outcome of the process by which gametes
(sperm and egg cells) are formed from gamete-producing cells in the
reproductive organs. This process is known as meiosis, and consequently Mendel's law is said to describe the outcome of "normal"
meiosis (see diagram of stages of meiosis, next page). During normal
meiosis, the total hereditary material of the gamete-producing cells
is fractioned through a series of cell divisions in such a way that
each resulting gamete receives a complementary half of the original
hereditary material. To describe the process in a bit more detail,
the hereditary material of the gamete-producing and other nongametic
cells comes in pairs of morphologically similar, or "homologous" chromosomes. The genes reside linearly along each chromosome; the two
genes that lie opposite one another when the homologous chromosomes
pair during meiosis are said to occupy the same chromosomal "locus"
(in the diagram, B and b occupy the same locus). During meiosis, the
chromosome pairs double and then divide twice so that each of the four
gametes formed receives one chromosome from each pair. Hence, each
gamete receives one gene from each locus. Moreover, the probability
that a gamete will receive one particular gene at a locus is one half.
Thus, a common brief formulation of Mendel's law is:

> Each of the two genes at a locus has a probability of one
> half of being the single gene at that locus carried by a
> particular gamete. (Edwards 1977, p. 3).

On the basis of this law, and an additional assumption to the
effect that gametes combine randomly with respect to their genetic
types, one can calculate the familiar genotypic frequencies of offspring born to parents with specified single-locus gene combinations.
For instance, a cross between two Bb-type parents will produce, in the
long run, 1 BB : 2 Bb : 1 bb ratio of offspring. If we add additional
assumptions about the dominance and recessiveness of the genes in
question, we can also calculate the phenotypic frequencies of the offspring. For instance, if B is completely dominant to b--i.e., if the
phenotype of a Bb-type organism is the same in the relevant respects
to the phenotype of a BB-type--then a cross between two Bb types will
produce the familiar 3:1 phenotypic ratio.

But the purpose of this brief excursion into genetics is not so much
to see what comes from normal meiosis, but to see where normal meiosis
comes from. We have seen that normal meiosis is a cellular process.

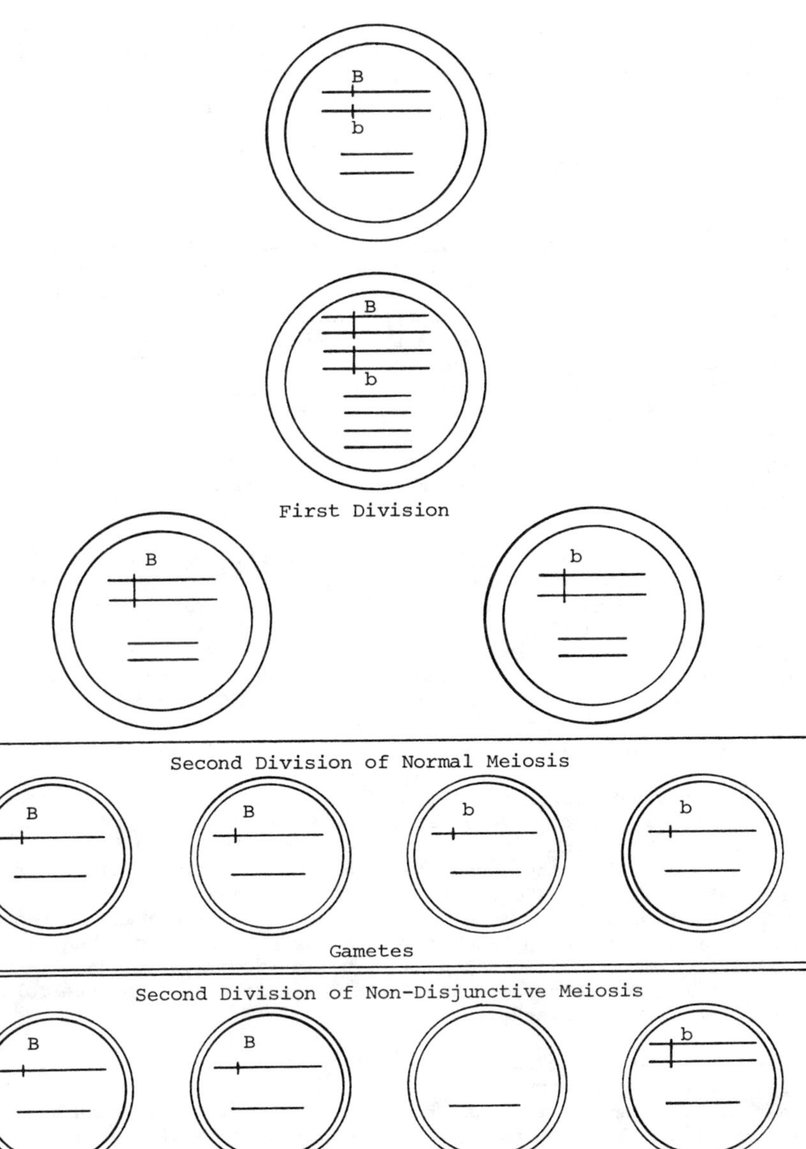
Stages of Meiosis
B
b
First Division
B
b
Second Division of Normal Meiosis
B
B
b
b
Gametes
Second Division of Non-Disjunctive Meiosis
B
B
b
Gametes

Now consider that this cellular process, like so many others, is _itself_ genetically controlled by genes at specific locations along the chromosomes. This has become more and more clear with the discoveries of more and more genetic mutations that affect meiosis. Two kinds of genetic variation in the meiotic mechanism have received considerable attention: a variation known as "non-disjunction" and a variation known as "meiotic drive".

Non-disjunction occurs when homologous chromosomes do not disjoin or separate during meiosis. As a result, some of the gametes formed contain both or neither of the homologous chromosomes--meaning that those gametes contain both or neither of the genes at each locus of the non-disjoining chromosomes (see the bottom of the preceding diagram). This exceptional phenomenon was recognized microscopically as early as 1913, but its genetic basis was not established until 1933 when Gowen reported his experimental findings that, "the processes through which the chromosomes pass during the meiotic divisions are in part, at least, subject to the same specific gene regulation that guides other bodily development and inheritance." (1933, p. 83). Through breeding experiments, Gowen determined that cases of the non-disjunction of one or all four pairs of chromosomes of _Drosophila melanogaster_ were influenced by a single gene locus that he located on the third chromosome pair (for more on the genetic basis of non-disjunction, see White 1973).

Meiotic drive, another acknowledged exception to normal meiosis, occurs when organisms with different genes at a locus (say, B and b) produce more than fifty percent of one kind of gamete (say B), on account of genetic differences between the homologous chromosomes. Perhaps the most well-known case of meiotic drive involves the so-called "t" gene in house mice, which is expressed in the form of taillessness. As much as ninety-five percent of the sperm of males of the Tt type may be of the t-type (Dunn 1953; for more on meiotic drive see White 1973).

So normal meiosis is a genetically controlled process--the phenotype of certain genotypes. Just as "blue eyes" is the phenotype of certain genotypes and "brown eyes" the phenotype of others, so too "normal meiosis" is the phenotype of certain genotypes and "abnormal meiosis" of others. The generality of 50:50 gamete ratios, as described by Mendel's law, is a consequence of the generality of the normal meiotic phenotype and its genotypes. Presumably, then, an explanation of Mendelian genetics in terms of molecular genetics would consist of a molecular account of the generality of this particular phenotype and its genotypes.

How can the generalizations of molecular genetics account for the generality of the normal meiotic phenotype and its genotype? First what are the generalizations of molecular genetics? I think a brief sketch will do. Among the generalizations of molecular genetics are:

(1) hypotheses concerning the structure of DNA, the various RNAs,

and proteins;

(2) hypotheses about how DNA replicates; and

(3) hypotheses about how DNA carries the instructions for the for-
     mation of other building blocks of life and the direction of
     other life processes--i.e., how DNA carries the information
     for the formation of RNA, how RNA carries the information for
     the formation of proteins, etc.

Now how can such generalizations explain the generality of the
normal meiotic phenotype and its genotype? The answer is that, alone,
they cannot. But let us get clear about what we are asking. Following van Fraassen (1980, pp. 126-129), let us consider an explanation
as an answer to a why question of the form,

Why P in contrast to X?

P is the phenomenon to be explained, the "presupposition" of the
question. The question only arises, or is in order, on the assumption of P. X is the "contrast class", the set of possible alternatives to P. When we ask for an explanation of P we want to know why
P is the case rather than some alternative to P.

With regard to the question at issue, P is the generality of normal
meiosis, and X is the set of abnormal meiotic processes. We want to
know,

> Why is "normal" meiosis general while alternative
> forms of meiosis are not?

3. The Evolutionary Perspective

We may someday know the answer to that question, but the answer will
not be strictly molecular-genetic. The generalizations of molecular
genetics alone--generalizations about the structure of DNA and its
replicative and informational properties--shed no more light on the
generality of normal vs abnormal meiosis than they do on any single
case of meiosis. And, alone, they no more explain any single case of
normal vs abnormal meiosis than they explain why I have blue eyes vs
brown. Normal meiosis vs abnormal, like blue eyes vs brown, requires
the "right" kind of DNA sequence--not just any DNA sequence. So an
account of the generality of normal vs abnormal meiosis in terms of
molecular genetics would have to be supplemented by a claim concerning
the generality of the right kinds of DNA sequences.

Now would any such supplemented explanation count toward the explanation of Mendelian in terms of molecular genetics? No, since the
supplemental generalization is no more part of molecular genetics than
are claims about the frequencies of any other kinds of DNA sequences.
More importantly, however, the supplemented explanation is not the most
appropriate explanation for understanding the phenomenon in question.

We need a different kind of explanation. In Mayr's (1961) terms, an "ultimate" rather than a "proximate" explanation is called for.

Proximate explanations concern the processing of environmental and genetic information--the causal pathways from environment and genotype to phenotype. Ultimate explanations, on the other hand, concern the history of environmental and genetic information--the causal pathways that lead from frequencies of genotypes and their phenotypes in earlier generations to their frequencies in later generations. Physiology and developmental biology are proximate sciences, while evolutionary biology is an ultimate science. Molecular genetics is also a proximate science--population genetics would be an ultimate science.[3]

It is especially important to recognize that proximate and ultimate explanations are generally not rival alternatives to the same questions; where one kind of explanation is appropriate, the other kind generally is not. In other words, some questions ask for proximate, others ultimate answers. Substantial confusion has resulted from confounding the roles of the two kinds of accounts. For instance, Mayr has made infamous the confusion of the early 20th-century physiologist Loeb, who complained that, "The earlier writers explained the growth of legs in the tadpole of the frog or toad as a case of adaptation to life on land. We know through Gudernatsch that the growth of the legs can be produced at any time even in the youngest tadpole, which is unable to live on land, by feeding the animal with the thyroid gland" (Loeb 1916, quoted in Mayr 1961, p. 1503). Thyroid hormone would be appropriately invoked to account for the growth of legs in an individual frog, but is not as appropriately invoked to account for the presence of legs as a general characteristic of frogs. An evolutionary account, presumably in terms of the differential reproductive success of legged vs nonlegged ancestors of frogs, would be more appropriate in the latter case.

Similarly, when we ask why normal vs abnormal meiosis is so general, we want an ultimate, evolutionary answer. As was explained earlier, the generality of normal meiosis rests upon the generality of the right kinds of DNA sequences. And the predominance of those sequences simply is not illuminated by a proximate explanation. Nothing (that we know of) that has to do with the processing of genetic information rules out the predominance of genotypes for alternative forms of meiosis. Evolutionary agents, on the other hand, could rule them out. For instance, it might be the case--we do not really know--that of the various forms of meiosis that have arisen through mutation, the normal form has always been selected. At least the latter explanation is an appropriate answer.

Molecular genetic considerations would be more appropriate for understanding why a certain kind of DNA sequence results in normal meiosis rather than abnormal. That question is no less interesting or important, but just a different question from why that kind of DNA sequence and its normal meiotic phenotype are so common. In order to differentiate more clearly these two types of questions, we need to focus on the "Why P" part of our general question form "Why P in contrast to X?". Depending on how we elaborate "Why P", the question may be one that calls for a molecular-genetic, or some other kind of proximate response, or one that calls for an ultimate, evolutionary response. The question about normal meiosis, as I have posed it, is an instance of a more general form:

(1)   Why is phenotype Φ and/or genotype G general in contrast to alternatives?

        (Why is/are normal meiosis and/or its underlying genotypes general in contrast to alternatives?)

Questions of form 1 call for ultimate, evolutionary answers--evolutionary considerations are most relevant to ruling out the generality of alternative phenotypes and genotypes. On the other hand, "Why P" questions of other forms call for proximate answers. Questions of form 2, for instance, are clearly more appropriately answered in molecular-genetic terms:

(2)   Why, given genotype G and environment E does phenotype Φ result in contrast to alternatives?

        (Why given a particular genotype does normal meiosis result in contrast to abnormal meiosis?)

In other words, molecular-genetic considerations may be relevant to ruling out alternative phenotypic outcomes. The point here is that the general question form "Why P in contrast to X?" does not automatically rule out the appropriateness of molecular-genetic responses. Whether such a response is appropriate depends on how one spells out "Why P?". The molecular-genetic approach has its place, to be sure, but its place is limited.

The case I have discussed--the bearing of evolutionary vs molecular perspectives on Mendelian genetics--is by no means an unusual case. In his recent, insightful analysis of theory structure in the biomedical sciences, Schaffner (1980) discusses four "representative" biomedical theories and the bearing of evolutionary theory on each. From these cases he generalizes the notion of "the metatheoretical function of evolutionary theory", which he summarizes as follows:

    We have seen in several of the examples mentioned above that evolutionary theory often functions as background providing a degree of intelligibility for components of a theory. Recall, for example, that the universality of the genetic code was explained evolutionarily, that the <u>lac</u> operon was conceived to have evolutionary advantages, that the "sharpness" of the clonal selection theory was justified evolutionarily, and that cell-mediated immunity was seen to confer important

> survival benefits as a defense against neoplasms in multi-
> cellular organisms. Evolutionary theory also allows us to
> understand why there is subtle variation in the organisms:
> Variation due to meiosis, mutation, and genetic drift, for
> example, predict that this type of variation should occur
> most frequently in evolving populations where strong selec-
> tion pressures towards sharpness (involving lethal variations)
> are not present. Thus evolutionary theory at a very general
> level explains some of the specific and general features of
> other theories in the biomedical sciences. (Schaffner 1980,
> p. 76)

Schaffner is not explicitly concerned here with the issue of the appropriateness of evolutionary perspectives vs molecular-genetic, or of
ultimate perspectives vs proximate. Nevertheless, a lesson to be
learned from his analysis is that the extent of generality in the
living world is a problem that calls for an ultimate, evolutionary
solution rather than a proximate one.

4. More on the Evolutionary Perspective

So far I have acknowledged a restricted role for molecular genetics
in the understanding of biological generalities. Molecular genetics
is restricted to questions concerning the causal pathways from particular genotypes and environments to their phenotypes--i.e., to
questions of form 2. But there is more to the restriction than that,
and it is worth exploring further the nature of the restriction. To
that end, let us consider a striking case in which a biological generality called for and received a molecular-genetic account. Having
paid this powerful molecular-genetic account its due respects, I will
then be in a better position to explain what I meant earlier when I
said that the molecular-genetic accounts that are brought to bear
upon biological generalities are often themselves subject to evolutionary scrutiny <u>in the final analysis</u>.

As early as 1922, the great geneticist Muller appealed for a molecular approach to genetics, and further suggested a criterion of adequacy that would have to be satisfied by a molecular account. That is,
a satisfactory solution of the molecular constitution of the genetic
material should illuminate the ability of the genetic material to
"autocatalyze", or replicate, even after it had mutated. In other
words, molecular genetics should address a question similar to form 2:

> Why, given the structure of the genetical material, is
> "mutable autocatalysis" possible rather than not?

In Muller's own words:

> The fact that the genes have this autocatalytic power is
> in itself sufficiently striking, for they are undoubtedly
> complex substances, and it is difficult to understand by
> what strange coincidence of chemistry a gene can happen

to have just that very special series of physico-chemical
effects upon its surroundings which produces--of all possible
end products--just this particular one, which is identical with
its own complex structure. But the most remarkable feature of
the situation is not this oft-noted autocatalytic action in
itself--it is the fact that, when the structure of the gene
becomes changed, through some "chance variation," the catalytic
property of the gene may become correspondingly changed, in
such a way as to leave it still autocatalytic. In other words,
the change in gene structure--accidental though it was--has
somehow resulted in a change of exactly appropriate nature in
the catalytic reactions, so that the new reactions are now
accurately adapted to produce more material just like that in
the new changed gene itself.

What sort of structure must the gene possess to permit it to
mutate in this way? Since, through change after change in the
gene, this same phenomenon persists, it is evident that it must
depend upon some general feature of gene construction--common to
all genes--which gives each one a general autocatalytic power--
a "carte blanche"--to build material of whatever specific sort
it itself happens to be composed of... . [This] question as to
what the general principle of gene construction is, that permits
this phenomenon of mutable autocatalysis, is the most fundamental question of genetics. (Muller 1922, pp. 34-35)

Muller's call for a molecular-genetic account of a biological
(specifically genetic) generality was taken up by Watson and Crick,
whose search for the structure of DNA was a search for a chemical structure with the appropriate biological properties. In other words, in
Watson's and Crick's eyes, if DNA were really the genetic material,
then a good chemical model of DNA would have to explain not only the
data of DNA's physical and chemical analysis, but also general properties of the genetic material--like replicability even after mutation.
As Watson later recollected in his textbook Molecular Biology of the
Gene, the "excitement" of solving the structure of DNA "came not merely
from the fact that the structure was solved, but also from the nature
of the solution. Before the answer was known, there had always been
the mild fear that it would turn out to be dull, and reveal nothing
about how genes replicate and function. Fortunately, however, the
answer was immensely exciting." (1965, p. 66).

In keeping with this goal, Watson and Crick triumphantly announced
their discovery in 1953: "We wish to suggest a structure for the salt of
deoxyribose nucleic acid (D.N.A.). This structure has novel features
which are of considerable biological interest." (1953a, p. 737). As is
now well known, their structure consisted of two intertwined helices,
each of which consisted of a linear arrangement of the bases adenine,
thymine, guanine, and cytosine. Moreover, the two helices were complementary in the sense that adenine on one helix was always paired with
thymine on the other, and guanine was always paired with cytosine.
After briefly describing the essentials of their structure, they con-

350

cluded, "It has not escaped our notice that the specific pairing we have
postulated immediately suggests a possible copying mechanism for the
genetic material." (1953a, p. 737). Their idea was that the intertwined
helices would, for purposes of replication, unwind and act as templates
upon which complementary helices would be constructed. As they explained in their second article on DNA, published five weeks after the
first:

> Now our model for deoxyribose nucleic acid is, in effect, a
> pair of templates, each of which is complementary to the other.
> We imagine that prior to duplication the hydrogen bonds [con-
> necting the corresponding base pairs of the two helices] are
> broken, and the two chains unwind and separate. Each chain then
> acts as a template for the formation on to itself of a new com-
> panion chain, so that eventually, we shall have two pairs of
> chains where we only had one before. Moreover, the sequence
> of the pairs of bases will have been duplicated exactly.
> (1953b, p. 966).

Mutational permutations in the order of the bases are similarly replicated. Muller's question about the possibility of mutable autocatalysis was thus answered in the molecular terms he sought.

Here we have a case where a biological generality called for and
received a strictly molecular-genetic account. The account seems
to be entirely independent of evolutionary considerations, but in an
interesting sense it is not. For the molecular account itself raises
further evolutionary questions. Consider that such an account of the
"mutable autocatalysis" of the genetic material in terms of the structure of DNA requires an identification of DNA as the genetic material.
We might want to add that claim to the generalizations of molecular
genetics, or we might not--that is not the point at issue. The point
is that the identification of DNA as the genetic material is a biological generality that calls for an evolutionary perspective. That is,
the fact, if it is a fact, that the genetic material is in all cases
DNA is a fact that is best understood in terms of an ultimate account
of the history of the genetic material, rather than wholly in terms of
a proximate account of the processing of genetic information. Such an
account would presumably concern the origin of life in a single common
ancestor. At least, that would be an appropriate account.

We might conclude, then, that molecular-genetic accounts are appropriate for understanding generalizable relations between the genetic
material and the other building blocks and processes of life, but the
general occurrence of any particular kinds of genetic material, and
any other building blocks and processes, ultimately calls for evolutionary understanding. In that sense, then, molecular-genetic generalities may account for biological generalities, but those molecular-
genetic generalities may very well be subject to evolutionary scrutiny
in the final analysis.

5. Difficulties and Conclusion

Exceptions to my general thesis are imaginable, however. As I excused myself earlier, I am in somewhat the position of the natural historian who recognizes that all possible things happen sometimes, but who emphasizes "only sometimes". Sometimes even questions of form 1 are better answered in terms of proximate rather than ultimate reasons. For instance, the question <u>why so many</u> trees of a certain population are charred and defoliated would be most appropriately answered in terms of circumstances that intervened in the processing of their environmental and genetic information, independently of any evolutionary considerations.

It would be nice if we could characterize this class of counterinstances generally and a priori instead of just citing instances of them a posteriori. It would be especially nice if we could tell simply from the vocabulary and/or grammar of a biological question whether it called for a proximate or an ultimate answer. I have suggested in effect that grammar and vocabulary are clues--that questions of form 1 should clue us to consider evolutionary answers, and questions of form 2 proximate answers. But to try to make the distinction completely in terms of language would be to overestimate seriously the power of philosophy. Whether a proximate or an ultimate explanation is most appropriate can only be determined by scientific investigation. For instance, there is no way to tell whether the generality of defoliation just discussed is the result of proximate or ultimate processes without "looking to see".

To be sure, it would be an error to overlook such exceptions as were just discussed. But it is a more common error, I think, to overlook the bearing of evolutionary perspectives on biological generalities, and to overemphasize the importance of the molecular-genetic perspective. We can at least say that the endeavor to explain all biological generalizations in molecular-genetic terms is a fundamentally misguided one, as is the endeavor to so explain any biological generalization if the sole rationale for trying to do so is the assumption that the deepest explanations in biology are molecular-genetic. Ultimately, evolution matters too.[4]

Notes

[1] Thanks to Lindley Darden, Ernst Mayr, Nancy Maull, Alexander Rosenberg, Mary Williams, and especially Philip Kitcher for valuable help with earlier drafts of this paper.

[2] See Schaffner (1969, 1974) and Ruse (1973, Chapter 10) on the possibility of explaining Mendelian genetics wholly in terms of molecular genetics. For an up-to-date review of other discussions concerning the bearing of molecular genetics on Mendelian genetics, see Hull (1982).

[3] Shapere (1974a, 1974b) has elaborated a related distinction between

"evolutionary" and "compositional" accounts. Briefly, an evolutionary
theory is appropriately invoked to solve problems that "call for
answers in terms of the development of the individuals making up the
domain [of the theory]," while a compositional theory is appropriately
invoked to solve a problem that "calls for an answer in terms of
constituent parts of the individuals making up the domain and the laws
governing the behavior of those parts." (1974a, p. 534). As examples
of evolutionary theories, Shapere considers, besides biological evolutionary theories, the nebular hypothesis, stellar evolutionary theories,
theories of the evolution of chemical elements, etc. And as examples
of compositional theories, he considers Dalton's law of combining pro-
portions, the periodic table, etc.

I say Shapere's distinction is "related" to Mayr's, while at the same
time I recognize important differences. For instance, as 'evolutionary'
and 'compositional' are defined above, developmental biology may be both
evolutionary and compositional. As Mayr characterizes proximate and ultimate accounts, however, developmental biology is proximate only. The
various differences between Shapere's and Mayr's distinctions are worth
exploring, though I cannot do that here. In what follows, I will rely
solely on Mayr's distinction--a distinction tailored to differences
among biological points of view.

[4]Williams (manuscript) has independently raised evolutionary objections to the explanation of Mendelian genetics in terms of molecular
genetics. She does so, however, within a different context--a context
well worth considering, if only briefly. The problem that concerns
Williams can be motivated in the following way. Remember the opening
quotation, attributed to Dobzhansky and championed by evolutionists
competing for funds with molecular biologists: "Nothing in biology
makes sense except in the light of evolution." There is, embodied in
that view, a problem that may have occurred to you already. That is,
evolutionary biology is part of biology. If nothing in biology makes
sense except in the light of evolutionary biology, then neither does
evolutionary biology make sense except in the light of evolutionary
biology. But how can evolutionary biology shed light on itself?

Let me put the problem in a slightly more technical way. Consider
the modern, so-called "synthetic theory" of evolution. The synthesis,
to which the name refers, was in large part a matter of incorporating
Mendelian genetics into evolutionary theory. The result was a genetic-
evolutionary theory, well exemplified by the so-called "Hardy-Weinberg
law", which describes the conditions for evolutionary equilibrium in
the following terms:

Within a breeding group, as long as gametes combine randomly, and
no forces affect gene frequencies from the time of fertilization to
the time of gamete formation, gene frequencies will remain constant.

This law is as important to the synthetic theory of evolution as
Newton's first law is to Newtonian mechanics (Ruse 1973, pp. 37-38).
Newton's law, recall, states that if there is no force on a body, its
momentum will remain constant. Each law allows investigators to

recognize the presence in nature of unbalanced forces against an otherwise stable background: change in the momentum of an object indicates
unbalanced physical forces acting on that object, while changes in the
gene frequencies of a breeding group indicate unbalanced evolutionary
forces acting on that group. Among those forces are natural selection,
mutation, and migration.

The Hardy-Weinberg law <u>seems</u>, from a "principle" point of view of
theories, to be the central principle of the synthetic theory. Ruse
refers to the law and its derivatives as the "core" of evolutionary
theory, in the sense that it is "presupposed by all other evolutionary
studies." (1973, p. 48). Munson (1975) and Schaffner (1980) concur with
Ruse in this regard.

But to demand such centrality for the Hardy-Weinberg law is to restrict the range of evolutionary theory in an interesting way. Consider
what happens when we ask for an account of the Hardy-Weinberg law--why
is it so universal? As it happens, the law is derived entirely from
Mendel's law discussed earlier (see Ruse 1973, Chapter 3 for a derivation). So in response to our query as to the generality of the central
principle of modern genetic-evolutionary theory, we get another generalization that can, in turn, only be understood in evolutionary terms. We
are thus in the position of having to admit that there are evolutionary
phenomena for which modern genetic-evolutionary theory cannot account,
namely the generality of the genetic mechanisms upon which the theory
rests. Another step has to be taken, and to a higher level of generality.

That is, as I see it, the why and whereto of Williams' particular
structuring of evolutionary theory (Williams 1970, 1973). Her axiomi-
zation of Darwinian evolutionary theory has been criticized on the
grounds that it "succeeds...only by avoiding all mention of genetics."
(Ruse 1973, p. 50, and echoed by Schaffner 1980, p. 63). The important
point, however, is not just that she is trying to formulate a nongenetic
theory of evolution, but that she is trying to formulate a theory at a
higher level of generality than classical Mendelian genetics (see also
Rosenberg 1979 and Beatty 1981). That she is trying to formulate a
higher level <u>evolutionary</u> theory is the point of most interest with regard to this paper. The direction of deeper explanation in this case,
as in so many others, is toward ultimate, not proximate, understanding.

References

Beatty, J. (1981). "What's Wrong with the Received View of Evolutionary Theory?" In _PSA 1980, Volume Two_. Edited by P.D. Asquith and R.N. Giere. East Lansing, Michigan: Philosophy of Science Association. Pages 397-426.

Dobzhansky, T. (1973). "Nothing in Biology Makes Sense Except in the Light of Evolution." _American Biology Teacher_ 35: 125-129.

Dunn, L.C. (1953). "Variations in the Segregation Ratio as Causes of Variations of Gene Frequency." _Acta Genetica et Statistica Medica_ 4: 139-147.

----------. (1957). "Evidence of Evolutionary Forces Leading to the Spread of Lethal Genes in Wild Populations of House Mice." _Proceedings of the National Academy of Science_ 43: 158-163.

Edwards, A.W.F. (1977). _Foundations of Mathematical Genetics._ Cambridge: Cambridge University Press.

Gould, S.J. and Lewontin, R.C. (1979). "The Spandrels of San Marco and the Panglossian Paradigm: A Critique of the Adaptationist Programme." _Proceedings of the Royal Society of London_ B205: 581-598.

Gowen, J.W. (1933). "Meiosis as a Genetic Character in _Drosophila Melanogaster._" _Journal of Experimental Zoology_ 65: 83-106.

Hull, D.L. (1982). "Philosophy and Biology." In _Contemporary Philosophy: A New Survey, Volume Two_. Hague: Nijhoff. Pages 281-316.

Loeb, J. (1916). _The Organism as a Whole._ New York: Putnam.

Mayr, E. (1961). "Cause and Effect in Biology." _Science_ 134: 1501-1506.

Muller, H.J. (1922). "Variation Due to Change in the Individual Gene." _American Naturalist_ 56: 32-50.

Munson, R. (1975). "Is Biology a Provincial Science?" _Philosophy of Science_ 42: 428-447.

Rosenberg, A. (1979). "Genetics and the Theory of Natural Selection: Synthesis or Sustinance?" _Nature and System_ 1: 3-15.

Ruse, M. (1973). _The Philosophy of Biology._ London: Hutchinson.

Schaffner, K.F. (1969). "The Watson-Crick Model and Reductionism." _British Journal for the Philosophy of Science_ 20: 325-348.

---------------. (1974). "The Peripherality of Reductionism in the Development of Molecular Biology." _Journal of the History of Biology_ 7: 111-139.

---------------. (1980). "Theory Structure in the Biomedical Sciences." _Journal of Medicine and Philosophy_ 5: 57-97.

Shapere, D. (1974a). "Scientific Theories and Their Domains." In _The Structure of Scientific Theories._ Edited by F. Suppe. Urbana: University of Illinois Press. Pages 518-565.

-----------. (1974b). "On the Relations Between Compositional and Evolutionary Theories." In _Studies in the Philosophy of Biology._ Edited by F.J. Ayala and T. Dobzhansky. Berkeley: University of California Press. Pages 187-204.

van Fraassen, B.C. (1980). _The Scientific Image._ Oxford: Clarendon.

Watson, J.D. and Crick, F.H.C. (1953a). "Molecular Structure of Nucleic Acids." _Nature_ 171: 737-738.

-------------------------------. (1953b). "Genetical Implications of the Structure of Deoxyribonucleic Acid." _Nature_ 171: 964-967.

------------. (1965). _Molecular Biology of the Gene._ New York: Benjamin.

------------. (1968). _The Double Helix._ New York: Atheneum.

White, M.J.D. (1973). _Animal Cytology and Evolution._ London: Clowes.

Williams, M.B. (1970). "Deducing the Consequences of Evolution: A Mathematical Model." _Journal of Theoretical Biology_ 29: 343-385.

---------------. (1973). "The Logical Status of Natural Selection and Other Evolutionary Controversies." In _The Methodological Unity of Science._ Edited by M. Bunge. Dordrecht: Reidel. Pages 84-102.

---------------. (Manuscript). _The Structure of Evolutionary Theory._

CAN <u>DARWINIAN</u> <u>INHERITANCE</u> <u>BE</u> <u>EXTENDED</u> <u>FROM</u> <u>BIOLOGY</u> <u>TO</u> <u>EPISTEMOLOGY</u>?

Carla E. Kary

The University of Illinois at Chicago Circle

1. Introduction

Despite continuing controversy over the detailed structure of the biological theory of evolution, and the scientific credentials of some of its central principles (see Goudge (1961), Peters (1976), Smart (1963), Williams (1970)), its striking success in explaining the development of living things has frequently inspired attempts to apply it beyond the biological domain (for example, Mach (1910) and White (1949)). Recently attempts have been made to explain the historical development of human concepts and ideas in terms of a Darwinian evolutionary theory. Various theories of "evolutionary epistemology" have been proposed by Stephen Toulmin (1972), Donald Campbell (1974), and Nicholas Rescher (1977), to name a few. In order to assess these theories adequately it would be useful to know in a general fashion under what circumstances and in what form a Darwinian evolutionary theory should be employed within non-biological domains. One way to gain this knowledge is to analyze evolutionary theory in its biological formulation in order to produce a generalized theoretical framework. Such a framework would contain a set of conditions which proper <u>Darwinian</u> <u>evolutionary</u> <u>theories</u> would have to fulfill. With a framework like this in hand, one would not only find it easier to go about constructing such theories for non-biological domains, but one would also be in a position to assess the adequacy of evolutionary theories already proposed.

In this essay I will perform only a small piece of this kind of analysis, and based on the results, assess the adequacy of a part of one theory of evolutionary epistemology.[1] The feature of biological evolutionary theory which my analysis will focus upon is the mechanism of inheritance. And the theory of evolutionary epistemology which I will later examine is contained in the book <u>Human</u> <u>Understanding</u> by Stephen Toulmin (1972). In the next section, I begin the analysis with an overview of the current evolutionary explanation of the development of

---

<u>PSA 1982</u>, Volume 1, **pp. 356-369**
Copyright ⓒ 1982 by the Philosophy of Science Association

species.

2. Fundamental Processes of Darwinian Evolution

Darwin's most striking accomplishment was an explanation of the origin of species that was based solely upon the features of organisms themselves. After Darwin one no longer needed to call upon the intervention of God, or gross environmental catastrophes, or spontaneous generation of species types in order to explain the origin of species. In an elegant manner, Darwin's theory of evolution restricted the explanation of the historical development of species to the domain of species themselves. Simply put, Darwin's theory showed that to find the cause of the existence of a particular species, one need look no further than the properties of some prior species.

Several authors (including both biologists and philosophers) have provided brief but informative descriptions of the current synthetic theory of evolution (Williams (1970), Lewontin (1970), Brandon (1978), Ruse (1973)). All seem to agree (though sometimes not explicitly) that three processes are fundamental to this theory:

    a) a process which generates traits (mutation);
    b) a process which transmits traits across generations (inheritance);
    c) a process which selects, or favors, organisms who exhibit advantageous traits (natural selection).

The obvious reason for including these three processes in a description of biological evolutionary theory is that they are viewed as jointly necessary for an answer to the fundamental Darwinian question, _viz_. How does one species give rise to a new species? Briefly, the answer provided by evolutionary theory is that the organisms of a species exhibit traits which can be advantageous in a particular environment. And that if they are in fact advantageous, then if they are represented in the genes of those organisms, they may be transmitted to future generations belonging to the species. The genetic theory, which is included in the evolutionary answer to the Darwinian question, links the production of new organisms by prior organisms to the very act of transmission of genes. That is, it asserts that genes possess both the information and causal efficacy for the production of a new organism, and that their separation and donation from a prior organism is the first necessary step in that production. Thus, the transmission of traits via genes is part and parcel of the causal act by which new organisms are produced.

Finally, according to genetic theory, if genes change (through mutation, recombination, etc.), then the traits they represent change also. When such changes are significant, the organism carrying these genes will exhibit new or greatly divergent traits with respect to its parent(s). Evolutionary theory goes on to say that, should these traits be advantageous, the organism will be favored relative to its co-organisms and as a result, it will be more likely to transmit its genes (which carry the new traits) to future generations.

The continual play of these three processes -- mutation, inheritance, natural selection -- permits the occurrence of fundamental changes in the state of a species. These changes may be sudden, within relatively few numbers of generations (as "punctuated equilibrists" maintain, Vrba (1980)), or they may be gradual, resulting from the accumulation of many small changes (phyletic evolution). However they occur, such changes mark the emergence of a new species. By means of genetic mutation, inheritance of traits, and selection of favored organisms, a population of distinctly new organisms may arise which either overwhelms its parent population (phyletic evolution) or else splits from its parent to inhabit a different niche (speciation). In this way, each of the processes listed above plays a critical role in the development of a new species from prior species.

3. The Structure of Darwinian Inheritance

I shall assume in my analysis that any generalized model for evolutionary theories must include these three processes. Put another way, I shall take it for granted that any application of Darwinian evolutionary theory in a non-biological domain must explain the development of that domain in terms of analogues to these processes. For my purposes this should not be controversial, as those who have advanced theories of evolutionary epistemology have said as much themselves. In any case, I intend to look only at the process of inheritance, and show how it should be generalized for non-biological applications. In order to do this, it will be especially useful to adopt some of the terminology introduced and defended by Hull (1978).

According to Hull, species should not be construed as classes, for they are historical entities. That is, they are tied to unique spatiotemporal periods, and so cannot form classes whose members are not spatiotemporally restricted. Hull argues convincingly that the historicity of species makes them unlikely candidates for typical classes. Further, Hull reminds us that the organisms which form a species do so in a manner different from the way in which members form classes. Organisms are included in a particular species not on the basis of similarity of traits, but "because they are part of the same chunk of the genealogical nexus." (p. 353). Thus, in order to decide which species an organism belongs to, one looks not to the similarities it bears to other organisms, but to the role it plays relative to other organisms around it, and to its historical origins. Organisms, as Hull views them, function within a population (and species) as parts function within a whole. They "belong" to that species in which they function as food gatherers, protectors, and reproducers. They do not belong to the species whose organisms they most resemble.

In line with the conclusions of Hull's argument then, species will be represented here as individuals, and organisms and populations of a species will be represented as parts of a whole individual. My strategy is to use this terminology as the vehicle for generalizing the

features required for the process of inheritance. It is clearly desir-
able, however, that the terminology be faithful to original Darwinian
theory.   Therefore I will begin by discussing the biological situation
of inheritance in some detail.

The function of inheritance is twofold.   First, it is the mechanism
by which prior organisms produce new, spatiotemporally distinct organisms.   Biologists call this event "reproduction".   Second, it is the
mechanism by which evolutionarily significant traits are passed from
one organism to the next.   Because of these functions, inheritance
plays a crucial explanatory role in evolutionary theory.   It accounts
for the appearance of organisms within a species, and the appearance of
traits in newly produced organisms.

To better understand the importance of this role, we might imagine
what would happen if these features were not present in the biological
domain.   What if, for example, spatiotemporally distinct organisms were
not generated? One curious result would be that the formation of a new
species (call it "B") could occur only by the donation of whole organisms belonging to some prior species (call it "A").   This would be the
only sense in which a prior species could be said to produce a new
species.   But the organisms lost to A would not be replaceable, as new
organisms are not generated in this imaginary situation.   Furthermore,
the organisms gained by B at the time of its formation would never
increase their numbers.   In effect, this "conservation of organisms"
would set a limit upon the number of new species which could exist at
any one time.   For the number of possible species which could exist at
a given time would be limited to the number of organisms which exist at
that time. And even more bizarre, as the number of species increased,
the number of organisms within each species would decrease.

Next, consider what would happen if the appearance of traits were
not accounted for in the way inheritance accounts for them.   One thing
is certain.   If the cause of the appearance of traits in one organism
could not be traced to another organism,   then we would be hard put to
explain how the forces of selection operating upon one organism, could
affect the appearance of traits in future organisms.   For unless we
knew to begin with that an organism could pass its traits to a future
organism, we could not argue that advantageous traits will appear in
greater frequency in future organisms.   And of course, if this could
not be argued, then the notion of evolution of species by natural
selection would be empty.   Thus, it is crucial in a Darwinian evolutionary explanation to have a mechanism by which traits are transmitted
from prior organisms to future organisms, and according to which the
frequency of the appearance of traits depends upon the selective advantages they confer.

I shall now go on to generalize this view of inheritance using the
part-whole terminology described earlier. Our discussion of the biological situation implies that there are several criteria which must be
met by the parts involved in inheritance. These are:

1. Prior parts must cause the production of numerically distinct (new) parts.
2. Parts must transmit traits to the new parts they produce.
3. Prior parts must remain distinct from the new parts they produce.

These criteria ensure that parts are not conserved, i.e., the generation of new parts does not simply mean that original parts are modified in some way.

In order to facilitate the discussion from this point and avoid irrelevant side issues, let us assume that we have available a reasonable criterion of "personal identity" for the parts comprising our individuals. The focus of our attention will be upon the principle of individuation for the individuals which are comprised of these parts. This is, in fact, what occurs in the biological situation. There it is assumed that a principle of identity for organisms is in effect and commonly accepted, and the focus of concentration is upon issues involving the distinctness and sameness of species.

With this assumption behind us, the following principle of individuation for Darwinian individuals can be introduced. Individuals of the Darwinian domain will be distinguished both by the traits of their parts and in virtue of the numerical discreteness of their parts. This permits us to say that A and B are the same individual if their parts have the same traits, even though their parts may not be the same. Two distinct individuals exist only when both their parts and the traits their parts exhibit are different. In biology, this situation is reflected by a single species which has different populations existing in different locations at the same time. It is also reflected by the history of a species, which is composed of numerically discrete populations which nevertheless possess the same identifying traits. Such species remain the same individual even though their parts change by the introduction of spatiotemporally new organisms.

In addition to the criteria which Darwinian parts must meet, and the principle of individuation discussed above, there is one last term which is needed to represent the process of inheritance. We have seen that in biology the production of spatiotemporally new organisms is joined with the transmission of traits. And that this is accomplished by the donation from a parent of something which its body produces and which is included in its structure, _viz._, its genes. To reflect this in the general terminology adopted here I introduce the term "parental causal center". A parental causal center is that which a part relinquishes in order to produce a new part. Like genes, parental causal centers contain the directions and the causal efficacy for producing new parts. They also carry the traits which will be exhibited by a new part and subjected to evolutionary forces.

The purpose of this section was to introduce a partial theoretical framework for evolutionary theories (see the appendix for a complete description of the framework). We have examined the biological

situation of inheritance in some detail, and developed a general terminology for representing it. The only other requirement is that the
representation be general enough so that it can range over non-biological domains. Using the terms "individual", "part", and "parental
causal center", the results of the analysis I have done can be put into
a set of conditions which represent inheritance in this framework.
They are offered as conditions which any Darwinian evolutionary theory
must meet.

<u>A</u> <u>GENERAL</u> <u>FRAMEWORK</u> <u>FOR</u> <u>EVOLUTIONARY</u> <u>THEORIES</u> (<u>Part</u> <u>II</u>)[2]

II. Parts of individuals have the following properties:
   A. Parts exhibit the traits which are subject to
      selection pressure from the environment.
   B. New parts[3] are produced from prior parts.
   C. When prior parts produce new parts they transmit
      some piece of themselves -- a parental causal cen-
      ter -- to new parts.
   D. Parental causal centers cause the traits of parent
      parts to appear in the new parts they produce.
   E. The transmission of a parental causal center to a
      new part is a necessary condition for the exis-
      tence of that new part.
   F. Prior parts are not identical with the parental
      causal centers which they transmit to the new
      parts they produce.

It is important to realize that, according to these conditions, new
individuals are produced only by both the production of new parts and
the transmission of new traits to new parts. In this way the conditions do provide the underlying structure for individuals such that the
parts of individuals can exhibit varying traits to a selective environment, and then transmit these traits to new parts. In addition the
continual production of new parts with differing traits, which the conditions permit, allows for the production of new and different <u>individ-</u>
<u>uals</u> from prior <u>individuals</u>. And this result warrants acceptance of
these conditions as an adequate representation of the Darwinian mode of
evolutionary development.

4. The Analogy to Darwinian Inheritance in Toulmin's Theory

The time has come now to use the conditions defended above as a
standard by which we can measure the success, at least in part, of some
theory of evolutionary epistemology. Because Stephen Toulmin states
explicitly in <u>Human</u> <u>Understanding</u> (1972) that his theory should and
does include a mechanism of inheritance analogous to that of the biological theory, his work is a good example for us to consider. The
theory of conceptual evolution which Toulmin describes will be summarized primarily from this book, where he intends to discuss "The Collective Use and Evolution of Concepts." Though some of his earlier
work also deals with evolutionary epistemology (1961, 1967), this book
appears to be the current and most complete statement of his position.

Therefore, all page references which follow will be from <u>Human</u> <u>Under-</u><u>standing</u>.

The first section (Section A) is devoted to criticism of non-evolutionary theories of conceptual development, particularly theories of scientific development. Toulmin attempts to show why these theories are inadequate, and consequently why an evolutionary interpretation ought to be attempted. In Section B Toulmin argues that any domain which is populated by "historical entities" is open to an evolutionary interpretation. He defines these entities as "comprising successive sets of constituent elements which maintain a recognizable unity despite their changing content." (p. 337) As applied to the conceptual development of science, historical entities are identified with scientific disciplines. As it happens, Toulmin restricts most if not all of the details of the analogy between conceptual develpment and organic development to the conceptual domain of science. Therefore I will use his characterization of scientific development as the best example of the analogy he believes exists between biological evolution and conceptual development.

Scientific disciplines, viewed as historical entities, are the things which evolve in Toulmin's account. And the "changing elements" of a scientific discipline are its concepts. These are not to be construed as private, subjective ideas, but as entities which are either currently being shared by the community of professionals within a discipline, or else have the potential for intelligibility in this common, public realm. Further, concepts are the entities which serve as solutions to problems encountered by a theory.

Disciplines may be characterized at any particular moment in time by a representative set of concepts. Though the role of theories in Toulmin's scheme is not clear, we may infer (pp. 200, 227, 266) that they provide a form for the organized sets of concepts he calls "representative". Yet, a representative set must include more than theories (p. 200), for it also contains the concepts which function as explanatory methods, theoretical problems, and the ideals and goals of the discipline.

Toulmin's analogy to the biological domain supposes that intellectual disciplines function as species do. The role he provides for concepts however is not so simple. On the one hand, they function like genes do in the biological domain. Not only are concepts organized into genealogies within a discipline, they are also the carriers of innovation in a population. They can mutate or change, and they are perpetuated through time. Thus, the history of a concept in a given discipline is supposed to be analogous to the history of a gene in a species (p. 200). The only exception is that concepts mutate in a coupled fashion, while genes mutate randomly, or in an uncoupled fashion. Toulmin's discussion of this point (p. 338) indicates that he does not view this discrepancy as significant.

On the other hand, Toulmin assigns an organismal role to concepts. Like organisms in the biological domain, concepts are the entities which adapt to environments and colonize ecological niches. Professional forums comprise the environment into which conceptual innovations are thrust. If they can adapt to these environments, i.e., solve the problems encountered by the theory, they colonize the niche offered by the environment. In effect, they are the units of selection in Toulmin's scheme, just as organisms are the units of selection in the biological theory. Concepts thus seem to be analogous to both genes and organisms of biological theory.

Let us look specifically now at the genetic role concepts play as the units of inheritance in the epistemological domain. Toulmin allows that a mechanism of inheritance is vital to the intelligibility of his theory (p. 208-209). Yet his description of such a mechanism is vague at best, though he does not appear to recognize this:

> Conceptual change in a science can proceed effectively, only where transient innovations do not automatically die with their creators. Certainly, individuals must be impelled to reflect about the relevant domain; but their private thoughts will by themselves, have as little long-term effect as transient organic variations, unless they are taken up by others and so become the raw material of collective evolution. Only then do they become 'available' in the relevant pool of conceptual novelties. While the final, indispensable source of conceptual variations lies in the curiosity and reflectiveness of individual men, these will be without effect unless other circumstances are favourable. While the 'genetics' of intellectual novelty on the individual level may be no great mystery, therefore, the occurrence of full-scale conceptual variations in collective disciplines depends also on those communal factors without which the original ideas of individuals may never get into professional circulation.
>
> A condition for the availability of genuine conceptual variants is, thus, the existence of suitable professional 'forums' of discussion. (p. 209)

While Toulmin goes on to describe the nature of professional forums, he does not try to say more clearly just how concepts, as carriers of information, are passed through time in an ordered fashion. Apparently, we are to think of concepts as being inherited in the following way: A new concept is generated by an individual human being who belongs to a particular profession which is part of (or wholly comprises) an intellectual discipline. The innovative concept (or conceptual variant) is placed before the professional forum. If it is selected by this forum, it becomes their concept as well as that of the human who created it, and takes a place in the theory. The theory, and hence its constituent concepts, is then passed down through generations of professional forums. When a professional forum no longer finds a particular concept suitable, it may modify it or reject it. This is assumed to be analogous to genes which arise through mutation or recombination, and then if selected are passed from organism to organism over generations.

However, there are serious difficulties with this analogy. Recall that the conditions offered in the last section all govern the mechanism of inheritance, by which parts are perpetuated. Put another way, the conditions illustrate the relation of descent which exists among parts. And they do this in terms of the transmission of parental causal centers from a parent part to a new part. Now Toulmin also emphasizes the need for descent among the parts of an evolutionary domain. He agrees that the parts which exhibit traits must be connected through time by descent, and that the perpetuation of traits must occur through descent:

> Given the whole range of historical precedents, the proper course is to treat the word 'evolution' as a general term, covering all historical processes in which a compact but changing 'population' is represented by successive sets of elements related by descent. On this definition, organic change, cultural change, social, conceptual, and linguistic change will be so many different varieties of historical 'evolution', all of which involve genealogical relationships between species, cultures, societies, and so on. (p. 339)

Unfortunately, Toulmin is extremely vague in his explanation of how descent occurs in the epistemological domain. In the conditions of the framework the relation of descent is captured by the idea of a parental causal center which every part must transmit in the production of a new part. But concepts, which we have seen are analogous to organisms (as well as genes) and so should function as the parts of a discipline, do not contain causal centers or anything resembling them. Indeed, in Toulmin's evolutionary account concepts do not produce new concepts at all. Instead they are either modified through time, or else abandoned. And the source of new concepts is not located in other concepts, but in the mental activity of some human being. Consequently, the way in which concepts are "inherited" is very unlike the way in which genes are inherited. Genes replicate themselves, but concepts must be "replicated" by human beings. Organisms contain the causal mechanism of inheritance, concepts do not. The term "part" as we have described it does represent organisms, but cannot represent concepts as characterized by Toulmin.

And so we should hardly expect the situation in which Toulmin's concepts exist to yield proper Darwinian results. For in the absence of some causal link among these concepts which would account for the production of new concepts by prior concepts through descent, we cannot expect new individuals (disciplines) to be produced from the parts (concepts) of prior individuals. And in the account Toulmin gives, since new concepts appear to be produced by the action of human agents who are external to the domain of concepts, we might expect new <u>disciplines</u> to be produced in this way as well. But this expectation is not at all in line with the Darwinian mode of evolutionary explanation whose structure is captured by the conditions given earlier. For the

Darwinian theory explains the development of individuals of the domain
solely in terms of the properties of the parts of those individuals.
There is no reliance upon the intervention of either God or human
agents.

Perhaps the conceptual development which Toulmin describes would be
better thought of as an example of continual modification of the same
parts of an individual, where the introduction of new parts is accomplished through the activity of external agents. There would be no
descent in this development, for the development would not proceed by
the production of new parts out of prior parts. This view of Toulmin's
theory might be more appropriate since the concepts he describes simply
do not have the requisite structure for the Darwinian descent relation.
We do not say that the adult is 'descended' from the child from which
it develops. Just so, we seem unable to say of Toulmin's concepts that
they are 'descended' from the concepts which preceeded them.

Given the failure of concepts to meet the conditions of the framework, we should not expect them, and the disciplines they compose, to
be governed by the individuating principle provided in Section 3. For
one thing, concepts do not seem to produce other, numerically discrete
concepts. And this makes it impossible to distinguish individuals on
the basis of the sameness of their parts, which is one of the criteria
of individuation in the principle. I have already argued that this
principle of individuation suitably reflects the status of individuals
in the biological domain. But since it cannot be applied to the epistemological domain described by Toulmin, he leaves us without a principle with which to distinguish the individuals that are supposed to
evolve. How <u>do</u> we know for any individual whether it is a newly
evolved discipline or simply an original discipline which has been modified? Obviously this is a critical question, for unless it is
answered we cannot be sure whether Darwinian evolution occurs at all in
a domain.

A final and related difficulty with Toulmin's analogy is his failure
to link the appearance of traits with inheritance. He tries to show
that concepts are selected for their advantageous traits. But he does
not seem to realize that selection has evolutionary significance only
when coupled with a mechanism of inheritance -- only when the traits
selected at one time and place are transmitted to future times and
places so they can <u>continue</u> to confer advantages upon the parts that
bear them. For Toulmin, concepts do not pass traits on, and so if one
particular concept has advantageous traits which are selected, this
tells us nothing about whether these traits may appear in future concepts. Consequently, the selection of a particular concept at a particular time does not indicate at all that the traits it exhibits will
persist. The effects of selection thus seem to be local. It is not
clear that selective forces shape the content of new disciplines, since
their effects are not transmitted to new concepts over time. However,
we saw earlier that unless the appearance of traits can be explained by
their selection in previous parts, it is senseless to try to explain
the appearance of new <u>individuals</u> by an appeal to the forces of

selection.

5. Conclusion

My analysis began with an overview of Darwinian evolutionary theory in biology. The primary insight of this theory is that the origin of a new species can be explained in terms of the properties of some prior species. A successful evolutionary explanation was shown to require, among other things, a suitable mechanism of inheritance to account for this insight. The structure of the parts of evolving individuals was generalized to display the process of inheritance.

I have argued that the conditions of this structure are not met by Toulmin's evolutionary theory of conceptual development. Although he claims that a mechanism of inheritance plays a role in his theory which is analogous to the role it plays in biology, he fails to show how new concepts can be produced by prior concepts, and how advantageous traits can be passed on to new concepts.

If the structure of the process of inheritance given in the conditions of the theoretical framework of Section 3 is a necessary feature of any Darwinian evolutionary explanation, then Toulmin's theory clearly falls short of being Darwinian. I have tried to show that the structure provided by these conditions is necessary.

Notes

[1]I have attempted a full analysis of biological evolutionary theory and assessed the epistemological theories of Toulmin and Campbell in my Ph.D. dissertation at the University of Illinois.

[2]See the appendix for the complete theoretical framework I have developed.

[3]Remember that a "new" part must be numerically discrete with respect to its parent.

<u>Appendix</u>

The following set of conditions is designed to form a complete theoretical framework for Darwinian evolutionary theories. All of the conditions must be met by a theory in order for it to be an instance of a Darwinian evolutionary theory.

A GENERAL FRAMEWORK FOR EVOLUTIONARY THEORIES

I.   The individuals of an evolutionary domain have the following properties:
   A.   Each individual has parts.
   B.   Every part is a part of at most one individual.
   C.   Populations of individuals are temporal groups of their parts.
   D.   Each individual arises from some other individual through the action of the parts of their populations.

II.   The parts of individuals have the following properties:
   A.   Parts exhibit the traits that are subject to selection pressure from the environment.
   B.   New parts are produced from prior parts.
   C.   When prior parts produce new parts they transmit some piece of themselves -- a parental causal center -- to new parts.
   D.   Parental causal centers cause the traits of parent parts to appear in the new parts they produce.
   E.   The transmission of a parental causal center to a new part is a necessary condition for the existence of that new part.
   F.   Prior parts are distinct entities from the parental causal centers which they transmit to the new parts they produce.

III.   Individuals interact with an environment that has the following properties:
   A.   Changes in the environment favor some traits and not others.
   B.   The environment remains constant at least until parts exhibiting favored traits within an individual are able to produce a new individual.

IV.   Individuals are distinguished according to the traits of their parts and in virtue of the numerical discreteness of their parts.

References

Brandon, Richard. (1978). "Adaptation and Evolutionary Theory." *Studies in the History and Philosophy of Science* 9: 181-206.

Campbell, Donald. (1974). "Evolutionary Epistemology." In *The Philosophy of Karl Popper*, Volume I. Edited by Paul Schilpp. Illinois: Open Court Press. Pages 413-463.

Cole, L.C. (1960). "Competitive Exclusion." *Science* 132: 348-349.

Goudge, Thomas. (1961). *The Ascent of Life*. Toronto: University of Toronto Press.

Hull, David. (1978). "A Matter of Individuality." *Philosophy of Science* 45: 335-360.

Lewontin, Richard. (1970). "The Units of Selection." *The Annual Review of Ecology and Systematics* 1: 1-18.

Mach, Ernst. (1910). "Die Leitgedanken meiner naturwissenschaftlichen Erkenntnislehre und ihre Aufnahme durch die Zeitgenossen." *Physikalische Zeitschrift* 11: 599-606. (As reprinted as "My Scientific Theory of Knowledge and its Reception by my Contemporaries." (trans.) A. Toulmin. In *Physical Reality*. Edited by Stephen Toulmin. New York: Harper and Row, 1970. Pages 30-31.)

Peters, R.H. (1976). "Tautology in Evolution and Ecology." *The American Naturalist* 110: 1-12.

Rescher, Nicholas. (1977). *Methodological Pragmatism*. Oxford: Basil Blackwell.

Ruse, Michael. (1973). *The Philosophy of Biology*. London: Hutchinson University Library.

Smart, J.J.C. (1963). *Philosophy and Scientific Realism*. London: Routledge and Kegan Paul.

Toulmin, Stephen. (1961). *Foresight and Understanding*. Indiana: Indiana University Press.

——————————. (1967). "The Evolutionary Development of Natural Science." *American Scientist* 55: 456-471.

——————————. (1972). *Human Understanding*. Volume I. Princeton: Princeton University Press.

Vrba, Elizabeth. (1980). "Evolution, Species and Fossils: How Does Life Evolve?" *South African Journal of Science* 76: 61-84.

White, Leslie. (1949). <u>The Science of Culture.</u> New York: Grove
    Press.

Williams, Mary. (1970). "Deducing the Consequences of Evolution: A
    Mathematical Model." <u>The Journal of Theoretical Biology</u> 29:
    343-385.

Part IX

Philosophy of Science,
Past and Future; Metaphor and Play

<u>Recovering Philosophy from Rorty</u>

Steve Fuller

Dept. of Hist. & Phil. of Science
University of Pittsburgh

There have been many responses to Richard Rorty's <u>Philosophy and the
Mirror of Nature</u>, but none of them have said why philosophers should be
assigned a unique set of offices within the halls of academe rather than
simply be distributed among the offices already provided for practicion-
ers of the special disciplines (which, after the German <u>Wissenschaften</u>,
will henceforth be referred to as the "sciences")--with logicians situated among the mathematicians, epistemologists among the psychologists,
and so forth. And ultimately the challenge of Rorty's book boils down
to this bureaucratic question. We may not be convinced by his story of
how philosophers, starting with Descartes and Locke, unwittingly came to
raise the sciences of their day to the status of metaphysics. Nor may we
find very illuminating his suggestion that philosophers now join the
ranks of cultural critics. Yet whatever his failures in diagnosing and
curing, Rorty has certainly isolated a disease peculiar to philosophy.
At first approximation, we may say that philosophy suffers from what
Habermas would call a "legitimation crisis". However, to stop here
would not do justice to the depth of Rorty's analysis, for he is not
arguing that philosophy is, as it were, "between paradigms"--no longer
able to justify itself as it once did (namely, as "queen of the sciences") and hence in search of some other means of justification. Too
many readers of Rorty, I fear, interpret his preference for continental-
style hermeneutics over Anglo-American analysis in just this manner: a
paradigm shift. However, the legitimation crisis that Rorty divines in
the history of modern philosophy more resembles the gradual withering
away of the state according to Marx: Here is a discipline that slowly
but surely loses its problems to the sciences, such that by the end of
the twentieth century philosophers are at a loss to specify what it is
that they do that nobody else does. In other words, philosophy suffers
from a radical <u>identity crisis</u>. It is not a matter of whether we should
do philosophy in one way rather than in another, but whether we are
doing anything in particular when we do "philosophy".

It is easy to see what Rorty finds attractive in continental philoso-

---

<u>PSA 1982</u>, Volume 1, pp. 373-383

phy: not so much its actual practice (which he all too eagerly reduces
to his home-grown pragmatism) but   the history of philosophy that
those practices presuppose--notoriously thematized as "the end of
philosophy" by Nietzsche and Heidegger. The Greeks started to philo-
sophize when they conceived of single problem whose solution would
encompass the very nature of "Being", or reality understood as the
integration of fact and value, a hint of which is still preserved by
the ambiguous interrogative "Why?" But since Galileo, the modern
period has been marked by the attempt to tackle the big "Why?" by
breaking it down into little "Hows", methods designed for studying
specific kinds of entities; hence, the rise of the special sciences.
In the midst of this cognitive division of labor and proliferation of
scientific methods, the big problem got lost (or "forgotten", as Heidegger would say). In fact, it has gotten so lost that it is now impossible to pose the problem intelligibly. Heidegger conceived of his
task as uprooting centuries of philosophical discourse to discover just
exactly what that problem was--_not_ in order to reassert the uniqueness
of philosophical inquiry, but to show just how alien the big problem is
to the current scientific temper, which Heidegger saw as irreversible.
Rorty himself is somewhat more cheerful--more Nietzschean perhaps--in
that he has no regrets about the passing away of philosophy. As a
pragmatist, he agrees with his more positivistic antagonists that big
questions arise only when our imaginations outstrip our techniques,
such that there is no way that our minds can be laid to rest, no definitive test to which our speculations can be put. In effect, philosophy is the disease of the Western mind for which science is the cure.

But even given the above account of the history of modern philosophy , philosophical problems are pathological only if we assume that
"normal" problems are defined by their solubility. Perhaps this is
not too much to ask. However, I disagree, for I would like to think
that the moral of Rorty's history is that philosophy has little to do
with _solving_ problems but a lot to do with _creating_ them. More importantly, the latter is (at least) as "wholesome" as the former. Logical positivism and computer science notwithstanding, a "well-formed"
philosophical problem does not suggest a means toward its own solution,
but, rather, deprives us of the means that we would most naturally turn
to in attempting a solution. And here we should recall how, in denying
the adequacy of conventional answers to apparently simple questions,
Socrates generated the deep problems of metaphysics. Today there are
philosophers who continue to cultivate a sense of the problematic--but
in monologue rather than dialogue. I am thinking especially of those
willfully obscure stylists Jacques Derrida and Stanley Cavell. Why
they write as they do will provide the first clue in recovering philosophical inquiry from Rorty's critique.

You do not have to be a literary critic to find Of Grammatology and
The Claim of Reason unusually discursive and unmotivated. After a few
moments with either text, were I to ask for your impression of the person who wrote it, you would perhaps say that the author is someone who
has _nothing_ in particular to talk about, but nevertheless feels that
he must talk about _something_, and so he ends up talking about _every-_

<u>thing</u>. Indeed, if the author took Heidegger's view of the history of
philosophy seriously, then that would seem to be the only way he could
legitimately write. Now the fate of my argument turns on the weight
that the word "legitimately" can support. At first glance, style seems
to be one of the few things that a writer has total control over, and
we quickly become impatient with an obscure writer precisely because it
is within his power to write more clearly. But this intuition rests on
a naive aesthetic, namely, that there is an ideal prose style--the one
whose anatomy is dissected in writing manuals--able to accomodate anything worth saying and to which any right-minded author should aspire.
Such a style clearly reveals the structure of the text so that the
reader may easily identify its arguments in order to see how adequately
they account for the phenomena in question. Writing manuals consequently drive home the point that the stylistically acceptable article in
philosophy differs from one in, say, history or psychology only in terms
of subject-matter and not in terms of genre. I call this aesthetic
"naive" because it fails to recognize that the question of style arises
not in trying to package a finished product of thought, but, rather, in
actually producing the thought. Indeed, a method is nothing more than
a canonical style, and a style no more than a personal method. (Needless
to say, there is more similarity between schools of painting and schools
of philosophy than meets the eye.) And a parallel relation can be drawn
between what I shall call "legitimacy" and "authenticity". A method is
"legitimate" if it accounts for its own presence; hence, the Cartesian
method is legitimate insofar as part of practicing it involves showing
why it is able to succeed as it does where other methods have failed.
Likewise, a style is "authentic" if it reveals the presence of the
author; hence, Descartes' style is authentic insofar as it offers some
clues as to the uniqueness of his enterprise, or why his text should be
written. (Notice that we have not taken sides on whether legitimacy or
authenticity is ever fully realized.)

When a scientist senses the need to make his method legitimate (or
his style authentic), he is responding to a distinctly <u>philosophical</u>
impulse, for as a scientist his ability to reveal the structure of some
well-defined set of phenomena is not problematic but expected. And he
displays this competence insofar as his observations are amenable to
the ideal prose style. Admittedly, the scientist may turn out to be
guilty of error, but there is no a priori reason to think that he will
be: The presumption of innocence is the presumption of access to reality. In principle, science can reflect the structure of phenomena without distortion. The fact that one scientist rather than another may
have the last word is irrelevant. What Wittgenstein claimed of philosophy can certainly be claimed of science's self-image. It leaves the
world as it is. We shall explore in more detail how philosophical concerns for legitimacy transform this notion, as we now try to draw the
ideal line between linguistics and the philosophy of language.

At first glance, the genesis of the philosophy of language at Oxford
in the 1940s seems to fit Rorty's thesis that philosophy is proto-
science. All of Ryle's and Austin's careful talk about "what we would

say" in certain discursive contexts mirrors the structure of our actual
linguistic practices. And indeed, the analytical clarity with which
they pursued their studies makes them exemplars for the writing manuals. Why then shouldn't we just regard these Oxonian efforts as exer-
cises in amateur linguistics? In fact, John Searle's subsequent reformulation of ordinary language philosophy as "speech-act theory" is
often considered a branch of linguistics. However, I think we can
begin to understand why the philosophy of language should be kept distinct by reflecting on a defense Ryle and Austin often invoked on be-
half of their meticulous analyses. Contrary to what even the speakers
themselves may think, the distinctions made in ordinary language are
far subtler than those drawn by metaphysicians. The irony here is
significant. If we assume that a science deals with a specified range
of phenomena, while philosophy--if it does anything at all--deals with
reality in general, then as a linguist I want to catalogue utterances,
perhaps with the goal of constructing a grammar. In other words, I
want to save the _linguistic_ phenomena. But as a philosopher I want to
save _all_ the phenomena, and so when I turn to language, not only do I
see the manifold structure of utterances, but I also see that the ut-
terers themselves do not realize the complex nature of that structure.
Thus, I need to account for the difference between the utterer's articulated and tacit knowledge of his language, and why I am able to articulate what is only tacit for him. Furthermore, I need to consider
whether making the tacit articulate changes the nature of what is
known. This is the question of whether analysis really leaves the
world as it is. Only philosophically minded linguists--often those of
a Marxist persuasion--have in fact taken up these issues. They require
reflection on what it means to know something "adverbially", as Ryle
would put it: to know detachedly, to know involvedly, as well as to
know tacitly and to know articulately.

The fact that _Marxist_ linguists have taken on these traditionally
philosophical concerns as their own is itself interesting. I believe
this has happened because such problems more easily permit the Marxist
to vent his practical commitment, which is _not_ to leave the world as
it is, but to change it. Consider what distinguishes Austin's articulated knowledge of ordinary language from the ordinary man's tacit
knowledge of it. The difference is that Austin is able to define the
limits of correct linguistic practice, while the ordinary man is only
able to speak correctly; the former can mention what the latter can
only use. Whenever Austin intuits this difference, he passes from
linguistics to the philosophy of language. The philosopher recognizes
that the scientist has reduced something actual--a world whose limits
are not known because one is immersed in it--to something possible--
a world whose limits are known because one, in some sense, has stepped
outside it. The obvious advantage of knowing something as a possibility is that it can be considered in relation to other possibilities.
Once we know the rules of a language-game,we know what needs to be
changed (the rules) in order to play another game. In fact, I would
go so far as to assert that if there is a strategy unique to philosophy, it is just this reduction of the actual to the possible. Another

way of putting it is to say that philosophy--rather than making room for faith, as Kant thought--makes room for doubt. It treats as merely one among many alternatives something that would otherwise be treated as having no alternatives whatsoever.

But while it is clear that Marxists would find such a strategy attractive, it is not clear that it requires a distinctly "philosophical turn" (pace Rorty). It could be argued that placing a system of linguistic practices against the background of socio-economic conditions suffices to show the ideological limitations of those practices without making any of the modal moves described above. In effect then, the linguist would simply be integrating phenomena from his science with those of other social sciences. And indeed, if the Marxist--or any scientist for that matter--were solely interested in providing a <u>description</u> of the society (as many anthropology monographs in fact seem to be), no philosophical turn would then be made. However, gaining a more comprehensive picture of a society does not ipso facto imply revealing the <u>limits</u> of, or <u>constraints</u> on, that picture. As a scientist I can systematically describe the pursuit of profit in a capitalist society without mentioning another activity that is prohibited in virtue of such a pursuit. But of course, such mentionings form the basis of the ideological critique so characteristic of Marxist methods. Moreover, such mentionings arise in science whenever explanations are requested for the phenomena being described. In other words, the move from description to explanation occurs when the scientist stops talking about just "x" and starts talking about "x and not y". Thus, "explaining 'x'" involves showing why "y" did not happen, which requires that the actual "x" and the non-actual "y" somehow be made comparable. In scientific practice, this comparability is accomplished through thought-experiments and counterfactual reasoning, which presuppose that the "actual x" can be reduced to a "possibly not x". Then, the scientist is in a position to consider why "x" and "y" are not possible together. Methodologically, we could say, after Husserl, that "x" has now been "bracketed"; stylistically, we could just as well say that "x" has been "put in scare quotes". This account accords with the view that explanations are needed whenever some phenomenon is said to be "necessary", which is to say, exist in virtue of excluding all other relevant possibilities. In short, Rorty is stood on his head. The desire for explanation expresses science's latent <u>philosophical</u> aspiration. Furthermore, in the course of satisfying this desire, the scientist is in a position to think through the ramifications of there being an alternative state-of-affairs, thereby making room for the Marxist, whose explanation of "x" may involve an implicit endorsement of "y".

Given these considerations, let us now gradually return to our earlier claim concerning the legitimacy of Derrida's and Cavell's obscure styles. First, it is easy to see why explaining a phenomenon involves questions of legitimacy, while describing it does not. Simply put, an implicit reference to the author is made during an explanation, insofar as he supplies the alternative states-of-affairs in virtue of whose prohibition the actual state-of-affairs obtains. Thus, in reading the explanation, we learn about the explainer as well as about the explained by seeing what he takes to be significant exclusions. And competing explanations may be evaluated on the basis of the relative signifi- cance of the states-of-affairs they make impossible. And although it remains to be shown how a science orders the "relative significance" of what are really nonevents, I would offer the following as a clue in that direction. Explaining why Caesar crossed the Rubicon <u>and</u> did not instead cross a nearby river is, ceteris paribus, better than explain- ing why Caesar crossed the Rubicon <u>and</u> did not instead cross the Eu- phrates. Given our knowledge of the circumstances surrounding Caesar's crossing of the Rubicon (including the minimum amount of time it could have taken him to travel from Western Europe to the Near East), the former nonevent was a much more viable possibility than the latter. The guiding intuition here is that to explain an event is to show how, if some highly probable nonevent occurred instead, other known events would probably not have occurred. We become better explainers as our explanations raise the probabilities of <u>both</u> the nonevent that could have happened <u>and</u> the noneventuality of known events that would have followed. Thus, to wax Peircean, explanations ultimately aim at show- ing how the slightest shift in the actual world would have universal (though not necessarily major) consequences. I suspect that this view appears deterministic only because explanations are usually of past events, which, given a realist view of time, could not (now) have been otherwise. But my tenseless formulation offers the possibility that small changes (in the future) could make big differences--so there is still hope for the Marxist.

These points go some way toward explicating what it means for two radically divergent paradigms to be within one science. For example, the grammar of economics is uniform enough so that it would be mislead- ing to say that Marxists and Neoclassicists <u>describe</u> economic phenomena radically differently. Indeed, recalling the ideal prose style of the writing manuals, insofar as the two schools mirror the structure of the phenomena in largely the same way, they are practicing a common science. And in virtue of that fact they need only identify their explanations as "economic" ones, rather than, say, "Marxo-economic" ones--as if the prefix indicated that the economic phenomena in question were discerni- ble to only Marxist eyes. However, even in the most mainstream profes- sional journal, where party-line rhetoric is kept to a minimum, the astute reader will nevertheless be able to sort out the schools to which the authors belong simply by noticing what nonevents form the background of their texts' explanatory strategies. Evaluating the explanations in this case is more difficult than weighing the counterfactuals to Caesar crossing the Rubicon "only" because these nonevents are signs of radi- cally divergent paradigms, and so not easily ordered according to their relative significance. But the guiding intuition as to what counts as a better explanation still applies.

As we have just seen, the first step in what may be called a "herme- neutics of legitimacy" involves establishing the narrative perspective, or "voice", of the explainer in a text by determining the things that didn't happen which are of interest to him--either because they were expected or desired. But as all literary critics know, the text may

speak with several voices at once in an effort to convey a comprehensive picture that incorporates not merely a description of the phenomena, which on our account do not speak for themselves but are instead
spoken for; nor do the voices necessarily cease with that of the omni-
scient narrator who, like Leibniz's God, situates things-as-they-are
against the backdrop of things-as-they-could-be in order to explain
the former in terms of the latter. In addition, there may be the voice
who speaks from knowledge that the narrator is not <u>actually</u> the omni-
scient supplier of possibilities who stands outside the phenomena he
explains, although it is a voice he <u>must</u> adopt in order to continue
narrating. This "metavoice," as it were, makes the explainer part of
what needs to be explained because ultimately his explanation cannot
be evaluated unless the reader understands the sense in which what the
explainer offers as constraints on the phenomena in fact reflect constraints on his own possibilities for explaining the phenomena.

Although the voices of a text may be analytically distinguished
along the lines we have just suggested, their textual interdependency
is also evident. A text with many voices will indeed be obscure if it
is read as having only one voice. In that case, the one voice will be
characterized as "equivocal", "paradoxical", "contradictory", and by
all the other adjectives at the command of the reader to express the
author's apparent duplicity. However, the reader is trading on a confusion that because a text is written by one author, it ought to have one
voice, or at least that there ought to be one of perhaps many textual
voices that clearly expresses the author's stance toward the subject-
matter. Clarity would then come from the reader's ability to latch on
to this one voice as it narrates and comments on the subject-matter.
(I suspect that this confusion arises partly as a result of children
no longer having to grapple with a difficult text like the Bible when
learning how to read; but this requires another Rorty-style history
that is beyond the scope of this paper.) In fact, the persuasiveness
of Rorty's argument for the end of philosophy capitalizes on our normally reading an explicitly nonfictional text as if it had only one voice,
which is to say, our judging the text against the standards of the ideal
prose style. I say this because were the reader instead to engage in
the hermeneutical exercise of sorting out the voices of the text, he
would be forced to take the philosophical turn, insofar as the different
narrative stances express the different modalities in which the subject-
matter is being regarded.

Literary critics have discerned many finer distinctions in perspective, usually subsumed under the category of rhetoric known as "tropes",
which could greatly enhance philosophical thinking. For example, the
many shades of irony--the trope whereby something that is not x is regarded as if it were x--may at first glance seem little more than part
of the poet's bag of tricks. But consider how one traditional issue in
epistemology--the significance of illusions for the possibility of veri-
dical perception--can be handled much more subtlely if illusion, which
satisfies the above definition of irony, is understood not as a particular subject-matter (one of our "mental contents") but as a particular
perspective from which anything may be regarded. In that case, we would

no longer ask what kind of _thing_ is an illusion, but, rather, in what
kind of _relation_ does one need to stand to something in order to say
that it is an illusion. It can be easily seen that the problematics
which follow from these two opening queries are radically different.
But notice too that only the latter query preserves the integrity of
the philosophical by acknowledging Rorty's point that philosophy has
no unique subject-matter.

This observation also suggests an answer to why certain philosophical
problems--especially "metaphysical" ones--are so difficult. The received view is that difficulty arises from an intractable subject-
matter, one that does not readily yield to analytical methods and clear
argument structures. Since metaphysical problems often seem to demand
that some defining feature of reality be encompassed within a single
proposition, their insolubility has usually been attributed to the
impossibility of arriving at a synoptic vision through the modest
methods of discursive reasoning. As a result, and as we have seen,
modern intellectual history has been marked by the breakdown of reality
into the more manageable pieces that are now the respective phenomena
of the sciences. But while this conception of philosophical difficulty
plays right into Rorty's hands, I would argue that there is good reason
to suspect that it misses the mark entirely, namely, that our actual
philosophical difficulties are not illuminated by regarding them as the
products of an intractable subject-matter. Rorty may render philosophical difficulties insignificant, but he goes no way toward addressing
them on their own terms. Intuitively, the symptoms of philosophical
difficulty do not so much involve trying to get a sense of the vast
range of phenomena under consideration as trying to find an _approach_
that will allow the phenomena to be considered. Once again, the problem _is_ setting up the correct combination of perspectives on which a
narrative structure can be built.

Consider the typically metaphysical dialectic between idealist and
realist. The idealist argues that only a paradox would allow the
realist's view to be tested--namely, that we would somehow "see" that
the world was not dependent on our seeing it, while the realist argues
that the correctness of the idealist's view implies that our commonsense understanding of the world is systematically misleading--a highly
improbable hypothesis. Because of the curious way in which this dialectic stalemates, Kant would situate it among his "antinomies of reason". It will be recalled that Kant assigns to reason a non-constitu-
tive role in cognition--which is to say, the difference between realism
and idealism would not be manifested simply in terms of different
descriptions of the world but in terms of different explanations, which,
as we have seen, are distinguished by their reference to counterfactual
possibilities. Is causality an intrinsically physical relation that
would obtain without the presence of minds? If so, then it is to be
explained by physics; if not, then by psychology. But since Kant
believed that nothing determinate could be said about counterfactuals,
it makes no sense to _argue_ about the relative merits of the two posi-
tions--although there may be some value in conducting empirical inquiry
_as_ _if_ one or the other were true. Rorty shares Kant's sensibility as

to how disagreements should be resolved--namely, by appealing to whether one saw the right thing, and <u>not</u> whether he saw it from the right perspective. Since metaphysical disputes do not fit the former category, they are difficult; and historically speaking, since Kant they have become disagreeably difficult, such that today it is taken for granted that differences in perspective cannot be evaluated. But given the current rejection of any simple-minded empiricism, we should <u>also</u> say that the Kantian intuition behind counterfactuals having indeterminate truth-values--namely, the impossibility of there ever being sufficiently "direct" evidence either to confirm or disconfirm them--is at least questionable. Indeed, we should consider what we need to see <u>not</u> happening in order to "merely" describe the impact of one ball on another, to explain it as something psychological, and to explain it as something physical. We would see then that what happens and what does not happen are equally "direct" and of equal evidentiary worth--and, as we have suggested, this is all in virtue of the explainer's stance toward the phenomena being integral to his explanation.

And again we see how the literary critic may help the philosopher, insofar as the former is a "strategist of difficulty", an expert in making judgments about what combination of voices will yield a narrative structure more or less capable of accomodating certain plots, addressing certain themes, and so forth. Posing a problem can be fruitfully regarded as "setting the stage" for an explanation, a metaphor that the critic Kenneth Burke has in fact used as a reduction language for reconstituting and evaluating metaphysical systems. The fact that he uses this metaphor for <u>both</u> constitutive <u>and</u> evaluative purposes reveals an interesting anti-Kantian intuition that deserves to be cultivated in philosophy. The intuition, common to many species of idealism, is that if you believe everything in the universe to be causally or intentionally interconnected, then it is reasonable to conclude that something epitomizes these connections so as to clearly reflect the structure of reality. Burke's microcosm is the drama; for many theists it is man himself ("the image and likeness of God"). Modern science has been drawn between the machine and the organism. But although these two metaphors also mix constitutive and evaluative concerns, most philosophers of science would argue that they function only in a pseudo-constitutive capacity. In other words, a metaphor models the phenomena until it <u>gets evaluated</u> as a literally false description. Still, the evaluative relation may be reversed, with conformity to or enrichment of the metaphor being the basis on which to judge the <u>relevance</u>, rather than the <u>truth</u>, of a given phenomenon. In other words, something that does not fit the metaphor well is not very significant.

But how is one to <u>argue</u> about significance? Before succumbing to Rorty's counsel of self-negation, I suggest that philosophers look at the very conceptual apparatus of contemporary literary criticism--semiotics and rhetoric--that Peirce, a century ago, said formed (with logic) the cornerstone of philosophical thinking. It might well be that we will need to start conceiving of argumentation in a new light, and here I can only offer a clue. Suppose I propose that "p" is necessary for "q", yet everyone else seems to think that "q" can obtain without "p" (and they even offer counterexamples to that effect). Has my attempt at a transcendental argument been foiled? Not if I can set up an explanation such that it makes sense for me to see that "q" requires "p" _and_ for my opponents to see what they see instead, given how I have posed the problem. Thus, it is not that they misunderstood my claim or argued fallaciously against me--or even that they are just plain wrong, but, rather, that they have in fact seen what I have seen --only from a systematically distorted perspective. And the burden is then on me to reconstitute the _significance_ of their perspective in light of my own.

383

References

Burke, Kenneth. (1967). <u>A Grammar of Motives.</u> Berkeley: University of California Press.

Cavell, Stanley. (1979). <u>The Claim of Reason.</u> New York: Oxford University Press.

Derrida, Jacques. (1967). <u>De la Grammatologie.</u> Paris: Éditions de Minuit. (As reprinted as <u>Of Grammatology.</u> (trans.) G. Spivak. Baltimore: Johns Hopkins University Press, 1976.)

Rorty, Richard. (ed.). (1967). <u>The Linguistic Turn.</u> Chicago: University of Chicago Press.

----------------. (1979). <u>Philosophy and the Mirror of Nature.</u> Princeton: Princeton University Press.

<u>Was Carnap a Complete Verificationist in the Aufbau?</u>

Richard Creath

Arizona State University

Was Carnap a complete verificationist in the <u>Aufbau</u>? The question
is asked in the precise sense which has been standard since 1936 in the
work of such classical writers as Carnap, Hempel and Scheffler.[1] It is
not an idle question because many, if not most, writers including Carnap himself have thought that he was such a verificationist in the <u>Auf-
bau</u>.[2] Moreover, it is generally agreed that verificationism in this
sense is a serious mistake. I propose to argue that the <u>Aufbau</u> does
not in fact embrace such a verificationism and to base my argument on
evidence directly from the <u>Aufbau</u> and surrounding documents.

Since 1936 'verificationism' has been standardly used in the sense
of a theory of empirical meaningfulness according to which a statement
is meaningful if and only if it is possible that there be evidence that
we might actually obtain which would conclusively establish the statement as true. Sometimes the word 'verificationism' is still used in a
wider sense, so some writers have chosen the name 'complete verificationism' for the view under investigation here. For reasons of brev-
ity I shall use the term 'verificationism' in the restricted sense in
which it is equivalent to 'complete verificationism'. It has also
been thought that evidence is restricted to observation sentences
(atomic in form), that we might directly establish at most a finite
number of observation sentences, and that a set of observation sentences can conclusively establish some statement only if that set is
consistent and logically implies the statement. The concept of verifiability thus gets analyzed by Hempel as the view that "A sentence
has empirical meaning if and only if it is not analytic and follows
logically from some finite and logically consistent class of observation sentences." (Hempel 1965, p. 104).

Whether this is an adequate analysis of either the original or the
ordinary concept of verifiability is beside the point here. It should
be emphasized again, however, that this is the sense in which the
charge against the <u>Aufbau</u> has repeatedly been made. Hence, the selec-

---

PSA 1982, Volume 1, 384-393

tion of this sense of the concept of verificationism is by no means
arbitrary, and rescuing the _Aufbau_ from the charge would not be a
trivial exercise in any way.

Those who would accuse the _Aufbau_ of verificationism are not without evidence to support their position. That evidence is of two sorts.
First, there are Carnap's autobiographical remarks. In the case of
someone as sober-minded and honest as Carnap these are to be taken
quite seriously. Second, there is one passage from the _Aufbau_ itself
which seems to embrace verificationism in its strongest form. I cannot argue that the passage is weaker than it seems. Instead I shall
produce evidence that it is unrepresentative of the _Aufbau's_ considered view.

Carnap's imputation of verificationism to the _Aufbau_ is quite
explicit. (For future convenience let us call this passage A.)

> According to the original conception, the system of knowledge, al-
> though growing constantly more comprehensive, was regarded as a
> closed system in the following sense. We assumed that there was
> a certain rock bottom of knowledge, the knowledge of the immedi-
> ately given, which was indubitable. Every other kind of knowledge
> was supposed to be firmly supported by this basis and therefore
> likewise decidable with certainty. This was the picture which I
> had given in the _Logischer Aufbau_; it was supported by the in-
> fluence of Mach's doctrine of the sensations as the elements of
> all knowledge, by Russell's atomism, and finally by Wittgen-
> stein's thesis that all propositions are truth-functions of the
> elementary propositions. This conception led to Wittgenstein's
> principle of verifiability which says that it is in principle
> possible to obtain either a definite verification or a definite
> refutation for any meaningful sentence. (Carnap 1963a, p. 57).

The last sentence of A, which demands only the possibility of either
"a definite verification of a definite refutation," seems to require
less than verificationism in our sense. Since, however, the apparently
disjunctive requirement was to follow from "Wittgenstein's thesis that
all propositions are truth-functions of the elementary propositions"
and was to imply, as Carnap goes on to indicate, that physical laws do
not satisfy the requirement, we have little choice but to interpret
the condition as being at least as strong as our verificationism. This
interpretation is supported by Carnap's own statement from the autobiography: ". . . Wittgenstein's principle of verifiability . . .
says . . . that a sentence is meaningful if and only if it is in principle verifiable, that is, if there are possible, not necessarily
actual, circumstances which, if they did occur, would definitely
establish the truth of the sentence." (Carnap 1963a, p. 45).

This charge of verificationism is supported by a very strong statement from the _Aufbau_ itself (call this passage B):

> Now, if it is the case that a genuine question is posed, . . .
> the sentence which was given when the question was posed . . .
> can in a step by step way be so transformed that it expresses
> a definite (formal and extensional) state of affairs relative
> to the basic relation. In keeping with the tenets of construction
> theory, we presuppose that it is in principle possible to recog-
> nize whether or not a given basic relation holds between two
> given elementary experiences. Now, the state of affairs in ques-
> tion is composed of nothing but such individual relation state-
> ments, where the number of elements which are connected through
> the basic relation, namely of elementary experiences, is finite.
> From this it follows that it is in principle possible to ascertain
> in a finite number of steps whether or not the state of affairs
> in question obtains and hence that the posed question can in
> principle be answered. (Carnap 1928a, pp. 291-2).

There is no room for interpretation here; this _is_ verificationism.
To make my case I need to show that B is unrepresentative of the _Auf-
bau_, that the balance of evidence weighs against taking Carnap to have
been a verificationist there.

Schematically my argument is this: First, verificationism is not
intrinsic to the main purpose of the _Aufbau_, i.e., construction theory.
Next, verificationism is not added as an incidental demand. Finally,
Carnap's outright rejection of verificationism in works appearing at
the same time as the _Aufbau_ counts heavily against the claim that he
had accepted it in the _Aufbau_ either.

Developing the notion of a constructional system is the whole purpose of the _Aufbau_, and every other issue is subordinate to it. If
one had to describe a constructional system in five words or less, one
would call it a system of definitions. That system is to be a rational reconstruction, an explication, of all the legitimate aspects of
our cognitive universe. Indeed, the legitimacy of a concept, object,
or sentence is defined in terms of the _possibility_ of placing it within a constructional system.

All of this would be uninteresting were it not for the conviction
that the whole structure is to be established on a very slender base
and built with rather restricted techniques. Carnap took as the base
for his own attempt only elementary experiences as the primitive individuals and the relation of remembered similarity as the only primitive concept. He was hopeful that this base would be sufficient, but
he was willing to be flexible. After all, he wanted only to convince
the reader that some such construction is possible. Though he was to
relax over the years, Carnap indicated no corresponding flexibility
about constructional techniques which were to consist solely of explicit and contextual definitions. All of the concepts, objects, and
logical connectives were of course to be construed extensionally.

Another way of summarizing a constructional system is to say that
it is a system of rules for translating every meaningful statement

about any object whatsoever into a statement about the basic objects. The only constraint on the translation is that the results be guaranteed to be materially equivalent to the original. "A <u>second thesis</u> of construction theory asserts that each <u>scientific statement is</u>, <u>in the final analysis</u>, <u>a statement about the basic relation(s)</u>; more precisely, each statement can be transformed into another statement which (besides logical constants) contains only the basic relation(s), where the logical value (although not the epistemic value) is retained." (Carnap 1928a, p. 187).

Since the elementary experiences are named from the beginning rather than introduced via construction procedures, Carnap might also have allowed that the statements into which scientific statements are to be transformed can contain names of the basic entities as well.

The truth value of every meaningful statement would depend on the truths at the basic level, but it would be a mistake to conclude from this that every meaningful sentence would be implied by some consistent class of basic sentences. Not even the class of all true basic sentences would imply some true universally quantified sentence. One would need to know in addition this <u>was</u> the class of all true basic sentences, and a general claim of this sort could not be expressed as a basic sentence.

Carnap's construction theory would guarantee verificationism only if he imposed either of two additional restrictions on his theory. First, he might prohibit the use of unrestricted (Russell-Whitehead) quantifiers, insisting instead on the restricted quantifiers, of language I of <u>The Logical Syntax of Language</u> (LSL) (Carnap 1934, pp. 21, 30). Every use of these latter quantifiers can be replaced with a <u>finite</u> conjunction or disjunction of their substitution instances. The idea that Carnap would have so restricted his quantifiers six years before LSL without announcing the fact is not only implausible, it is refuted by the text: "Propositional functions (§28): if $f\underline{x}$ is a propositional function, then $(\underline{x}) \cdot f\underline{x}$ means: '$f\underline{x}$ holds for every $\underline{x}$'; $(\exists \underline{x}) \cdot f\underline{x}$ means 'there is an $\underline{x}$ for which $f\underline{x}$ holds.'" (Carnap 1928a, p. 155).

There is a second way in which construction theory could involve verificationism. If the number of basic individuals could be guaranteed to be finite, and if, further, for some finite stock of names one happened to know that it contained a name for each object (a bit of knowledge which could not be expressed as a basic sentence) then plausible adjustments could be made in the rules for implication so that for any consistent meaningful sentence (including universally quantified ones) there would be a finite consistent class of basic sentences which would imply it. Whether this would change the meaning of the quantifiers can be left undecided. In any case it depends on the guarantee and not just the possibility that the domain is finite (as well as the guarantee that one's list of basic names is complete.)[3]

It would seem that Carnap did entertain the possibility that the

388

domain of basic individuals is finite. There is no evidence that he
was prepared to insist dogmatically on this. He does say "all objects of knowledge, inasmuch as they are objects of conceptual knowledge, are somehow designated or, at least, can in principle be
designated." (Carnap 1928a, p. 34). This does not say that we can
individually designate all of the objects, but only that for any object we can designate that one. In addition there is always the possibility of general reference via quantifiers as opposed to individual
references via names, etc.

Carnap comes close to speaking of the number of basic individuals
in still another place: "In choosing as basic elements the elementary experiences, we do not assume that the stream of experience is
composed of determinate, discrete elements. We only presuppose that
statements can be made about certain places in the stream of experience, to the effect that one such place stands in a certain relation to another place, etc. But we do not assert that the stream of
experience can be uniquely analyzed into such places." (Carnap 1928a,
p. 109). This does not deny the finitude of the number of basic individuals, but he seems definitely to resist asserting it. If Carnap
had believed that the number of basic objects is finite, then surely
this would have been the place to say so. That he said nothing,
apparently deliberately so, suggests that he wanted the question left
very much open.4

There is one final line of reasoning which seeks to show that
Carnap's interpretation of quantifiers and/or his views on the number
of basic entities would commit him to verificationism. This argument
is that since Carnap was very much influenced by Wittgenstein's
Tractatus, and since Wittgenstein there seems to adopt a non-standard
interpretation of quantifiers as well as some form of finitism, Carnap
may well have adopted similar views in the Aufbau. Without detailing
Wittgenstein's views or their undoubtedly important impact on Carnap
we can rule out this line of reasoning from the very start. Wittgenstein was able to abandon the standard (Russell-Whitehead) interpretation of quantifiers and to entertain finitism only because he was
also heroically willing to abandon the Russell-Whitehead construction
of classical mathematics itself. Carnap was not willing to do this.
In the Aufbau he explicitly embraced the system of Principia Mathe-
matica (Carnap 1928a, pp. 23, 61, 177) and explicitly adopted the
Frege-Russell definition of cardinal numbers (Carnap 1928a, p. 68).
If, as his discussion indicated, basic objects are the objects of
which the cardinal numbers are classes of equinumerous classes, then
the infinitude of cardinal numbers would specifically commit Carnap to
the infinitude of basic objects. There may be ways in which this conclusion could be avoided, but clearly Carnap took no steps to do so.

It would, therefore, seem that Carnap does not commit himself to
the additional assumptions that would make verificationism intrinsic
to constructionalism.

While verificationism might be beside his main purpose, Carnap may

have intended to impose it anyway. The textual evidence is that he did
not. The issue is raised several times in the Aufbau in contexts which
do not mention constructability. In each case, Carnap uses a criterion
of meaningfulness which is vastly weaker. Even more strikingly, in
Pseudoproblems in Philosophy, he explicitly considers alternative criteria and explicitly chooses the more liberal.

In one place the Aufbau reads: ". . . various contradictory solutions have been proferred between which no (conceivable) experience
could decide. Hence essence problems must be transferred from science
to metaphysics." (Carnap 1928a, p. 44). Obviously this condition
would be satisfied if each "solution" implied one basic sentence such
that the other solution implied the negation of that sentence.

Carnap touches the issue in another passage and again does not
demand a verification: "A decision which has been made through empa-
thy or otherwise, which cannot in principle be subjected to a rational
test through conceptual criteria would forfeit every claim to scientific status." (Carnap 1928a, pp. 82-3).

On two occasions (1928a, pp. 257, 289) Carnap uses the word 'verifiable' but in each case he means statements whose objects are con-
structable: "For only the construction formula of the object--as a
rule of translation of statements about it into statements about the
basic objects, namely, about relations between elementary experiences
--gives a verifiable meaning to such statements, for verification means
testing on the basis of experience." (1928a, p. 289). If Carnap had
in fact demanded verifiability in our sense, then he is being extra-
ordinarily reluctant to say so in the very passages which call for such
a comment. The claim is not that these passages are inconsistent with
verificationism; the claim is rather that Carnap is repeatedly not
espousing verificationism when supposedly he should be.

Any residual doubts, however, are dispelled in Pseudoproblems,
which was published in the same year as the Aufbau and did not an-
nounce any change of position. There Carnap distinguished between
testable statements and those merely having factual content in this
way: "A statement p is said to be "testable" if conditions can be
indicated under which an experience E would occur which supports p
or the contradictory of p. A statement p is said to have "factual
content", if experiences which would support p or the contradictory
of p are at least conceivable, and if their characterstics can be
indicated." (Carnap 1928b, p. 327). Testability implies having
factual content, but not vice versa. To make matters even clearer
some examples are given.

The statement "in the next room is a three-legged table" is
testable; for one can indicate under what circumstances (going
there and looking) a perceptual experience of a certain kind
would occur which would support the statement. Hence this statement has factual content. The statement "there is a certain red
color whose sight causes terror" is not testable, for we do not

know how to find an experience which would support this statement. Nevertheless, this statement has factual content, for we can think and describe the characteristics of an experience through which this statement would be supported. Such an experience would have to contain the visual perception of a red color and at the same time the feeling of terror about this color. (Carnap 1928b, p. 327).

Given this distinction, Carnap clearly states his criterion of scientific meaningfulness:

> . . . each statement which is to be considered meaningful in any one of these fields (i.e., which is either considered true or false or which is posed as a question) either goes directly back to experience, that is, the content of experiences, or it is at least indirectly connected with experience in such a way that it can be indicated which possible experience would confirm or refute it; that is to say, it is testable, or it at least has factual content... . We have not taken the strict viewpoint which requires of each statement that it should be supported or testable; rather we consider statements meaningful even if they merely have factual content, but are neither supported nor testable. Hence we are using as liberal a criterion of meaningfulness as the most liberal minded physicist or historian would use within his own science. (Carnap 1928b, p. 328).

I do not know how much more clearly Carnap could have rejected the verificationism attributed to the _Aufbau_ in the autobiography. The fact that it is rejected coupled with the fact that _Pseudoproblems_ announced no change of position from the _Aufbau_ which had been written so shortly before strongly suggests that the earlier book was not verificationist either.

If we look back over our discussion, it is quite clear that any charge of verificationism is unfounded. Constructionalism itself can be construed as a verificationist demand only if all quantifiers are restricted or if the number of basic individuals is known to be finite. Neither of these conditions is met. While verificationist demands might have been imposed in addition to constructionalism, the evidence is that they were not. Except for B, every time that the subject is broached, weaker demands are announced. This would be hard to explain if we thought that Carnap was a verificationist, and it strongly suggests that B is an ill-considered and un-representative remark. This suggestion becomes almost irresistible when we consider the fact that in _Pseudoproblems_, published at about the same time and with no announced change of position, verificationism is explicitly rejected.

It might be interesting to speculate as to how Carnap could have come in the intervening thirty or so years to so underestimate the _Aufbau_. It would also be interesting to trace Carnap's very real flirtation with verificationism in the years immediately following that book. Both of these tasks would carry us well beyond the scope of the present paper. In short, the best guide to what the Aufbau means is still the Aufbau itself. In that the preponderance of evidence points away from verificationism.

Notes

[1] See especially (Carnap 1936-37), (Hempel 1950, 1951, 1965), and (Scheffler 1963).

[2] In particular see (Carnap 1963a), (Jørgensen 1951), (Popper 1963). See also (Carnap 1963b, 1963c) and (Laudan).

[3] I suspect that this is as close as one can come to making passage B plausible. Even so the necessity of knowing that one's list of basic names, and hence of basic sentences, is complete undermines the support it offers verificationism.

[4] On this issue Nelson Goodman has suggested that the number of elementary experiences (he calls them erlebs) is finite and that therefore the notion of dimensionality, including Carnap's, is in need of revision. He says: ". . . according to ordinary mathematical definitions of dimensionality--and Carnap's definition of dimension number (DZ) does not appear to differ in this respect--any finite array is zero-dimensional. Since we are here concerned with... a finite set of quality classes, a sense class can thus hardly be multidimensional according to ordinary mathematical usage." (Goodman 1951, pp. 179-80).

At the risk of urging the well-known doctrine that one man's modus ponens is another man's modus tollens, it must be remarked that the difficulty over dimensionality can be construed as evidence that Carnap was not exclusively committed to a finite number of elementary experiences, and that he did not always have the finite case in mind when he was framing his definitions. In any case there is still a problem about dimensionality, for even on my reading of the Aufbau, Carnap wanted to leave the possibility of a finite number of elementary experiences an open question, and thus he would still need a definition of dimensionality that could accommodate such a case.

References

Carnap, Rudolf. (1928a). Der Logische Aufbau der Welt. Berlin-
     Schlachtensee: Weltkreis-Verlag. (As reprinted as The Logical
     Structure of the World. (trans.) R.A. George. (Published with
     Pseudoproblems in Philosophy.) Berkeley: University of
     California Press, 1967.)

------------. (1928b). Scheinprobleme in der Philosophie: Das
     Fremdpsychische und der Realismusstreit. Berlin-Schlachtensee:
     Weltkreis-Verlag. (As reprinted as Pseudoproblems in Philosophy.
     (trans.) R.A. George. (Published with The Logical Structure of
     the World.) Berkeley: University of California Press, 1967.)

------------. (1934). Logische Syntax der Sprache. Wien: Verlag
     Julius Springer. (As reprinted with revisions as The Logical
     Syntax of Language. (trans.) Amethe Smeaton. London: Kegan
     Paul Trench, Truber & Co., 1937.)

------------. (1936-37). "Testability and Meaning." Philosophy of
     Science 3: 419-471, 4: 1-40

------------. (1963a). "Intellectual Autobiography." In Schilpp
     (1963). Pages 3-84.

------------. (1963b). "K.R. Popper on the Demarcation Between
     Science and Metaphysics." In Schilpp (1963). Pages 877-881.

------------. (1963c). "Paul Henle on Meaning and Verifiability."
     In Schilpp (1963). Pages 874-877.

Goodman, Nelson. (1951). The Structure of Appearance. Cambridge:
     Harvard University Press.

Hempel, Carl G. (1950). "Problems and Changes in the Empiricist
     Criteria of Meaning." Revue International de Philosophie II.
     41-66.

------------. (1951). "The Concept of Cognitive Significance: A
     Reconsideration." Proceedings of the American Academy of Arts
     and Sciences 80: 61-77.

------------. (1965). "Empiricist Criteria of Cognitive
     Significance: Problems and Changes." In Aspects of Scientific
     Explanation and Other Essays in the Philosophy of Science. New
     York: The Free Press. Pages 101-119.

Jørgensen, Jørgan. (1951). The Development of Logical Empiricism.
     (International Encyclopedia of Unified Science. Volume II. No.
     9. Edited by Otto Neurath, et al.) Chicago: University of
     Chicago Press.

Laudan, Laurens. (n.d.) "From Testability of Meaning (and Back Again)." Unpublished Manuscript.

Popper, Karl R. (1963). "The Demarcation Between Science and Metaphysics." In Schilpp (1963). Pages 183-226.

Scheffler, Israel. (1963). _The Anatomy of Inquiry._ New York: Alfred A. Knopf.

Schilpp, Paul Arthur. (ed.). (1963). _The Philosophy of Rudolf Carnap._ LaSalle, IL: Open Court.

<u>The Creation of Similarity: A Discussion of Metaphor</u>
<u>in Light of Tversky's Theory of Similarity</u>

Eva Feder Kittay

SUNY, Stony Brook

The cognitive interest in metaphor and simile derives from a conceptual detour: through these figures we regard one thing in terms of another, and in so doing our understanding of the first is modified in light of perspective gained by the second.[1]

But why, in terms of cognitive gain, engage in such conceptual detours? That is, what motivates our use of metaphor and simile? The motivation of metaphor has often been thought to be the assertion of a similarity amongst things generally not thought to be similar. However, as Max Black pointed out in criticizing the "comparison theory of metaphor", metaphors are not simple comparisons of "objectively given" similarities, wherein to the question "is A like B in respect of P?" we might have a definite answer. Instead, he writes, "It would be more illuminating in some of these cases to say that the metaphor creates the similarity rather than records a similarity antecedently existing." (1954, p.37). Consider Wallace Steven's metaphor of a garden as a "slum of bloom" ("Banal Sojourn"): what objectively given similarity is there between a garden and a slum of bloom prior to the formation of the metaphor itself?

It is this "creation" of similarities which I take to be the prime conceptual gain of and motivation for the use of metaphor. But we need to understand how metaphors and similes can "create similarities". On a first view, it would appear that things are either similar by virtue of properties they share or they fail to share common properties and so are not similar. As long as similarity is regarded only as relating "objectively given features" of the world, we cannot explain the metaphoric "creation of similarities". We need an account which ties similarity more closely to the judgments we make concerning the world than to its "objectively given" features.

Amos Tversky (1977) has, I believe, put forth such a view,

PSA 1982, Volume 1, pp. 394-405

accompanied by intriguing experimental results. Moreover, his account allows us to understand metaphor as a special case, a sort of limiting case of category formation. Metaphor thereby gains its force in being grounded in a conceptual activity which is pervasive in cognition.

Of particular importance is the account he gives of the relation between context and similarity, but of interest as well is his claim that similarity is not necessarily a symmetrical relation. Each point bears on the question of the way in which similarity functions in the formation of categories and consequently on the question of metaphor.

Tversky's foil is the geometric model in which similarity is a measure of proximity between two objects characterized as points in a coordinate space; dissimilarity is a measure of their metric distance. On this model, similarity must always be symmetrical, i.e., if a is similar to b, then b is similar to a. Further, given objects a, b, c, d, e in a given coordinate space, if the distance between a and b is judged to be less than the distance between a and c, that measure should remain indifferent as to whether d is replaced by another object, e. Tversky demonstrates that the two assumptions, symmetry and context independence, are belied by common usage and by experimental results. For each assumption, I will first, briefly present Tversky's position, discuss it in more philosophical terms and apply the findings to the question of metaphor.

1. Symmetry

Tversky suggests that objects be construed not as points in a geometric space, but as collections of features[2], and that similarity be construed as a set-theoretic process of feature matching. As such, similarity may, but need not be, symmetrical. Unlike the symmetrical judgments we find when, for instance, we speak of the similarity between two triangles, many similarity judgments are asymmetrical. For example, we say "the son is like the father." Were we to say "the father is like the son", we would mean something quite different. Again, we are more likely to say that an ellipse is similar to a circle or to claim that North Korea is more similar to Red China than Red China is to North Korea than the reverse. Tversky notes that directionality and asymmetry are especially prominent in figurative similarity, as in metaphors and similes: compare "My love is as deep as the ocean" with "The ocean is as deep as my love." (1977, p.328).

To the extent that similarity is an asymmetrical relation we need terms to differentiate the relata. In statements of the sort "a is similar to b", Tversky calls 'a' the 'subject' and 'b' the 'referent'. Because 'referent' is used quite differently in most philosophical discussions I propose, instead, to introduce two terms based on the verb 'to confer', whose etymological derivation, 'to bring together', bears on the process of making similarity judgments. The two objects named by 'a' and 'b' I will call 'conferee' and 'conferer' respectively, although I will also speak of the conferee as the subject, since it so appropriately conveys the role of the conferee.

396

The "contrast model" of similarity which Tversky develops is able
to accommodate both the directionality of some similarity statements
and judgments as well as the symmetry of others. On this model,
similarity is represented as a weighted difference of common and
distinct features of compared objects:

$$S(a,b) = \theta\, f(A\ B) - \alpha f(A-B) - \beta f(B-A), \text{ where } \theta, \alpha, \beta, \geq 0.$$

S and f are interval scales, S is a similarity scale, and A and B are
the sets of features of the objects a and b, respectively.

Symmetry is obtained when $\alpha = \beta$ (e.g., in the task: "assess the
degree to which a and b are similar."). When more weight is given
to the distinctive features of the conferee than to the conferer, $\alpha > \beta$,
(e.g., in the task: "assess the degree to which a is similar to b
"), we are focusing on the former. When we focus on the conferee we
take the conferer to be a prototype or to have the more salient
features by virtue of which the former is compared to it. This
Tversky calls "the focusing hypothesis".

The notions of prototype, variant and salience which Tversky uses
are based largely on the work of Rosch (1973,1975) and associates
(Rosch and Mervis 1975, Rosch et al. 1976). According to them, the
members of a category share prominent features of the prototypes and
are related to one another in the manner characterized by Wittgenstein
as family resemblance. The relative saliency of prototype and variant
leads us to make comparative judgments of similarity in such a way
that the variant is thought to be more similar to the prototype than
the prototype is to the variant, hence North Korea is thought to be
more similar to Red China than Red China is thought to be similar to
North Korea; and it leads us to choose the more salient stimulus, or
prototype as conferer, and the less salient stimulus as subject.

The striking asymmetry of metaphors and similies is related to the
fact that the conferer is chosen particularly for the saliency of
certain features while the focus is always on the subject, the
conferee. It is the salient features of the conferer rather than the
distinctive features of the subject which are what count toward the
judgment that the subject is similar to the conferer and not vice
versa. For example, in the metaphor "Juliet is the sun", we focus on
Juliet so that her distinctive features are more heavily weighted
against a similarity with the sun than the distinctive features of the
sun count against the similarity with Juliet.

2. Classification

Features are salient in what Tversky calls a 'diagnostic' sense
insofar as they are the basis upon which classifications are made.
(The other factor determining saliency is 'intensity', i.e. "the
signal to noise ratio.") If we take into consideration the category
implicit in the use of the term prototype (given a Roschian account of
these terms), we can give Tversky's account of similarity a somewhat
different rendering which will also bring it closer to the relation
between metaphor and the process of classifying or category formation.

The geometric model of similarity has certain intuitive plausibility as long as we think of similarity as a partial identity -- not identity between two objects _per se_, but between two objects considered as exhibiting certain attributes or properties by virtue of which they are being compared. Objects which are more similar to one another are identical in more regards than those which are less similar. Identity, of course, is symmetrical, and similarity, as a partial identity, would appear necessarily to be symmetrical as well.

The asymmetry to which Tversky points is largely the result of the attention paid to the distinctive features of conferee and conferer. Where the differences between conferee and conferer are well defined and even conventionalized, as in the case of the similarity between two triangles whose sides, while different in length, are nonetheless proportional, the differences can be ignored and similarity can be treated as a partial identity. But considerations that Tversky brings to bear indicate that similarity is more often not a modified, i.e., partial, identity between two objects but a modified, i.e., implicit, classification of one object, the conferee, as suggested by its shared features with another object, the conferer.[3] When we classify objects, we do so not only according to shared features but also according to distinctive features.[4]

A classificatory statement is one in which we claim that some _a_ is a _C_. In many asymmetric similarity statements, where the conferer is relatively more prototypical than the conferee, there is no explicit mention of the category _C_; rather there exists the implicit understanding that the conferer _b_ is representative of some category[5] _C_ into which we want to include and place the object _a_. To say that _a_ is similar to _b_, on this account, is to say that where _b_ is understood as a prototype of _C_, _a_ is to be included in category _C_ insofar as it shares the features of _b_ which qualify _b_ as a member of _C_. If we understand Red China as a prototype of Communist Asian nations, and we say "North Korea is like Red China", then we are saying that by virtue of the features North Korea shares with Red China, insofar as Red China is paradigmatic of an Asian Communist nation, North Korea is an Asian Communist nation. Unless the context is quite clear, we are unlikly to say "Red China is like North Korea" because we are less clear of what category of things North Korea is likely to epitomize.

Since _a_'s membership in _C_ is, epistemically, mediated by its sharing of some of _b_'s features by which _b_ claims membership in _C_, the distinctive features of _b_ which _a_ does not share but which are pertinent to its prototypical status (or at least to its inclusion within _C_), are therefore less significant in counting against the similarity judgment than those features of _a_ which are distinctive and which count against its inclusion in category _C_.

As we have noted, the category of which the conferer _b_ is a paradigm or prototype need not be explicitly mentioned. Indeed, there may be no such named category, although we may have a conception of _b_

-like things. Or the category may be so abstract or general as to convey too little information to be useful in a given situation. If I were to say of a woman who cared for and looked after me that "She was like a mother to me", I would be saying that she partook of the qualities of a nurturing parent. We might say that she, like my mother, belonged to the class of nurturing caring persons in their relation to me; but this is far more abstract than simply using the paradigmatic "mother" for a person who bears such a relationship to me. We can say that a suitable conferer when effectively used can better provide what Boyd (1979) has called "epistemic access" (as well, we can add, as affective access) to the relevant category.

One can glean from this example that in the case of metaphor and simile, the conferer will be relatively prototypical (or paradigmatic) of some category to which the conferee is ordinarily denied membership. In this sense, metaphors and similes do not record "antecedently pre-existing similarities". I have expressed this claim elsewhere by saying that the entities brought together in a metaphor must come from two distinct semantic fields.[6] What I will try to show in the remainder of the paper is that by virtue of the bringing together of a given conferee and conferer, certain features of the latter are made salient which may not have been salient before, and in terms of these it serves as a prototype for a newly formed category. This "newly" formed category need not be new in that it has never before been conceived, only in that it does not conform to the network of categories responsible for the prior classifications in which the conferee and the conferer belong to two distinct categories or semantic fields.

3. Context Dependency

To understand how the bringing together of two objects can make salient features which were not so prior to the grouping we have to understand the extent to which context plays a role in the relation between similarity and classification. That context figures prominently in literal comparison statements can already be gathered from the fact that the category for which the conferer serves as a prototype is often not explicitly mentioned but is implicit in the context in which the similarity judgment is made. Tversky provides provocative experimental results indicating that judgments of similarity are significantly influenced by the object-set under consideration and that the salience of a feature which contributes to the similarity between objects is partly determined by the set of objects under consideration.

In speaking of the saliency of a feature, we are concerned here with "diagnosticity", or the importance of a given feature in determining the classification of the object. The saliency of diagnostic features is highly sensitive to the context or the object-set (unlike intensive features which are relatively context-independent). For example, if we are considering a group of animals, all of which actually exist, the feature 'real' has little diagnostic value. If we add to this group some imaginary animals, this feature assumes significant diagnostic value. Considerations such as these lead Tversky to formulate the "diagnosticity principle" (henceforth DP).

DP begins with the understanding that when we are faced with a set of objects we tend to cluster them into groupings such that within each group there is maximal similarity among the objects and each group is maximally distinct from every other group. If some objects are added and/or others deleted from the set, then the cluster may change, altering the diagnostic value of the features on which the clusters, old and new, are based. The change in diagnostic value of the features will also result in a change in the similarity judgments which are made among the objects of the set. DP refers to such relations between groupings and similarity.

A number of experiments were designed to test DP. In one, each of two groups of subjects were presented with a different set containing four schematic faces. The sets differed only in one item: set 1 contained faces _a_, _b_, _c_, _p_ and set 2 contained faces _a_, _b_, _c_, _q_. Both groups were asked to judge which of the remaining three faces was most similar to the target face, _a_. One group rated _b_ as most similar to _a_ while the other group rated _c_ most similar to _a_. The presence of _p_ or _q_ significantly influenced subjects in the choice of the other faces as most similar. See Figure 1.

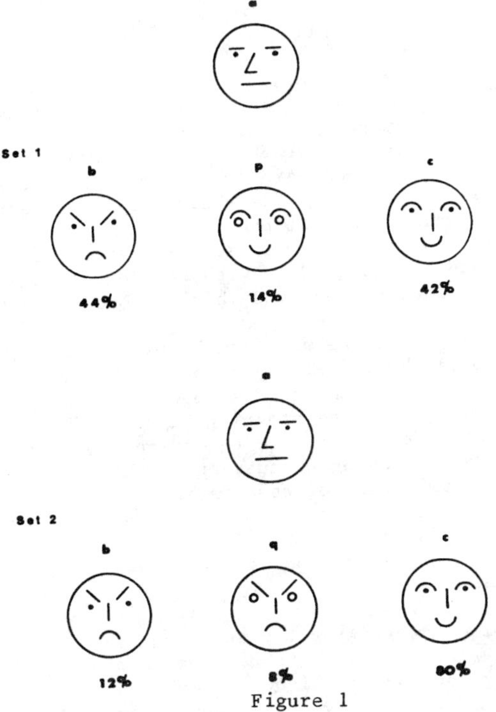

Percent numbers indicate % of subjects who chose each object from the set as most similar to the target.
(Taken from Tversky 1977, p.341).

Figure 1

400

The results can be explained by the fact that the subjects cluster-
ed the items in the two sets differently by virtue of the odd element
and then made their judgments of similarity to the target based on
these clusterings. For example, faces p and c clearly share the
common features of a smiling face: the upturned mouth and pleasantly
curved eyebrows. In relation to this grouping, a and b appear to
be non-smiling faces and are grouped together. The judgment that b
is most similar to a is based on this grouping. In the set excluding p but including q, b and q clearly share the features of a
frowning face: the downturned mouth and raised eyebrows. In this setting a gets grouped with the smiling face as an instance of a non-
frowning face, and based on this grouping c is judged most similar
to a. The clusterings are dialectically related to the objects of
the set and the extent to which some clusterings appear more immediately evident than others. In this situation, the distinctly smiling and
frowning faces determined the way in which the other faces were
grouped. The indifferent face, i. e., the one in which both eyebrows and mouth are drawn with straight lines, could be interpreted as
non-smiling or as non-frowning. Some groupings appear natural, others
more forced, depending perhaps, on the number of shared and jointly
distinctive features and, on the importance of the shared features for
specified or assumed purposes.

4. Metaphor and the Diagnosticity Principle

In the case of metaphor we have a grouping of objects thrust upon
us by the form of the metaphor or simile. The tenor/vehicle, or more
recently the subject/vehicle of metaphor coincides with the conferee/
conferer of similarity judgments.[7] In metaphor as well, there is an
object set against which the common and jointly distinctive features
of conferee and conferer take on saliency: the vehicle is introjected
into a context dominated by elements from the semantic field of the
subject. However, through the metaphoric identification, or the
figurative comparison, the subject is grouped with the vehicle and it
is their shared and jointly distinctive features which become salient.

Tversky sums up his findings on the relation between similarity,
classification, and context thus: "The effects of context on similarity, therefore are treated as changes in the diagnostic value of features induced by respective changes in the grouping of the objects....
It is generally assumed that classifications are determined by similarities among the objects. The preceding discussion supports the converse hypothesis: that the similarity of objects is modified by the
manner in which they are classified. Thus similarity has two faces:
causal and derivative. It serves as the basis for the classification
of objects, but it is also influenced by the adopted classification."
(1977, p. 344).

In metaphor, understood as bringing together objects and hence
causing new groupings, similarity takes on its derivative aspect; previously given similarities are foregone in light of features made salient through the new grouping. We might say that metaphor exploits this

cognitive process to "create" similarities or call attention to features in the subject which are obscured in its usual classifications.

That obscurity can again be related back to the functioning of the diagnosticity principle, for an important factor in the diagnostic value of a feature has to do with whether or not it is shared by all the members of a given object set. From DP we gather that those features which are shared by all the members of the object set have no diagnostic value. This alone indicates some of the motivation for using metaphor. If I wanted to describe the callous attitude and careless ways surgeon Smith has with regard to his patients, it may be quite inadequate to compare him to other poor surgeons. Their many commonalities can obscure the point of the comparison. However, to call Smith "a plumber" is to force the hearer to draw on those features of being a plumber which differ from those of the class of surgeons or the semantic field of surgery, but which coincide with those of the particular surgeon named Smith.

Classification has traditionally been regarded as based on features objects held in common -- at least those features which are salient. According to DP, the commonality of features _per se_ is not significant in classification except where the commonality is rendered salient through its contrast with other objects which do not share these features. Metaphor may be seen to proceed along the very same basis of classification, if the metaphorical process is viewed correctly.

In ordinary judgments of similarity, we chose the conferer so as to maximize those common features which simultaneously distinguish the conferee and conferer from other objects in the object set. In metaphorical judgments of similarity, we chose the conferer (the vehicle) so as to maximize the features which distinguish it from the conferee, except for the few features which are held in common and which are meant to be underscored. It is precisely by virtue of the few commonly held features that the saliency of the features increases, making metaphor so effective in bringing to light those aspects of the subject.

Furthermore, not all the features held in common between subject and vehicle are salient, only those which are also distinctive to subject and vehicle in contrast to other objects in the semantic field of the subject. In the metaphor "Smith is a plumber", where Smith is not an actual plumber but a surgeon, Smith and plumber share the feature of being human. But because all surgeons share this with plumbers it has no diagnostic value in this object set.[8]

Metaphors, then, involve a classification across semantic fields where the aim is not to construct a "superfield" consisting of the original two fields and constructed by virtue of their common features, but to reorganize the subject's field in light of the new diagnostic value of some of the features made salient in the subject by virtue of its being grouped with the vehicle.

If it is the case that features shared by all objects under consideration cannot be used to classify these objects and are devoid of diagnostic value, then metaphor -- in bringing together two <u>sets</u> of objects (i.e., the elements of the two semantic fields, those of the subject and the vehicle) which have very few, and then only very general features in common -- substantially increases the diagnostic value of those remaining features shared by just the subject and the vehicle; indeed it maximizes the saliency of those shared features by providing a context so sparse in shared features and so plentiful in distinct features. Insofar as the diagnostic value of features is the basis for classification, the forced intensification of the diagnostic value of shared features in metaphor provides the ground for a possible, though often tenuous category.

In the case of ordinary similarity statements, we tend to chose a conferer which is more prototypical than the subject, but which is some element of the same semantic field as that of the subject. (Prototypicality may be nothing but an object's epistemic priority for a given audience.) We tend to use metaphor and simile precisely when the usual classification of that subject provides no prototype within that subject's own semantic field for the pertinent features which we want to display or attribute to the subject. With this consideration we move still closer to understanding the motivation for using metaphor. One significant reason no such prototype exists is that often the features we may be interested in exhibiting are relational rather than substantive. It is generally the substantive features which are shared by the elements of a semantic field. The relational features tend to define the often unique position an element within a field holds vis-à-vis other elements of the field. To underscore these relational features, or to suggest a possible shift in the relations one object might bear to another, we have to look toward another semantic field for the prototypical relational features.

When Socrates wants to explain his role as educator to Theaetetus he cannot simply liken himself to an educator of a more paradigmatic sort such as Protagoras, for instance. There is no paradigm within the field of education or philosophy for the sort of relationship Socrates bears to his student, and hence for the sort of educating Socrates does. Thus no literal statement of similarity will do the job of highlighting the distinctive features of Socrates' capacity as an educator. To describe and justify his activities, Socrates moves to a distant field, that of human procreation and speaks of himself as a midwife and of his student as one who is in labor. The field of philosophical education and the field of human procreation certainly share some common features -- for example, both are fields involving human agents. The common feature of midwife and philosophical educator, 'human', therefore has no diagnostic value. Since these two fields, particularly as constituted in the practice of Ancient Greece, are completely differentiated along sexual lines, there are few other substantive features which they may share. What (parts of) the two domains do share are a set of relational features, features which are brought to the fore precisely by virtue of the otherwise scarce and purely general similarity shared by the two fields. These relational features primarily concern the relations of the mother to her baby and to the midwife and that of the student to his ideas and to Socrates. By elaborating on these relations, the midwife metaphor is played out -- the relations which pertain to the field of procreation are borrowed and used to order and reorder suitable relations within the field of philosophical education. The relations of the former field serve as prototype or paradigm for the latter field, in much the same way that the circle serves as the prototype for the ellipse. But in the literal similarity prototype and variant come from the same field.

5. The Cognitive Motivation of Metaphor

We can now connect and apply to metaphor insights gained by considering asymmetrical similarity judgments with the impact of context on the diagnostic saliency of features. The grouping given by the metaphor is in contrast to the usual grouping of the subject, but this only serves, through DP, to make less salient those features of the subject which it shares with other elements of its field and more salient the features it shares with the vehicle. This, in turn, is "the creation of a similarity". This created similarity is, again in turn, the basis of a "new", generally transient, classification. Metaphors, in this way, form the basis for category formation, though generally these categories have no easy fit into our larger conceptual scheme. Insofar as these are categories which are created to answer to a particular occasion, and insofar as they do not easily fit into our larger conceptual scheme, they are unstable, often never acquiring an appropriate superordinate term which names them. They often dissolve as soon as <u>this</u> particular subject is no longer grouped with <u>this</u> particular vehicle. Categories are generally meant to extend beyond individual instances. Insofar as they are transient, literally <u>ad hoc</u>, they push the notion of category to its limit.

That metaphor creates similarities in such a way that its functioning is based on a principle which is operative in our considerations of groupings, similarities and classifications generally, allows us to see why some metaphors might, in fact, become more stable and generate categories which become entrenched in our conceptual schemes. As such, these may serve as scientific models and, in time, become a scientific theory which demands accommodation within our conceptual scheme. Metaphors, seen in this way, take their place, along with induction, in the 'logic of discovery' and give sense to how new categories and classifications can emerge -- both in everyday conceptualizing and in the scientific enterprise.

'Notes'

[1]Simile, figurative comparison, is no less a conceptual detour than metaphor. It differs, rather, in nontrivial linguistic considerations.

[2]Broadly construed, "object" includes entities of all sorts; "feature" includes attributes, functions, and relations.

[3]There is a parallel with statements of the form "_a_ is _b_" which do not always express identity. At times _a_ is said to belong to the class of things which are _b_, or "is _b_" is predicated of _a_.

[4]Notice here the advantage of regarding objects as sets of features rather than as points in a coordinate space.

[5]In some asymmetrical similarity judgments conferee and conferer are not related to one another as variant to prototype, (even a relativized prototype.) In "The son is like the father" or "The globe is like the earth", "father" and "earth" are rather "templates" which bear a direct or indirect causal relation to the conferee. Here there need not be a class of which conferee and conferer are both members, but instead a class of things which can be _generated_ by the conferer. Not only is a globe similar to the earth, but any representation of the earth which is similar to it will be similar in this asymmetric sense; not only is this particular son like his father, but his other offspring may also be similar to him. In these cases it is not a quasi-class inclusion which is the reason for the directionality but the fact that the conferer can generate a class of things such as the conferee may belong to, while the conferee does not generate a class (or the same class) of things to which the conferrer may belong.

[6]See Kittay and Lehrer (1981). The semantic field is a lexical field (a set of related words) as it covers a certain conceptual domain and its members bear certain specifiable semantic relations to one another. In metaphor a portion of a semantic field is transferred from one domain to another and imposes its structure on the recipient domain; when the lexical field from one conceptual domain is transferred to another, the semantic relationship between the lexemes remains, e.g., _hot_ and _cold_ are antonyms in the temperature domain; when transferred to athletics, a _hot_ player is one who scores and a _cold_ one does not. The antonymy of the pair is preserved. Moreover, other terms in the first domain are available for transfer: a _warm_ player scores less than a _hot_ one but better than a _cool_ one.

[7]For subject/vehicle terminology see Ortony (1979). When speaking of figurative similarity I will call the conferer, the vehicle.

[8]In the metaphor, "man is a wolf", aspects of a wolf, e.g., being a mammal, which he shares with man, are excluded from metaphoric considerations. Black and others have claimed that this is because being mammalian is part of the definition of wolf and metaphor plays on an "associational network", and not, they claim, on definition. But it is false that metaphors never depend on definitional features of the vehicle. Reference to DP allows us to explain why being a mammal does not figure here without recourse to the dubious claim concerning the role of definitional features.

References

Black, Max. (1954). "Metaphor". _Proceedings of the Aristotelian Society_ 55: 273-294. (As reprinted in _Models and Metaphors._ Ithaca, NY: Cornell University Press, 1962. Pages 25-47.)

Boyd, Richard. (1979). "Metaphor and Theory Change: What is 'Metaphor' a Metaphor for?" In Ortony (1979). Pages 356-408.

Kittay, Eva. (1978). _The Cognitive Force of Metaphor: A Theory of Metaphoric Meaning._ Unpublished Ph.D. Dissertaion, City University of New York. Xerox University Microfilms Publication Number 79-00789.

------------ and Lehrer, Adrienne. (1981). "Semantic Fields and the Structure of Metaphor." _Studies in Language_ 5: 31-63.

Lehrer, Adrienne. (1974). _Semantic Fields and Lexical Structure._ Amsterdam: North Holland.

Lyons, John. (1963). _Structural Semantics._ Oxford: Blackwell.

------------. (1977). _Semantics._ Cambridge: Cambridge University Press.

Ortony, A. (ed.). (1979). _Metaphor and Thought._ Cambridge: Cambridge University Press.

Rosch, E. (1973). "On the Internal Structure of Perceptual and Semantic Categories." In _Cognitive Development and the Acquisition of Language._ Edited by T.E. Moore. New York: Academic Press. Pages 111-144.

---------. (1975). "Cognitive Reference Points." _Cognitive Psychology_ 7: 532-547.

---------- and Mervis, C.B. (1975). "Family Resemblances: Studies in the Internal Structure of Categories." _Cognitive Psychology_ 7: 573-603.

---------- _et al._ (1976). "Basic Objects in Natural Categories." _Cognitive Psychology_ 8: 382-439.

Tversky, Amos. (1977). "Features of Similarity." _Psychological Review_ 84: 327-352.

--------------- and Gati, I. (1979). "Studies of Similarity." In _On the Nature and Principle of Formation of Categories._ Edited by E. Rosch and B. Lloyd. Hillsdale, NJ: Erlbaum. Pages 79-98.

Science and Play[1]

Michael Goldman

Miami University

Many philosophers interested in problems of scientific progress--
what it is and whether it is possible--have in recent years, and with
good reason, focused on the challenges directed by Kuhn (1962) and
Feyerabend (1975) to the traditional belief in progress by accretion.
I have argued (in response to Feyerabend's proliferation theory) that
the proper locus for a solution to these problems must be the social
(or what I prefer to call the "material") conditions in which proposed
competing theories must exist (Goldman 1980). Recently, Gonzalo
Munévar has proposed similar criteria in his book Radical Knowledge
(1981). Our approaches have much in common. Most importantly they
eschew the wholly "intellectual" criteria offered by almost all philosophers; that is, criteria which focus entirely on a theory's ability
to solve problems arising only in thought (whether they be called
'conceptual' or 'empirical' problems). Even Laudan, whose unwilling-
ness to separate the problems of science from the problems of metaphysics or indeed from problems arising in other areas of human intellectual concern provides a notable advance over more typical approaches, nevertheless insists upon considering science (and, one must assume, the other intellectual pursuits as well) wholly in isolation
from the social context with which it interacts (Laudan 1977).
Munévar's position, and my own, depart in similar ways from this constraint. Nevertheless, while I take my own account to be a refutation
of Feyerabend's anarchistic epistemology in general and proliferation
theory in particular, Munévar apparently sees his work as supportive
of Feyerabend's overall program (and Feyerabend seems to agree, as he
has written a Foreword to the book). This difference, I think, is instructive, and I would like to investigate the reasons for it in this
paper.

Munévar, following Toulmin (1972) and Scriven (1972), uses an
evolutionary model to generate his account of a "successful" scientific theory, an account which, informally, tells us "that a theory is
'better' than another if it allows us to 'get along' better in the

PSA 1982, Volume 1, pp. 406-414

universe." (Munévar 1981, p. 54). 'Getting along better' is glossed
as follows:

>   (a) dealing with greater ease with our environment (our 'niche');
>   (b) increasing the number and diversity of environments that we
>   can deal with (enlarging our 'niche');
>   (c) coping with a continuously changing environment (which puts a
>   premium on flexibility of response)...(Munévar 1981, p. 54).

   While I do not think that this account of 'getting along' is explicit enough to serve the purpose it is meant to serve,[2] I do think
that, to the extent it does, it is correct. It is condition (c) which
is especially crucial (and which I stressed in 1980), for it suggests
why it would be a fatal mistake to expect that any fixed mode of interaction with nature will have unlimited survival potential for human
beings. Briefly, the reason is that even when conditions do not
change dramatically we are faced with the pervasive problem of dealing
with a continually diminishing supply of any given resource (and of
crucial importance, of any known source of usable energy). Since it
is human creative and productive activity which defines a resource it
is clear that it is only continual and successful science, coupled
with the ability of a population to rapidly assimilate the practical
implications of that science that can assure the continued survival of
the human species. Munévar is correct to point out the importance of
being prepared for radical shifts in survival strategies. Such shifts
are necessary even when circumstances themselves do not shift rapidly
but where (as in the case of diminishing supplies of a critical resource) the shift is slow but inexorable.

   It is also condition (c) which leads Munévar to stress the importance of play, which, as Lorenz (1950, 1954), he associates with
curiosity, as a partial model for understanding science. It is this
playful curiosity, which often leads to the development of theories
with no immediate payoff, that best enables the human species to be
prepared for predictable as well as unpredictable changes, and hence
that maximizes our survival potential.

   It is clear how these considerations provide support for a proliferation theory of science. (As far as I can tell they provide no
support at all for an anarchistic epistemology.) The greater the
number of theories being actively pursued the greater the chances of
hitting upon some result whose application will be just the thing for
dealing with an unpredictable turn of events. Whether it be a re-
spectable, though esoteric, branch of theoretical mathematics, a somewhat disreputable astrology, or some border-line case like ESP, it
would be foolish to deny that it is at least possible that some worthwhile consequence will follow from its pursuit. Moreover, and perhaps
more importantly, it seems that even if we do not expect any particular useful outcome, there is probably something socially valuable
about experimentation and "curiosity" for its own sake: under the
best circumstances the quality of mind which is encouraged by such experimentation, by such openness, can be of value to a culture even if

some of the particulars it deals with are not. Such mental resources
in the population can often be harnessed to the solution of pressing
problems, even when their objects cannot be. Even when this is not
possible, the very existence of such mental qualities is probably in-
fectious enough to increase the average adaptability and flexibility
of the culture.

Having admitted all this, however, it is necessary to raise considerations which seriously undermine much of it. Where Munévar goes
wrong, I think, is in his failure to recognize the material circumstances which make such "play" possible. It is revealing that he
notes that the "play" of certain species ends when the animal reaches
maturity. Similarly we all recognize that play is more characteristic
of the behavior of human beings when they are still children, though
of course it is never fully extinguished in adults. What is pertinent
about these facts is that play characterizes the behavior of animals
only when their survival needs are not threatened; that is, when short
term needs have already been assured either by the previous production
of enough surplus to see them through the play period or by the labor
of others (parents, for instance). Summarizing the social scientific
work on play of recent years, D. E. Berlyne points out that "play is
usually abandoned promptly as soon as a clear biological emergency is
encountered." (1969, p. 815).

It is crucial to notice, then, that whatever its long-term significance, play is completely parasitical in the immediate and short term.
This or any culture must make the decision to _permit_ play, because
play does not provide the players with the necessary elements for
their immediate physical survival. Parents permit their children to
play not merely, or even primarily, by providing verbal cues but by
providing them with the essentials for physical existence. The age
at which play must end as characteristic of the individual's behavior
may be fixed in the case of animals but is obviously not fixed in the
case of human beings. There are class differences at any given time
and, over time, the age changes, with the general trend having been
toward higher ages. This is normally a cultural matter, contingent
in large part upon the ability of the working members of the culture
to provide enough surplus to support the players.

When I say that a culture must make the decision to allow some of
its members to engage in play (be it scientific or otherwise) it
should be understood that the mechanisms for doing this are as varied
as the social, economic, and political structures of those cultures.
With straightforwardly centralized economies the mechanisms are correspondingly straightforward. But even in nominally decentralized economies such mechanisms exist: more or less money is made available for
scholarships in certain academic areas; enrollment in certain programs
is opened up or closed down; laws prohibiting certain kinds of research (for instance, recombinant DNA, fetal experimentation) are
passed. Even more dramatically, a culture may make certain resources
completely unavailable. For instance, space research is possible only
to the extent that the culture is willing and able to produce the

necessary hardware. In general, any form of scientific research requires, implicitly, the cooperation of vast numbers of non-scientific personnel. It is always a social decision, though not always a self-conscious social decision, to provide that cooperation.

This is the locus of my objection to Munévar's position. It is a pre-condition of any culture's ability to permit its able-bodied (or able-minded) members to "play" that the culture have enough material surplus to support them and their activities. This in turn means that the culture has, at least in the present and for the foreseeable future, solved its survival problems. To the extent that some approach to science has contributed to that solution we can agree with Munévar that that approach is "better" than other contenders. But under what conditions can a culture continue to provide resources to any and all players of the science (or any other) "game"? This is a way of raising some of the very practical questions about the distribution of resources to science, questions which add a practical, critically pressing dimension to the philosopher's intellectual interest in criteria of scientific truth and scientific progress.

It is my contention that the kind of social munificence implicit in a call for the support of the widest variety of scientific approaches is possible only under the most fantastic, science-fiction-esque circumstances, namely when the culture is so very wealthy in terms both of resources and of available productive techniques that problems of survival are not even remotely pressing. In this case, of course, science will have already proved as successful as we can possibly hope it to be, and whatever science remains to be done will resemble play in its most basic sense.[3] We will expect no practical results from <u>this</u> science either in the short or long term because there will be no<u>thing</u> important left for it to do. Science will have approximately the same status as chess: there may be problems to solve, but they will be wholly self-contained problems. Nothing whatever will turn upon their solution.

All of this is a way of reiterating the fact that the "play" aspect of science is parasitical upon the already successful (and, perhaps, rather stodgy) approaches to science in so far as those approaches contribute to the production of the material surplus which makes such play possible. Given a society with limited resources-- financial, human, material--it is extremely important to make rational allocations of those resources to competing scientists (as well as to other worthy non-scientific endeavors). It would be entirely irrational (in the very practical sense of self-destructive) to squander resources on approaches which have either long ago proven fruitless (eg., astrology, witchcraft) or which cannot or do not even try to make a persuasive case that they will bear fruit. To be sure the notion of "bearing fruit" is unclear (I will have more to say about this below) but the main point I wish to accentuate here is that there are very real constraints upon what a culture can and cannot support in the way of scientific "playfulness" and that these constraints are

410

in part defined by the successes or failures of science itself.

Of course it is necessary to admit that the implementation of such
constraints may backfire: a genuinely valuable innovation may lurk
just around the astrological corner. But as a practical matter this
is irrelevant. No real culture can afford to squander its resources
in any and every direction, and once we admit (with Feyerabend) that
no methodology holds an eternally privileged position with respect
to survival potential, then we are always at risk.

In the absence of any guarantees, and when we are dealing with approaches to science which are radically different from already successful ones, it is very important to be able to identify the approaches which are worthy of support from those which are not. Does
the philosopher have anything to contribute to this very practical issue once it is admitted that none of the traditional philosophical
attempts to demarcate "true" science from "mere metaphysics" or
"pseudo-science" will do? Agreeing with Munévar that we can know, ex
post facto, which approaches were better than others is not very helpful in the practical case.

Two kinds of approach for such decision making come to mind: approaches which focus on the method by which such allocation decisions
should be made and approaches which focus on the content of such decisions.

Those who despair of finding universally applicable characteristics
of the content of successful science are, not surprisingly, inclined
to favor approaches which focus on the method by which allocation decisions should be made. Thus, Feyerabend's suggestion that direct
democratic procedures be used in such allocation decisions is an example of the methodological approach, and one can easily imagine
others. For instance, a Marxist might insist that only those approaches consistent with the interests of the most progressive class
of society ought to be supported; this claim would then have to be
supplemented by an account of what the interests of that class are and
how we might identify a given scientific approach as consistent or inconsistent with it. (This seems in part to be the editorial commitment of the journal Science for the People).[4]

Without passing judgment upon the validity of either of these (or
other possible) methodological approaches, I would like to suggest
that the content is more important than the method, and that there
are at least a few suggestions about the content of preferable approaches to science that the philosopher can make even in the absence
of the reassuring traditional dogmas. Let me emphasize that what
follows is highly speculative and is presented in large part to generate discussion about this most important issue.

Two preliminary precautions are in order. (1) It is probably not
possible to answer the kind of question posed here in the abstract.
That is, the specific approaches to science which are worthy of sup-

port will have to be decided in the context of the real historical circumstances in which the questions arise. At one time, for instance, alchemy might be a legitimate approach, while at another time it might not be. The reason for this difference will become clearer below. (2) The question before us concerns the allocation of what are unambiguously surplus resources. We must admit, at least if we accept Munévar's account of a successful science, that an approach which is generating "results" or has a recent history of generating results, deserves our support. We are concerned here only with resources available above and beyond what is necessary for that support.

My suggestion is that a useful starting point is to consider what it is that we assume when we assume that the systematic methods of science are going to be more helpful to us in our struggle for survival than are the unsystematic (even if true) rules of thumb, folklore, habits, family and tribal traditions, etc., that science (whatever its content or style) is supposed to replace. One assumption, it would seem, is that the universe is lawful in a more profound sense than is suggested by the mere regularities implicit in these other techniques. This in turn means acknowledging that these other "strategies" will work only to the extent that they are instances of limited, localized applications of more universal relationships. Given this assumption it is plausible to insist that the kind of mind most suitable for doing the science which will ultimately be useful to human beings in their struggle for survival is the kind of mind not overly impressed by localized applications or temporary successes. It is the kind of mind which "drives toward universality", which insists that local successes (whether spatially or temporally localized) be explained by more and more universal relationships. It is further plausible to insist, therefore, that our limited resources be allocated to those approaches which respect this drive toward universality, rather than those satisfied with mere parochial success. This is especially important because it is quite possible that this "quality of mind" is a quality not selected for by Darwinian mechanisms, precisely because in the short run and at the tribal or family level, the ability to operate with "rules of thumb" (and not be concerned with profounder implications of them) has survival potential. We no longer live in that kind of environment, however.

Munévar insists upon the importance of theoretical mathematics, and he does so on the (not unreasonable) ground that there may be unforeseen applications of even the purest mathematics. Plato also stressed the importance of mathematics, but he saw it as a pedagogical stepping stone to an understanding of universal truths. I think Plato's account is quite plausible if we understand it in the psychological sense I am suggesting here (and without Plato's ontological presuppositions). Mathematicians (for example) deserve support from the materially productive community because their way of thinking, of approaching problems, recognizes the kind of universality ultimately necessary for a practical understanding of a changing universe, a universe in which rules of thumb, traditions, etc., have very little survival value. It is impossible to be more specific than this because individual cases have to be judged on their individual merits (keeping this criterion in mind) and the kind of arguments to be made in individual cases may vary enormously. In addition, the picture is complicated by sociological factors. Whatever the intrinsic merits of an approach to science, some can and do become associated within a given culture with irrationality, with self-indulgence, with a kind of parochial, tribalized mentality. Certainly that has been the recent history of many "cult" sciences such as astrology. Under these circumstances support for such approaches is highly counter-productive, for such support tends merely to give credence to a cultish, parochial mentality so inimical to the universal outlook required of the scientist.

A second "quality of mind" that it makes sense to encourage when we assume that science can provide us with the theoretical understanding which makes possible survival techniques, is a sense of potency. If 'drive toward universality' is a vague concept, I have to admit that 'sense of potency' is at least equally vague. It is fairly easy to see what is meant here when we discuss technologies. Many people urge that certain technologies not be implemented (nuclear power, genetic engineering, are examples) either because the power implicit in the use of such technologies is greater than they feel human beings are capable of handling or because it is greater than they feel human beings should handle. (This is not to say that there are not other reasons for caution in the use of these techniques; that is not the issue here.) I would assess these attitudes as a fear of human potency, and would argue that human survival is not possible until such fear is overcome. It is more difficult to see how this applies to the case of theory-creation, but certainly it would imply that support be denied to "magical" forms of science such as voodoo, witchcraft, etc., which define a realm of occult power over which human beings have little or no control. (This does not mean that we shouldn't try to find out why these approaches "work" within limited populations, but, as I said above, our explanatory base must be one which is more universally applicable: physiology, psychology, etc.)

My claim, then, is that a "drive toward universality" and a sense of potency are two characteristics necessary to a culture which has come to depend upon science for a significant contribution to its survival techniques. I certainly do not claim to have made a decisive case for these two characteristics, and it was not my intention to do so. Rather, my intention was to suggest that there are criteria quite different from the ones typically proposed in philosophical literature for the allocation of resources to one or another approach to science, and that these criteria can be discerned only if we recognize, with Munévar, that science plays a significant role not merely in our understanding of the world, but in our living in the world.

My argument can be summarized as follows:

(1) Theoretical science is parasitical upon the materially productive sectors of society.

(2) Such science can contribute to human survival by providing the theoretical tools useful in the creation and application of survival techniques under changing environmental circumstances.

(3) The limited material resources of a culture define corresponding limits to the extent and diversity of approaches to science which the culture can support.

(4) Therefore, uncontrolled proliferation is not possible, and decisions have to be made about the distribution of such surplus as is available for science.

(5) First priority must go to those approaches currently proving fruitful, or which can plausibly be supposed to do so shortly.

(6) Further surplus must go to those approaches likely to promote in the population the psychological attitudes most appropriate to the use of scientific techniques for survival; among these are an appreciation of universality and a sense of potency.

Notes

[1] This paper was written while I was studying under a National Endowment for the Humanities Residential Fellowship for College Teachers. I am grateful to my colleagues in that program for their suggestions, and to Peter Schuller, my colleague at Miami University, with whom I have discussed these issues.

[2] In "Against Feyerabend: The Meaning of Progress in Science"(1980) I tried to give a criterion of scientific progress which makes use of the concept of social reproduction, and offered a quantifiable criterion of social reproduction.

[3] Some may object that there is an important difference between that science which leads to technological advance and that science which provides understanding. This seems a dubious distinction at best, especially today when the interaction between science and technology is at its height. But even if it were so, my arguments would remain unchanged for it is the former kind of science which would, of necessity, take priority.

[4] Many Marxist critics have insisted that the choice of methodologies in the social sciences has always been dictated by political considerations. The claim that such influences have been felt in the physical sciences is also, if less frequently, made. See any early issue of the journal Fusion for examples.

References

Berlyne, D.E. (1969). "Laughter, Humor, and Play." In *Handbook of Social Psychology,* Volume III. Edited by Gardner Lindzey and Elliot Aronson. Reading, Mass.: Addison-Wesley. Pages 795-852.

Feyerabend, Paul. (1975). *Against Method.* London: New Left Books.

Goldman, Michael. (1980). "Against Feyerabend: The Meaning of Progress in Science." In *Research in Philosophy and Technology,* Volume III. Edited by Paul Durbin. Greenwich, Conn.: JAI Press. Pages 28-38.

Kuhn, Thomas. (1962). *The Structure of Scientific Revolutions.* Chicago: University of Chicago Press.

Laudan, Larry. (1977). *Progress and Its Problems.* Berkeley: University of California Press.

Lorenz, Konrad. (1950). "Ganzheit und Teil in der tierischen und menschlichen Gemeinschaft." *Studium Generale* 9: 455-499. (As reprinted as "Part and Parcel in Animal and Human Societies: A Methodological Discussion." (trans.) Robert Martin. In Lorenz (1971). Pages 115-195.)

----------------. (1954). "Psychologie und Stammesgeschichte." In *Die Evolution der Organismen.* 2nd ed. Edited by G. Heberer. Jena: G. Fischer. Pages 131-172. (As reprinted as "Psychology and Phylogeny." (trans.) Robert Martin. In Lorenz (1971). Pages 196-245.)

----------------. (1971). *Studies in Animal Behavior,* Volume II. Cambridge, Mass.: Harvard University Press.

Munévar, Gonzalo. (1981). *Radical Knowledge.* Indianapolis: Hackett Publishing Company.

Scriven, Michael. (1972). "The Concept of Comprehension: From Semantics to Software." In *Language Comprehension and the Acquisition of Knowledge.* Edited by Roy O. Freedle and John B. Carroll. Washington: V.H. Winston and Sons. Pages 31-39.

Toulmin, Stephen. (1972). *Human Understanding.* Princeton: Princeton University Press.